T5-AGS-367

DISCARDED

GEOGRAPHICAL INFORMATION SYSTEMS IN HYDROLOGY

Water Science and Technology Library

VOLUME 26

Editor-in-Chief
V. P. Singh, *Louisiana State University,*
Baton Rouge, U.S.A.

Editorial Advisory Board

M. Anderson, *Bristol, U.K.*
L. Bengtsson, *Lund, Sweden*
A. G. Bobba, *Burlington, Ontario, Canada*
S. Chandra, *New Delhi, India*
M. Fiorentino, *Potenza, Italy*
W. H. Hager, *Zürich, Switzerland*
N. Harmancioglu, *Izmir, Turkey*
A. R. Rao, *West Lafayette, Indiana, U.S.A.*
M. M. Sherif, *Giza, Egypt*
Shan Xu Wang, *Wuhan, Hubei, P.R. China*
D. Stephenson, *Johannesburg, South Africa*

The titles published in this series are listed at the end of this volume.

GEOGRAPHICAL INFORMATION SYSTEMS IN HYDROLOGY

edited by

VIJAY P. SINGH

Department of Civil Engineering,
Louisiana State University,
Baton Rouge, U.S.A.

and

M. FIORENTINO

Department of Environmental Engineering and Physics,
University of Basilicata,
Potenza, Italy

KLUWER ACADEMIC PUBLISHERS

DORDRECHT / BOSTON / LONDON

A C.I.P. Catalogue record for this book is available from the Library of Congress

ISBN 0-7923-4226-7

Published by Kluwer Academic Publishers,
P.O. Box 17, 3300 AA Dordrecht, The Netherlands.

Kluwer Academic Publishers incorporates
the publishing programmes of
D. Reidel, Martinus Nijhoff, Dr W. Junk and MTP Press.

Sold and distributed in the U.S.A. and Canada
by Kluwer Academic Publishers,
101 Philip Drive, Norwell, MA 02061, U.S.A.

In all other countries, sold and distributed
by Kluwer Academic Publishers Group,
P.O. Box 322, 3300 AH Dordrecht, The Netherlands.

Printed on acid-free paper

All Rights Reserved
© 1996 Kluwer Academic Publishers
No part of the material protected by this copyright notice may be reproduced or
utilized in any form or by any means, electronic or mechanical,
including photocopying, recording or by any information storage and
retrieval system, without written permission from the copyright owner.

Printed in the Netherlands

To our families:

Anita   Maria-Rosaria
Vinay
Arti

Table of Contents

Preface

The last few years have witnessed an enormous interest in application of GIS in hydrology and water resources. This is partly evidenced by organization of several national and international symposia or conferences under the sponsorship of various professional organizations. This increased interest is, in a large measure, in response to growing public sensitivity to environmental quality and management.

The GIS technology has the ability to capture, store, manipulate, analyze, and visualize the diverse sets of geo-referenced data. On the other hand, hydrology is inherently spatial and distributed hydrologic models have large data requirements. The integration of hydrology and GIS is therefore quite natural. The integration involves three major components: (1) spatial data construction, (2) integration of spatial model layers, and (3) GIS and model interface. GIS can assist in design, calibration, modification and comparison of models. This integration is spreading worldwide and is expected to accelerate in the foreseeable future. Substantial opportunities exist in integration of GIS and hydrology. We believe there are enough challenges in use of GIS for conceptualizing and modeling complex hydrologic processes and for globalization of hydrology. The motivation for this book grew out of the desire to provide under one cover a range of applications of GIS technology in hydrology. It is hoped that the book will stimulate others to write more comprehensive texts on this subject of growing importance.

Discussing the role of GIS, the introductory first chapter discusses the current needs in hydrologic modeling and the meeting of these needs with application of GIS, and is concluded with a brief reflection on the outlook for the future. Integration of remote sensing and GIS for hydrologic studies constitutes the subject matter of Chapter 2. Introducing the remote sensing system, the image processing system, and the global positioning system (GPS), the chapter goes on to discuss the interface amongst remote sensing, GIS, and GPS; and several applications in surface-water hydrology, groundwater hydrology, and water-quality hydrology. Chapter 3 presents hydrologic data development, including data needs, sources of data, conversion of data into appropriate format for GIS software, and maintenance of the database, with emphasis on quality control measures. Spatial models and data structures are presented in Chapter 4. Raster spatial data structures as well as vectorial data structures are described under spatial data structures. The spatial models include a discussion of categorical and object approaches, as well as a combination of these approaches. Chapter 5 presents methods for spatial analysis, including variogram models, trend surface models, and kriging. Also presented are some examples of these models and their integration into GIS software.

GIS needs and GIS software constitute the subject matter of Chapter 6. It briefly illustrates GIS capabilities, and disucsses uses of GIS software for information management, queries and mapping, spatial data preparation, modeling, and simulation needed for hydrological applications. Chapter 7 discusses digital terrain modeling. It includes a review of techniques for generating a digital terrain model (DTM), algorithms for calculating information relevant to hydrological modeling, and available software packages for generating DTM and hydrological parameters. GIS for distributed rainfall-runoff modeling is the topic of Chapter 8. It discusses applications of GIS in analyses of different runoff production mechanisms, and use of a distributed rainfall-runoff model, TOPMODEL, for simulating the hydrologic behavior of a gaged basin in southern Italy. Chapter 9 extends the discussion to large-scale watershed modeling, encompassing macroscale modeling, integration of a GIS with SLURP macroscale hydrological model, and appraisal of the present modeling situation and recommendations for future development of GIS. Lumped modeling and GIS in flood prediction is the subject of Chapter 10. It briefly reviews hydrologic models for flood prediction, and shows the GIS unit hydrograph models as distributed rainfall-runoff models, and makes an argument for application of GIS to facilitate the merging of deterministic and stochastic models into one unified modeling approach.

GIS in groundwater hydrology is discussed in Chapter 11. It presents the role of "loosely coupled" GIS-groundwater models with field examples. Chapter 12 discusses non- point source pollution modeling with GIS, with emphasis on hydrologic and water quality modeling. Soil erosion assessment using GIS is presented in Chapter 13. It discusses several soil erosion models, and demonstrates integration of erosion models with GIS with use of LISEM soil erosion model as an example.

Chapter 14 discusses application of the GIS technology to study of landslides. With a short review of techniques and examples of landslide hazard mapping, the chapter goes on to describe a GIS methodology based on the identification of 3-D hydro-geotechnical models, together with the difficulties of defining hazard asssessment from rainfall triggering mechanisms. It is concluded with a discussion of examples of GIS output for investigating the relationships between land use change and the distribution of landslides. Land-use hydrology with GIS constitutes the subject matter of Chapter 15. It reviews basic concepts in land-use systems from a hydrologic perspective, approaches employed for quantifying hydrologic effects of land-use changes, use of GIS in the modeling of processes, and specific management issues that may arise when GIS is used as a tool for land-use hydrology. The concluding Chapter 16 presents the design of GIS for hydrologic applications. It analyzes some of the main criteria for GIS design and considers a wide range of commercially available GIS packages.

The editors would like to express their gratitude to all the contributors who, despite their busy schedule, were generous to write the chapters. The book is the fruit of their labor. They also would like to express their appreciation to their families; the Singhs – wife Anita, son Vinay, and daughter Arti; and the Fiorentinos – wife Maria-Rosaria, for their love and support – without which the book would not have come to fruition.

V.P. SINGH
Baton Rouge, Louisiana, U.S.A.

M. FIORENTINO
Potenza, Italy

CHAPTER 1

Hydrologic modeling with GIS

V.P. Singh and M. Fiorentino

The last few years have witnessed an enormous interest in application of GIS in hydrology and water resources. The increased interest, in a large measure, is in response to the growing public sensitivity to environmental quality and management. The GIS technology has the ability to capture, store, manipulate, analyze, and visualize the diverse sets of georeferenced data. On the other hand hydrology is inherently spatial and distributed hydrologic models have large data requirements. The integration of hydrology and GIS is therefore quite natural. The integration involves three major components: (1) spatial data construction, (2) integration of spatial model layers, and (3) GIS and model interface. GIS can assist in design, calibration, modification and comparison of models. This integration is spreading worldwide and is expected to accelerate in the foreseable future. Substantial opportunities exist in integration of GIS and hydrology. We believe there are enough challenges in the use of GIS for conceptualizing and modeling complex hydrologic processes and for globalization of hydrology. This chapter introduces GIS and discusses a range of its applications in hydrology.

1.1. A Short Historical Perspective of Hydrologic Modeling

Prior to the advent of the unit hydrograph by Sherman (1932), hydrologic modeling was mostly empirical and based on limited data. Those days, graphs, tables, and simple analytical solutions were the standard models, and hand calculations, in conjunction with a sliderule, reflected the computing prowess. In the decades that followed, perhaps up to the end of 1950's, simple analytical models received a great deal of attention and a range of such models were limited, and data processing was easily handled manually.

In the decade of sixties, a new frontier in hydrologic modeling unfolded on account of digital revolution and easy access to the resulting massive computing capability. For the first time it became possible to synthesize the entire hydrologic cycle and process large quantities of data. A seminal and the first example of such

V. P. Singh and M. Fiorentino (eds.), Geographical Information Systems in Hydrology, 1–13.
© 1996 *Kluwer Academic Publishers. Printed in the Netherlands.*

a synthesis was the Stanford watershed model pioneered by Crawford and Linsley (1966). In subsequent years and decades, computing capability grew without leaps and bounds. Today's computing capability is virtually limitless and is still growing. In hydrologic arena, synthesis of the entire hydrologic cycle became popular, so popular that its computer models are too many to count. Virtually every agency in the United States, federal, state, or local, had either developed its own model or adopted one of the popular models. The private sector was not far behind. Of course, universities and federal research agencies such as the U.S. Army Corps of Engineers, U.S. Geological Survey, U.S. Department of Agriculture, and U.S. National Weather Service, took a lead in this area. Similar trends emerged in other countries. In the meantime, growing environmental and ecological concerns demanded greater information of these models. Thus, these models were expanded in their scope to encompass simulation of erosion and sediment transport, chemical transport, and the effect of land use changes due to human activities. Clearly, the data needs of such models multiplied, and so did the need for data processing. Models continued to expand, incorporating statistical analyses, risk and reliability, graphics, etc., all because more and more demands were imposed on these models. These days, the models are becoming more and more distributed, and many of them are being integrated with global circulation models on one hand and oceanic models on the other. The net result of all this modeling fury is that the data needs have become enormous and so has the need for processing them. The old ways of handling and managing the data are no longer adequate. Furthermore, technology exists that can handle and process massive quantities of data and such a technology is GIS. The need for a marriage between hydrology and GIS is, therefore, apparent.

1.2. Hydrology in Environmental and Ecological Continua

In the environmental continuum, water plays a central role and brings soil and air closer. Water shapes the landscape and turns into water vapor. It sustains the ecological continuum. Therefore, hydrology is the core of environmental and ecological continua. It is the environmental and ecological problems and growing public concern for their solution that placed increasing demands on hydrologic models. It is these demands that triggered the development of comprehensive hydrologic models, both surface water and groundwater, with details in up to 3 dimensions. Emergence of hydrology as a science in its own right may partly be ascribed to the role that it plays in addressing a range of environmental and ecological problems and to its being a critical component of the environmental and ecological continua. Furthermore, these problems have helped either integration or bringing closer of seemingly disparate areas. For example, there is growing acceptance that biological and chemical transport processes are as much a part of hydrology as physical processes. Similarly, geomorphologic processes underlying drainage networks are very much a part of hydrology. The areas that some years ago appeared remote or peripheral to hydrology are coalescing to form parts of hydrology. As a result, with continuing evolution, hydrology is expanding both horizontally and vertically. To that end, more data and better processing capability are clearly needed.

1.3. Current Needs in Hydrologic Modeling

Hydrologic modeling is increasingly becoming global, both in terms of spatial scale and depth of treatment. The globalization of hydrology is manifest in many ways. First. distributed watershed models are receiving a lot of attention these days. Second, coupling hydrologic models with global circulation models and oceanic models is being triggered by the need for quantifying the repercussion of the oft and much talked about climate change. These two types of modeling activities represent two ends of the spectrum. Third, water, whether on the surface or below it in vadose zone and/or saturated geologic formations, is being treated in the sense of a continuum. Artificial divisions of hydrology into surface water hydrology, vadose zone hydrology, and groundwater hydrology are being discarded and a unified body of scientific knowledge being emphasized. This globalization of hydrology is placing new and greater demands for data and more sophisticated techniques for managing and processing them. Fortunately, sophisticated remote sensing and satellite technologies are evolving which are capable of providing the needed data. Similarly, comprehensive GIS packages are becoming available that are capable of managing and processing massive quantities of data.

1.4. Role of GIS

1.4.1. WHAT IS GIS?

The term Geographic (or Geographical) Information System dates back to mid sixties and seems to be rooted in two separate applications, one in Canada and the other in the United States: (a) Management of the mapped information collected for the Canada Land Inventory and obtaining estimates of the area of land available for certain types of uses; and (2) accessing the many different types of data from large volumes of information for analysis and presentation of the results in map format (Coppock and Rhind, 1991) for use by the large scale transportation models in the United States. Nearly 30 years later, perhaps the same justifications can be advanced for the use of GIS in hydrology, and environmental and water resources.

GIS is a general-purpose computer-based technology for handling geographical data in digital form. It is designed to capture, store, manipulate, analyze and display diverse sets of spatial or geo-referenced data. They contain both geometry data (coordinates and topological information) and attribute data (i.e., information describing the properties of geometrical objects such as points, lines and areas). Goodchild (1993) states that the GIS technology has the ability to perform a variety of tasks some of which are: (1) preprocessing of data from large stores into a form suitable for analysis exemplified by reformatting, change of projection, resampling, and generalization; (2) supporting analysis and modeling: forms of analysis, calibration of models, forecasting, and prediction; and (3) postprocessing of results through reformatting, tabulation, report generation, and mapping. According to Mark and Gould (1991), in any operation involving GIS, the user defines the requirement and interacts with the system through a "user-friendly" intuitive interface that makes use of such contemporary concepts as graphic icons and desktop metaphors.

1.4.2. GEOGRAPHICAL DATA MODELING

A computer database is a finite, discrete store, whereas a geographical reality may be quite complex. For example, soil variables are inherently complex. The complex reality has to be converted into a finite number of database records or objects through the process of discretization. This process is referred to as geographical data modeling. Objects may be points, lines, or areas and may also be descriptive attributes. For instance, precipitation is sampled by precipitation gages which are point objects. Precipitation at these gages are object attributes.

Two types of geographical data models are distinguished in the GIS technology: (1) field models, and (2) object models. A field model represents the spatial variation of a single variable by a collection of discrete objects. A field may be associated with a variable measured on either continuous or discrete scales. A spatial database may comprise many such fields. Goodchild (1993) discusses 6 field models that are commonly used in GIS:

(1) Irregular point sampling: The database contains a set of triples $\langle x, y, z \rangle$ representing point values of the variable at a finite set of irregularly spaced locations. An example is precipitation measurement.

(2) Regular point sampling: The sampling points are regularly arrayed on a square or rectangular grid. This is similar to (1) except for regular spacing of measurement location. An example is a Digital Elevation Model (DEM).

(3) Contours : The database contains a set of lines, each represented by an ordered set of $\langle x, y \rangle$ pairs and an associated z value. The points in each set are assumed connected linearly. Examples are digitized contour data, and isohyetal map data.

(4) Polygons: The area is divided into a set of polygons such that every location falls into one polygon. Each polygon possesses a value that the variable is assumed to have for all locations within the polygon. The boundaries are defined by an ordered set of $\langle x, y \rangle$ pairs. Examples are the soil, the Thiessen polygons, etc.

(5) Cell grid: The area is divided into regular grid cells as, for example, a finite difference grid. Each cell contains a value and the variable is assumed to have value for all locations within the cell. An example is remotely sensed imagery.

(6) Triangular net: The area is divided into irregular triangles. The value of the variable is specified at each triangle vertex and is assumed to vary linearly

over the triangle. An example is the Triangular Irregular Network (TIN) model of elevation (Weibel and Heller, 1991).

Models (2) and (5) are frequently referred to as "raster" models and models (1), (3), (4), and (6) are "vector" models (Peuquet 1984, Goodchild 1993). Models (1), (2), and (3) yield the point definition of the value of the variable, whereas models (4), (5), and (6) explicitly yield the areal definition of the value of the variable. Thus, models (1), (2), and (3) must be supplemented by a spatial interpolation model to make them more general. In practice, model (6) is employed for elevation data. Models (1) and (3) are frequently encountered in hydrology. Models (4), (5), and (6) are for data storage and analysis. Sometimes the distinction between point samples and area averages is not important. Then, models (2) and (4) become indistinguishable. Through GIS it is possible to convert between different data models.

Object models characterize objects as points, lines, or areas. These are employed in cases where the objects are defined a priori, rather than as part of the modeling process. For example, the object, the Mississippi river, has an identity of its own which is independent of any discretization. In some implementations no distinction is made in the database between object and field models. For example, a set of lines may represent contours of piezometric head (field model) or geological formation (object model), each consisting of an ordered sets of $\langle x, y \rangle$ pairs and associated attributes.

Object models are commonly employed to represent man-made facilities, such as drainage pipes, channels, canals, highways, etc. which are naturally represented as linear objects. These models can capture aspect of human experience. For example, the concept of a "dam" may be important in building a database for agricultural and economic development forcing the database designer to represent it as a geographical object. At present, hydrologic data modeling is dominated by the field view and its spatially continuous variables, but the object view is clearly important in interpreting and reasoning geographical distributions.

Both field and object models are models of two dimensional variation. However, in hydrologic environments variables and parameters such as flow, elevation, width, roughness, slope, etc. may vary continuously over a watershed, and are well represented as homogeneous attributes of subwatersheds or channel reaches. These variables represent a class of geographical information continuously varying over the one dimensional space of a 2 dimensional watershed. The GIS technology needs to provide representation in this form.

1.4.3. APPLICATIONS

The most successful applications of GIS can be identified in four areas: (1) mapping, (2) data preprocessing, (3) modeling, and (4) policy formulation. GIS includes the capabilities of digital cartography in its input and output subsystem plus much more capabilities for storing and handling relationship between entities, and representing entities with multiple attributes. GIS can handle digital representation of continuous variables in numerous ways.

GIS can act as a data manager or data integrator. It includes facilities for input and output of data in different formats, ability to extract information in a user friendly window, scale and projection change, and resampling. In many applications, GIS is used to preprocess data, or make maps of input data or model outputs. However, models can be calibrated and run directly in GIS, using the GIS commands and language by matching the GIS data model with the hydrologic model. The data model must satisfy the needs of the hydrologic model.

The GIS technology has the capability to postprocess the model results, i.e., translate the results of distributed hydrologic modeling into a water resources policy. This translation may involve aggregation of results by administrative units, transformation of results consistent with social and economic data for comparison and correlation, displaying the results in a sensible manner, and / or development of alternative scenarios.

Goodchild (1993) lists the following most essential features the GIS should have to support environmental modeling: (1) Support for efficient methods of data input, (2) support for alternative data models, (3) ability to compute relationships between objects based on geometry and handle attributes of pairs of objects, (4) ability to carry out a range of standard geometric operations, (5) ability to generate new objects on request, (6) ability to assign new attributes to objects based on existing attributes and complex arithmetic and logical rules, and (7) support for transfer of data to and from analytic and modeling packages. Clearly it would be difficult to design a single GIS system integrating all of these support features and it may perhaps not be even desirable. However, it should provide appropriate linkages between different software components to act in union.

1.5. Hydrologic Modeling with GIS

Hydrologic models that lack a spatial component have no use for GIS. However, concern for resource management, and environmental quality requires application of distributed models. One characteristic of distributed-parameter models is that they are data intensive. Multiple data types such as hydrometeorology, topography, land use, soils, geology, streamflows, etc. are commonly required. Many of these data are often used to identify a hydrologic unit, because application of these models requires partitioning of the watershed into homogeneous units. This often proves to be cumbersome. Therefore, a spatial data analysis and manipulation tool is desirable and that tool is GIS. Furthermore, the use of computers in hydrologic modeling has become so widespread that the marriage between hydrology and GIS should be a logical step.

The early users of GIS in hydrology were municipalities and water distribution authorities and companies (Wallis, 1988). For example, the United Kingdom's Thames Water Authority (TWA) uses GIS for keeping an up-to-date inventory of the existing water supply network maintaining demographic and economic data-bases, planning, assessing flood and pollution damage, and real-time on-line fore-casting. The Connecticut Natural Resource Center in the United States is another example. Its database contains the statewide complete, updated set of the USGS digital line graph file for the USGS 7.5 minute quadrangle maps, and map layers displaying soil type, watershed boundary, 100 and 500 years flood plain zones. Sur-ficial and underground geology, land use, individual plot ownership, water quantity and quality information for both surface and groundwater, tax assessor's records growth and change etc.

Zhang et al. (1990) and Schultz (1993) presented overviews of hydrologic modeling with GIS, and DeVantier and Feldman (1993) reviewed GIS applications in hydrologic modeling, with particular reference to rainfall-runoff models, flood-plain management and forecasting, erosion prediction and control, water qual-ity prediction and control, and drainage utility implementation. Wallis (1988) re-viewed the GIS hydrology interface and commented on its future outlook. The in-tegration of GIS with hydrologic model is increasingly facilitating their design, calibration, modification, and comparison. A short review of hydrologic modeling with GIS is in order.

1.5.1. HYDROMETEOROLOGICAL FORECASTING

Shih (1990) applied satellite data and GIS for estimation of rainfall in Florida. Eleven convective clouds that encompassed 26 raingages were analyzed, where the cell size with raster format in the GIS was 48 km^2. He found high correlation between the rainfall estimated from satellite data and the rainfall observed on the ground and concluded that the satellite data would be useful in monitoring rainfall in an area having no raingage. Barrett (1993) presented an application of GIS in the Nile hydrometerorological forecast system in Egypt. A customized GIS was developed to show political boundaries, rivers, terrain elevation, and gaging net-work and was employed to develop hydrologic parameters for the basin and for multiple display features (maps and graphics).

1.5.2. STORMWATER MANAGEMENT

Meyer et al. (1993) coupled a GIS package with a physically based, spatially dis-tributed urban watershed model for storm-water management in Fort Collins, Col-orado. The GIS was used for preprocessing information for the model, and postpro-cessing the model output in order to examine consequences of various storm-water management strategies. Shea et al. (1993) integrated a GIS with hydrologic and hydraulic models for developing a drainage management plan for Polk County,. Florida. This area is undergoing rapid urbanization, resulting in rapid changes in land-use patterns. They examined different growth scenarios and their inpact. They were also able to identify areas where the growth would be detrimental to localized flooding, bridges, and other structures, and to local surface water quality.

1.5.3. WATERSHED MODELING

Perhaps the greatest application of GIS in hydrology has been in the area of watershed modeling. This is understandable because the computer models of watershed hydrology are highly data intensive and GIS is a natural technology to process volumes of data. Schumann (1993) discussed development of conceptual semi-distributed models and estimation of their parameters with the aid of GIS. Schultz (1994) reported on meso-scale modeling of runoff and water balances using remote sensing and other GIS data. Neumann et al. (1990) discussed a GIS as data base for distributed hydrological models and Wolfe (1992) presented a GIS-assisted input data set development for a finite-element storm hydrograph model. Wolfe noted that the input data set development procedure provided an efficient way for considering alternate land use and management scenarios in Florida. By integrating hydrologic modeling with GIS, Ross and Tara (1993) obtained increased detail in evaluation, minimized user subjectivity in parameter selection, and reduced costs of analysis due to significant time savings.

Warwick and Haness (1994) noted that the growing presence of detailed spatial information owing to GIS would assuredly foster continued development and implementation of more complex hydrologic models. In developing watershed models for two Sierra Nevada basins Jeton and Smith (1993) found that the GIS was an objective, efficient methodology for characterizing a watershed and for delineating hydrologic response units.

Bhaskar et al. (1992) employed GIS for estimating parameters of a hydrologic model WAHS (Singh and Aminian, 1985) for 12 watersheds in eastern Kentucky. In a similar vein, Mitchell et al. (1993) validated the agricultural nonpoint source (AGNPS) model by use of a GIS for predicting runoff and sediment delivery from small watersheds of mild topography. The integration of AGNPS with GIS facilitated development of input data, operation of the model, and interpretation of the results. Neumann and Schultz (1989) quantified the effect of catchment characteristics with use of GIS and a hydrological model. The influence of seasonally varying vegetation status on flood flows and the effect of land use changes over the years on runoff were investigated. Shih and Jordan (1993) assessed soil moisture by investigating Landsat data with GIS.

1.5.4. FLOOD PREDICTION

Chairat and Delleur (1993) coupled a GIS-GRASS with a physically-based hydrologic model TOPMODEL (Beven and Kirby, 1979) to predict peak flows and to quantify the effects of digital elevation model resolutions and contour lengths on the predicted peak flows. Muzik and Pomeroy (1990) coupled a GIS with a flood prediction software package based on the unit hydrograph theory for prediction of design flood hydrographs for selected watersheds in Alberta's Rocky Mountain foothills. In a similar vein, Hill et al. (1987) employed a GIS to prepare a computerized data base for flood prediction using WAHS (Singh and Aminian, 1985) in the Amite River basin in Louisiana.

1.5.5. GROUNDWATER MODELING

There is increasing trend for use of GIS in groundwater hydrology. Hudak et al.(1993) applied GIS to groundwater monitoring network design by examining spatial relationships between candidate sampling sites and aquifer zones suscept- ible to contamination. They demonstrated the role of GIS in a ranking approach to network design through application to a landfill. Hinaman (1993) employed a GIS to assemble input data sets for the USGS' MODFLOW (McDonald and Har- baugh,1988). He noted that the ability to change large sets of spatial data quickly and accurately with GIS enhanced the model-calibration process. In a similar vein, Orzol and McGrath (1993) modified MODFLOW so it could read and write files used by a GIS. Richards et al. (1993) integrated GIS with MODFLOW for ground- water resource assessment of the Floridian Aquifer in Santa Rosa County in Flor- ida. They used GIS for model calibration as well. Raaza et al. (1993) integrated GIS in groundwater applications using numerical modeling techniques to the Sand and Gravel Aquifer in Escambia County, Florida. The GIS integration of spatial topology, data attributes, and specialized tools provided an ideal environment for development of a digital model.

Flockhart et al. (1193) employed GIS for groundwater protection planning in Wellfleet and Hadley in Massachusetts, and Cortland County in New York. Baker (1993) used a GIS to develop an automated procedure for identifying the primary aquifers supplying groundwater to individual wells in eastern Arkansas. The Mis- sissippi River Valley alluvial aquifer was identified as the primary aquifer for 23,500 wells. Lackaff et al. (1993) developed and implemented a GIS-based con- taminant source inventory over the Spokane Aquifer in Spokane County, Wash- ington. Their study demonstrated the usefulness of GIS technology in the develop- ment, management, maintenance, and analysis of vast quantities of data associated with a local wellhead protection program. Along a similar line, Baker et al. (1993) coupled a GIS with the Rhode Island Department of Environmental Management (RIDEM) groundwater protection strategy to delineate wellhead protection areas and prepare their maps. These would then be used as the basis for wellhead pro- tection planning required under the Rhode Island protection program.

1.5.6. NON-POINT POLLUTION MODELING

Robinson and Ragan (1993) developed a GIS to support regional scale assessment of the spatial distribution of uncontrolled nonpoint source pollution, including ni- trogen, phosphorus, zinc, lead, BOD and sediment. The GIS was linked to a model developed by the Northern Virginia Planning District Commission to simulate the consequences of additional development or alternate management strategies for a 37000 acre area in the Montgomery County in Maryland. Along similar lines, Tim et al. (1992) coupled two water quality computer simulation models with a GIS to delineate critical areas of nonpoint source pollution at the watershed level. These models estimated soil erosion, sediment yield, and phosphorus loading from the

Nominal Creek watershed located in Westmoreland County, Virginia. Ventura and Kim (1993) employed GIS in the assessment of urban nonpoint source pollution and the development of a pollution control strategy. He et al. (1993) integrated GIS and a computer model to evaluate the impacts of agricultural runoff on water quality. Management scenarios were explored to minimize sedimentation and nutrient loading, and these included variations in crop cover, tillage methods, and other agricultural management practices. Heatwole and Shanholtz (1991) employed GIS to produce maps and corresponding tables of facilities ranked by animal waste pollution index for seven counties in Virginia. These rankings could be used to target pollution potential. Kalkoff (1993) determined the relation between stream quality and geology in the Roberts Creek watershed in Clayton County, Iowa, using a GIS. He found a weak but statistically significant relation between water temperature, pH, and nitrogen concentration in Roberts Creek and underlying geology during base flow conditions. De Roo et al. (1989) estimated soil erosion by coupling a GIS with the model ANSWERS (Beasley and Huggins, 1982).

1.5.7. WATER RESOURCES PLANNING

Leipnik et al. (1993) discussed issues related to implementation of GIS for water resources planning and management: initial decision to use a GIS; system selection, installation, and training, and up-to-date base development and production generation. They alerted to the critical choices to be made about each issue. McKinney et al. (1993) presented an expert GIS for Texas water planning, and showed its application to the existing water supply system for the Corpus Christi area in Texas. Battaglin et al. (1993) applied a GIS to assess water resources sensitivity to changes in climate in the Gunnison River basin. Juracek and Kenny (1993) presented an overview of the USGS' use of GIS technology to assist in the management and analysis of water use in Kansas. Griner (1993) developed a water supply protection model using the Southwest Florida Water Management District's GIS. He developed maps showing groundwater supply suitability, protection areas for surface-water supply, protection areas for major public supply wells, susceptibility to ground-water contamination, and recharge to the Floridian Aquifer. These were also combined into a final map to prioritize protection areas for water supply. Djokic and Maidment (1993) discussed the types of problems suitable for GIS-water network analyses.

1.6. Outlook for the Future

To reflect on the future of hydrologic modeling is nothing less than risky. Nevertheless. we share our opinion here. In the future, hydrologic modeling will be more global, more distributed, and more sophisticated. It will avail of the potential of remote sensing and satellite technology, GIS, computer graphics, and artificial intelligence. The models will be usable by those not trained in hydrology and interpretation of model results will not tax the ability of the user. They will be simple to use and explicitly state as to how good they are or where and under what conditions they should be used. The ultimate driving force for the models will be the user. The models will compete with each other and those which do not survive the competition will vanish. In a sense, the free market will determine the suitability of models. With such models, better informed decisions about the environment we live in will, hopefully, be made. The public will be better able to appreciate the role of hydrology in their lives and this, in turn, may further strengthen hydrology as a science and as a profession. At the moment, it is weak in both areas.

References

Baker, N.T., 1993. Utilization of a geographic information system to identify the primary aquifer providing ground-water to individual wells in eastern Arkansas. Water Resources Bulletin, Vol.29, N.3, pp.445-448.

Baker, C.P., Bradley, M.D. and Kaczor-Bobiak, S.M., 1993. Well head protection area delineation: linking flow model with GIS. Journal of Water Resources Planning and Management, Vol.119, N.2, pp.275-287.

Barrett, C.B, 1993. The development of the Nile hydrometeorological forecast system. Water Resources bulletin Vol.29 N.6 p.933.

Battaglin, W.A., Hay, L.E., Parker, R.S. and Leavesley, G.H., 1993. Application of a GIS for modeling the sensitivity of water resources to alterations in climate in the Gunnison River basin, Colorado. Water Resources Bulletin, Vol. 29, N.6, pp.1021- 1028.

Beasley, D.B. and Huggins, L.F., 1982. ANSWERS–user's manual. Department of Agricultulral Engineering, Purdue University, West Lafayette, Indiana.

Beven, K.J. and Kirby, M.J., 1979. A physically-based, variable contributoring area model of basin hydrology. Hydrological Sciences Journal, Vol. 24, N.1, pp.43-69.

Bhaskar, N.R., James, W.P. and Devulapalli, R.S., 1992. Hydrologic parameter estimation using geographic information system. Journal of Water Resuorces Planning and Management, Vol. 118, N.5, pp.492-512.

Chairat, S. and Delleur, J.W., 1993. Effects of the topographic index distribution on predicted runoff using GRASS. Water Resources Bulletin, Vol. 29, N.6, pp.1029-1034.

Coppock, J. T. and Rhind, D.W., 1991. The history of GIS. In Maguire, D. J., Goodchild, M.F. and Rhind, D.W., editors, Geographal Information Systems: Principles and Applications (2 volumes), London: Longman, and New York: Wiley, pp. 21-43.

Crawford, N.H. and Linsley, R.K., 1966. Digital simulation in hydrology: The Stanford Watershed Simulation Model IV. Technical Report No. 39, Department of Civil Engineering, Stanford University, Stanford, California.

DeRoo, A.P.J., Hazelhoff, L. amd Burrough, P.A., 1989. Soil erosion modeling using 'ANSWERS' and geographical information systems. Earth-Surface Processes and Land Forms, Vol.14, pp.517- 532.

DeVantier, B.A. and Feldman, A.D., 1993. Review of GIS applications in hydrologic modeling. Journal of Water Resources Planning and Management, Vol.119, N.2, pp. 246-261.

Djokic, D. and Maidment, D.R., 1993. Application of GIS network routines for water flow and transport. Journal of Water Resources Planiing and Management, Vol. 119, N.2, pp.229-245.

Flockhart, D.E., Sham , C.H. and Xiao, Y., 1993. Maximizing the value of information for groundwater protection: three test cases. Water Resources Bulletin, Vol.29, N.6, pp. 957- 964.

Goodchild, M.F. 1993. The State of GIS for environmental problem-solving. Chapter 2 in Enveronmental Modeling with GIS, edited by M.F.Goodchild, B.O. Parks, and L.T.Steyaert, Oxford University Press, 488 p., New Yok.

Griner, A.J., 1993. Development of a water supply protection model in a GIS. Water Resources Bulletin, Vol.29, N.6, pp. 973-979.

He, C., Riggs, J.F. and Tang, Y.T., 1993. Integration of geographic information systems and a computer model to evaluate impacts of agricultural runoff on water quality. Water Resources Bulletin, Vol.29, N.6, pp. 891-900.

Heatwole, C.D. and Shanholtz, V.O., 1991. Targeting animal waste pollution pootential using a geographic information sustem. Applied Engineering in Agriculture, Vol.7, pp.692-698.

Hill, J.M., Singh, V.P. and Aminian, H., 1987. A computerized data base for flood prediction modeling. Water Resources Bulletin, Vol.23, N.1 pp.21-27.

Hinaman, K.C., 1993. Use of a geographic information system to assemble input-data sets for a finite-difference model of ground-water flow. Water Resources Bulletin, Vol.29, N.3, pp. 401-448.

Hudak, P.F., Loaiciga, H.A. and Schoolmaster, F.A., 1993. Application of geographic information system to groundwater monitoring network design. Water Resources Bulletin, Vol.29, N.3, pp. 383-390.

Jeton, A.E. and Smith, J.L., 1993. Development of watershed models for two Sierra Nevada basins using a geographic information system . Water Resuorces Bulletin, Vol.29, N.6, pp.923- 932.

Juracek,K.E. and Kenny,J.F., 1993. Management and analysis of water-use data using a geographic information system. Water Resources Bulletin, Vol. 29, No. 6, pp. 973-979.

Kalkoff, S.J., 1993. Using a geographic information system to determine the relation between stream quality and geology in the Roberts Creek watershed, Clayton County, Iowa. Water Resources Bulletin, Vol.29, N.6, pp. 989-996.

Lackaff, B.B., Hunt, B.J. and Von Essen, I.E., 1993. The developement and implementation of a GIS-based contaminant source inventory over the Spokane Aquifer, Spokane County, Washington. Water Resources Bulletin, Vol.29, N.6, pp. 949-955.

Leipnik, M.R., Kemp, K.K. and Loaiciga, H.A., 1993. Implementation of GIS for water resources planning and management. Journal of Water Resources Planning and Management, Vol.119, N.2, pp.184-205.

McDonald, M.G. and Harbaugh, A.W., 1988. A modular three dimensional finite difference groundwater flow model. U.S. Geological Survey Techniques of Water Resources Investigations, Book 6, Chapter A1, 586 pp.

Mark, D.M. and Gould, M.D., 1991. Interacting with geographic information: A commentory. Photogrammetric Engineering and Remote Sensing, Vol.57, N,4, pp.1427-1430.

Meyer, S.P., Salem, T.H. and Labadie, J.W., 1993.Geographic information systems in urban storm.water management. Journal of Water Resources Planning and Management, Vol.119, N.2, pp.206-228.

McKinney, D.C. and Maidment, D.R., 1993. Expert geographic information system for Texas water planning. Journal of Water Resources Planning and Management, Vol.119, N.2, pp.170-183.

Mitchell, J.K., Engel, B.A., Srinivasan, R., and Wang, S.S.Y., 1993. Validation of AGNPS for small watersheds using an integrated AGNPS/GIS system. Water Resources Bulletin, Vol.29,N.5, pp.833-842.

Muzik, I. and Pommery, S.J., 1990. A geographic information system for prediction of design flood hydrographs. Canadian Journal of Civil Enginnering, Vol.17, pp.965-973.

Neumann, P., Fett, W. and Schultz, G.A., 1990. A geographic information system as data-base for distributed hydrological models. Proceedings, international Symposium on Remote Sensing and Water Resources.

Neumann, P.and Schultz, G.A., 1989. Hydrological effects of catchment characteristcs and land use changes determined by satellite imagery and GIS. IAHS Pubblication, N.186 , pp.169-176.

Orzol, L.L. and McGrath, T.S., 1993. Summary of modifications of the U.S. Geological Survey modular, finite- difference, ground-water flow model to read and write geographic information system files. Water Resources Bulletin, Vol.29, N.5, pp.843-846.

Peuquet, D.J., 1984. A conceptual framework and comparison of spatial data models. Cartographica, Vol.21, N.4, pp. 66-113.

Raaza, H., Raaza, R.M., and Wagner, J.R., 1993. Integrating geographic information systems in groundwater applications using numerical modeling techniques. Water Resources Bulletin, Vol.29, N.6, pp. 981-988.

Richards, C.J., Raaza, H. and Raaza, R.M., 1993. Integrating geographic information systems and MODEFLOW for groundwater resources assessments. Water Resources Bulletin, Vol.29, N.5, pp.847-853.

Robinson, K.J. and Ragan, R.M., 1993. Geographic information system based nonpoint pollution modeling. Water Resources Bulletin, Vol.29, N.6, pp. 1003-1008.

Ross, M.A. and Tara, P.D., 1993. Integrated hydrologic modeling with geographic information systems. Journal of Water Resuorces Planning and Management, Vol.119, N.2, pp.129-140.

Schultz, G.A., 1993. Application of GIS and remote sensing in hydrology. IAHS Pubblication, N. 211, pp.127-140

Schultz, G.A., 1994. Meso-scale modelling of runoff and water balances using remote sensing and other GIS data. Hydrological Sciences Journal, Vol. 39, N. 2, pp. 121-142.

Schumann, A.H., 1993. Development of conceptual semi- distributed hydrological models and estimation of their parameters with the aid of GIS. Hydrolocal Sciences Journal, Vol. 38, N. 6, pp.519-528.

Shea, C., Grayman, W., Darden, D., Males R.M., and Sushinsky, P., 1993. Integrated GIS and hydrologic modeling for countywide drainage study. Journal of Water Resources Planning and Management, Vol.119,N.2, pp.112-128

Sherman, L.K., 1932. Streamflow from rainfall by unit- graph method. Engineering News Record, Vol. 108, No. 4, pp. 501- 505.

Shih, S.F., 1989. Satellite data and geographic information system for rainfal estimation. Journal of Irrigation and Drainage Engineering, Vol.116, N.3, pp.319-331.

Shih, S.F. and Jordan, J.D., 1993. Use of landsat thermal- IR data and GIS in soil moisture assessment. Journal of Irrigation and Drainage Engineering, Vol.119, N.5, pp.868-879.

Singh, V.P., and Aminian, H., 1985. The watershed hydrology simulation model. Procceedings, International Workshop on Operational Applications of Mathematical Models (Surface Water) in Developing Countries, UNESCO, New Delhi, India.

Tim, U.S., Mostaghimi, S. and Shanholtz, V.O., 1992. Identification of critical nonpoint pollution source areas using geographic information systems and water quality modeling. Water Resources Bulletin, Vol.28, N.5, pp. 877-882.

Ventura, S.G. and Kim, K., 1993. Modeling urban nonpoint source pollution with a geographic information system. Water Resources Bulletin, Vol.29, N.2, pp. 189-198.

Wallis, J.R., 1988 the GIS/hydrology interface: The present and future. Environmental Software, Vol.3, N.4, pp. 171- 173.

Warwick, J.J. and Haness, S.J.,1994. Efficacy of ARC/INFO GIS application to hydrologic modeling. Journal of Water Resuorces Planning and Management, Vol.120,N.3, pp.366-381.

Weibel, R.and Heller, M.,1991. Digital terrain modeling. In Geographical Information Systems: Principles and Applications (2 Volumes), edited by D.J.Maiguire, M.F. Goodchild, and D.W. Rhind, Longman, London, pp. 269-297.

Wolfe, M.L.,1992. GIS-assisted input data set development for the finite element storm hydrograph model. Applied Engineering in Agriculture, Vol.8, N.2, pp.221-227.

Zhang, H., Haan, C.T. and Nofziger, D.L., 1990. Hydrologic modeling with GIS: An overview. Applied Engineering in Agriculture, ASAE, Vol.6, N.4 pp, 453-458.æ

CHAPTER 2

Integration of Remote Sensing and GIS for Hydrologic Studies

S.F. Shih

Abstract.   Recently, the importance of integration of remote sensing and geographic information system (GIS) to study hydrologic processes has been realized by water-resources workers. However, the interface between remotely sensed data and GIS is still weak and many problems must be solved before it becomes widely available. In the meantime, the public use of satellite data to manage water resources is still in its infancy, and more application techniques are urgently in need of development. Thus this chapter is separated into two major parts. The first part is to introduce general information about remote sensing systems, GIS, and the global positioning system (GPS). The second part is to exemplify successful applications of the integration of remote sensing and GIS in hydrologic studies such as land use/land cover classification, precipitation, soil moisture, evapotranspiration, water extent, groundwater, water quality, and runoff.

2.1. Introduction

The endless recirculation of water in the atmosphere-hydrosphere-lithosphere is known as the hydrologic cycle. The cycle can be studied according to the particular scale of reference i.e., global scale, basin scale, etc. From the hydrological point of view, the basin scale cycle is emphasized. The basin-scale cycle is considered to be a continuous circulation of water from water vapor to precipitation, stream flow, lakes, reservoirs, soil moisture, groundwater, and out of basin transfer, to the return to water vapor through evaporation and transpiration. Within a basin, the dynamics of the hydrologic processes are governed partially by the temporal and spatial characteristics of inputs and outputs and the land use/land cover conditions. Currently, little is known about the accuracy of the estimation of hydrologic parameters in a given area, primarily because of the wide variation of the distribution in

15

V. P. Singh and M. Fiorentino (eds.), Geographical Information Systems in Hydrology, 15–42.
© 1996 *Kluwer Academic Publishers. Printed in the Netherlands.*

space and time and the lack of information concerning the optimization of the conventional measurements. In general, the conventional methods used in hydraulic parameter estimation are expensive, time consuming, and difficult to update. While the conventional methods continue to be the principle technique at this moment to gather the point information at ground level, remote sensing could provide a more rapid and comprehensive overview of the distribution in a given area. However, satellite data have not been implemented widely by the public for managing earth resources. This could be due to the lack of knowledge and understanding of data applicability, availability, and accessibility. These matters all imply that the public use of satellite data to manage earth resources is still in its infancy, and that more techniques and programs for application, data access, publicity, and education are urgently needed.

Traditionally, the hydrologic data, map data, aerial photography, and tabular data describing tracts of land use/land cover have been stored separately, and updating the information has been cumbersome and expensive. Now, however, computer and remotely sensed data have given us the opportunity to merge data sets, and update the information by a low cost operation. In particular, a widely used technique called geographic information system (GIS) which can provide a means for merging spatial and attribute data into the computerized database systems allowing input, storage, retrieval, overlay, analysis, and tabulation of geographically referenced data.

Currently, however, the interface between remotely sensed data and GIS system is still weak, and many techniques must be developed before it becomes widely available for hydrologic studies. Thus, this chapter covers two major components. First, the existing remote sensing system which includes ground-based, airborne-based, and satellite-based sensors is introduced. Second, the successful applications of integration of the remotely sensed data and GIS for hydrologic studies are exemplified.

2.2. Remote Sensing System

Remote sensing is the gathering of information about an object without using an instrument in physical contact with the object. The human ears, eyes, and nose are typical natural sensors with limited capabilities. Today, based on scientific research and development, remote-sensing systems are being developed for ground-based, airborne-based, and satellite-based platforms.

2.2.1. GROUND-BASED SENSORS

2.2.1.1. *Radiometer*
Recent developments in solid-state sensors and microprocessors have led to radiometers that are hand-held and portable (Williams et al., 1984), and which can provide continuous reflectance information for the entire spectral range from visible light to the near-infrared wavelengths. Williams and Shih (1989) used a ra-

diometer to study spectral response differences among selected crops in the Everglades Agricultural Area of Florida.

2.2.1.2. *Infrared Thermometry*

The sun (average temperature = 6000 K) emits its maximum energy in the visible wavelength range (0.4–0.7 μm), while the earth (average temperature = 287 K) emits its maximum energy in the thermal-infrared wavelength range (8–14 μm). Therefore, a good way to sense temperature remotely is to use a device sensitive to thermal infrared wavelengths. Also, to avoid destructive action to the sensed surface, a passive infrared measuring device is best. An infrared thermometer, using a narrow field of view, measures radiation in a predetermined (commonly 7–16 μm) wavelength range. Using optics, radiation is then focused on an infrared detector, which converts the radiation measurement to an electrical signal. The signal is amplified and linearized to provide a voltage - temperature relationship. After further conditioning, the signal is converted to digital form and displayed. Using infrared thermometry to study soil moisture content has been conducted in sandy soil (Myhre and Shih, 1990a), organic soil (Jordan and Shih, 1993), pasture area (Shih et al., 1991), and alfalfa area (Idso et al., 1981).

2.2.1.3. *Lightning Detection System*

A state-of-the-art lightning detection network, the National Lightning Detection Network (NLDN), has operated with full coverage of the contiguous United States since 1989. The network provides lightning flash occurrence time, location, polarity, stroke count and signal amplitude in near real-time (Orville et al., 1983). Lightning data and rainfall relationship has been studied by several researchers (Piepgrass et al., 1982; Shih, 1988a; Tan and Shih, 1991a). However, for non-electrical researchers, the detailed occurrence time, polarity, signal amplitude, and multiplicity become burdensome and difficult to analyze. For example, a total of over 13.4 million ground flashes were recorded in the contiguous United States in 1989 by NLDN (Orville, 1991). Tremendous storage space is needed in order to archive the data. Although only a portion of the lightning data is collected by local users, the lightning information from the network still creates large data files. It becomes even more difficult when other related environmental factors (e.g., rainfall, land cover, land use) need to be analyzed with the data. Tan and Shih (1993a) found that GIS database management capability provided a convenient, yet powerful means for lightning data analysis.

2.2.1.4. *Weather Radar*

Gilman (1964) indicated that the ability to determine the areal distribution of precipitation intensity depends on the type of radar employed, and that the best all-purpose weather-search radar would have the following properties: (a) wavelength such that the effect of precipitation attenuation is minimized, (b) power and pulse length selected to ensure that the lowest significant amounts of precipitation are detected at maximum range, (c) correction for range, (d) beam width as narrow as possible, (e) antenna large enough to receive the weakest possible reflected energy. Miller and Thompson (1975) mentioned that the amount of backscatter from raindrops depends strongly on their size and the wavelength of the radio waves. For example, with a radio wavelength of close to 1 cm, raindrops and snow crystals having a diameter of 1 mm or more can be easily detected. The much smaller cloud droplets and ice crystals (diameter < 0.2 mm) scatter so little energy that only extremely powerful and sensitive radar systems can "see" them except at very short wavelengths (< 1 mm). But a sufficiently powerful radar system that generates and receives such extremely short wavelengths is very difficult to build.

2.2.1.5. *Ground-Penetrating Radar*

The ground-penetrating radar (GPR) system, which currently operates in the frequency range of 80 MHz to 1 GHz, has been designed specifically to penetrate earthen materials. It radiates repetitive, short-duration, electromagnetic pulses into the soil from a broad-bandwidth antenna. An impulse generator in the control unit transmits short pulses of energy into the ground from a transmitting antenna. When these pulses strike an interface having differing dielectric characteristics, a portion of the pulses' energy is reflected back to the receiving antenna. The receiving antenna amplifies these reflections and converts them into a wavelength of similar shape in the audio frequency range. Reflections are composite returns which have been averaged across that interface within the area of radiation. The variations of the reflected signals are displayed on a graphic recorder or are recorded on a tape recorder for future playback or processing. Towing the antenna along the ground produces a continuous profile of subsurface conditions on the graphic recorder. The GPR has been implemented for studying water-table depth (Shih et al., 1986), and salt-affected region assessment (Shih and Myhre, 1994).

2.2.2. AIRBORNE-BASED SENSORS

Aircraft, helicopter, drones (unmanned aircraft), and gliders are commonly used in airborne-based remote sensing systems. The airborne-based sensors, which have been frequently used in hydrologic studies, include aerial photographs, thermal scanner, airborne radar, and airborne multispectral scanner. Other sensors such as airborne lidar, airborne microwave system, airborne radiometer, and airborne gamma radiation system are also used occasionally.

2.2.2.1. *Aerial Photography*

Black-and-white film, color film, black-and-white infrared film, and color infrared film are commonly used in aerial photography. Aerial photograph techniques are commonly used for detection of a portion of the wavelength range from 0.3 to 0.9 μm. For instance, the sensitivity of wavelength range for main aerial photograph film are

- Black-and-white film: visible
- Color film: blue, green, and red
- Black-and-white infrared film: visible and near infrared
- Color infrared film: green, red, and near infrared

All aerial photograph films mentioned above require filters to remove unwanted radiation (haze). Recently, the color infrared photograph has been digitized as digital format data which can be used not only for simulating the satellite data appearance, but also for developing the imagery models to improve the satellite imagery interpretations. For instance, the green, red, and near-infrared digital data can be used to simulate the appearance of a "true color composite" which would be obtained using either Landsat TM (thematic mapper) bands 2, 3, and 4, or SPOT (Systeme Probatoire d'Observation de la Terre) (Systeme Probatoire d'Observation de la Terre) HRV (High Resolution Visible) bands 1, 2, and 3.

2.2.2.2. *Airborne Thermal Scanner*

Several airborne-based thermal scanners are currently in use. Most use wavelength ranges used from either 3-5 μm or 8-14 μm. The extremes and rates of temperature variation of a material are governed by the thermal conductivity, thermal capacity, and thermal inertia. Thermal inertia is an indicator of soil resistance to diurnal surface temperature fluctuation. It increases with an increase in material conductivity, thermal capacity, and density. In general, materials with high thermal inertia have a smaller range of diurnal temperature fluctuation than materials with low thermal inertia. For instance, water has a very high thermal capacity compared to other types of material. This implies that a soil with higher moisture content will have a smaller range of diurnal temperature fluctuation. Conversely, a dry soil will have a higher maximum (daytime) temperature than a wet soil. This theoretical relationship is in agreement with a study using Geostationary Operational Environmental Satellite (GOES) thermal infrared data (Shih and Chen, 1984) that reported a range of daytime temperature 2.4–3.2 °C higher in dry soil than in wet soil. However, it needs to be remembered that there are factors affecting the thermal imagery such as air temperature, wind, look angle, and time of measurement. Thermal imagery has been successfully used in many fields of hydrology such as assessing soil type and soil moisture, mapping land use/land cover, monitoring irrigation canal leaks, estimating evapotranspiration, and locating thermal plumes in lakes and rivers.

2.2.2.3. *Airborne Radar*

Side Looking Airborne Radar (SLAR) is commonly used in airborne radar systems. The SLAR has two advantages over optical photographic or scanning systems; i.e. it can produce imagery not only in the day and night, but also under both clear, and rainy and cloudy conditions. The common wavelength bands used in SLAR are K (8 to 24 mm), X (24 to 38 mm), C (38 to 75 mm), and L (150 to 300 mm). The spatial resolution of a radar system is mainly governed by the antenna size. A larger antenna will produce better spatial resolution. However, the main limitation of the SLAR system is the deterioration in resolution in the azimuth direction. Recently, the Synthetic Aperture Radar (SAR) has been designed to overcome the azimuth resolution limitation. The SLAR system has been used for mapping terrain and water resources including large rivers, unknown rivers, water supplies, timber inventory, canal systems, and sites suitable for agriculture.

2.2.2.4. *Airborne Multispectral Scanner*
The airborne-based Multispectral Scanner has been used increasingly because it has more channels covering the spectral range from 0.42 to 13 μm. Depending upon the flight height, the spatial resolution can be on the order of 1–2 m. The output of the MSS data is in digital format. These data can be used to improve the satellite imagery interpretation not only as mentioned in the aerial photography section but also by simulating the imagery which can be expected from future satellite systems due to it coverage of a wider spectral range than the aerial photograph.

2.2.3. SATELLITE-BASED SENSORS

Satellite-based sensor systems in water resources are primarily of two basic types.

2.2.3.1. *Earth-Resources Satellites*
Earth-resources satellites include the Landsat (USA), SPOT (French), ERS-1 (Earth Resources Satellite-1, European), JERS-1 (Japanese Earth Resources Satellite-1, Japan), IRS (Indian Remote Sensing Satellite, India), and Radarsat (Canada). The Landsat remote sensing system was first launched in 1972 and has been applied widely around the world for studying water resources. The type of sensor, band designation, spatial sensitivity range, and spatial resolution used in the Landsat system are listed in Table 2.1. As Table 2.1 shows, the existing Landsat system has two important sensors, i.e., TM and multispectral scanner (MSS). The Landsats 4 and 5 have 16-day repeat cycle with equatorial overpass time at 9:30 a.m.

Recently, the SPOT system has also been used. The SPOT has HRV imagery which includes three bands (0.5–0.6 μm, 0.6–0.7 μm, and 0.78–0.9 μm) with 20-m resolution color mode, and 10-m resolution panchromatic mode (0.51–0.73 μm). Nadir repeat coverage is every 26 days with equatorial overpass time at 10:30 a.m., but off-nadir repeat can be every 3–4 days.

The ERS-1, which has 10:30 a.m. equator crossing time with 3 day repeat cycle, contains active microwave instrument (AMI) and along track-scanning radiometer (ATSR). The AMI has C-band 5.3 GHz SAR with VV polarization at 30 m resolution. The scatterometer (wind mode) has 5 km resolution and the altimeter is Ku-band (12.5 GHz) with 10 cm precision for land and 40 cm for ocean. The ATSR has radiometer 3.7, 11 and 12 μm bands with 1 km resolution.

The JERS-1, which repeats coverage every 42 days, has an L-band (1.275 GHz) SAR system with HH polarization at 25 m resolution. The visible and near infrared radiometer (VNIR) contains four bands (0.45–0.52 μm, 0.52–0.60 μm,

TABLE 2.1. Comparisons of Type of Sensor, Band Designation, Spectral Sensitivity Range, and Spatial Resolution Used in Landsat Remote Sensing Systems.

Type of sensor (1)	Band (NASA designation) (2)	Spectral Sensitivity Range		Spatial Resolution (m) (5)
		Wavelength (μm) (3)	Color (4)	
(a) Landsat 1 and 2[a]				
RBV[b]	Band 1	0.475-0.575	Blue-green	76
	Band 2	0.580-0.680	Yellow-red	76
	Band 3	0.690-0.830	Red-infrared	76
MSS[c]	Band 4	0.5-0.6	Green	76
	Band 5	0.6-0.7	Red	76
	Band 6	0.7-0.8	Near infrared	76
	Band 7	0.8-1.1	Near infrared	76
(b) Landsat 3[a]				
RBV[b]	Two cameras	0.505-0.750	(Panchromatic)	40
MSS[c]	Band 4	0.5-0.6	Green	76
	Band 5	0.6-0.7	Red	76
	Band 6	0.7-0.8	Near infrared	76
	Band 7	0.8-1.1	Near infrared	76
	Band 8	10.4-12.6	Thermal infrared	234
(c) Landsat 4, 5, and 6[a]				
TM[d]	Band 1	0.45-0.52	Blue	30
	Band 2	0.53-0.61	Green	30
	Band 3	0.62-0.69	Red	30
	Band 4	0.78-0.91	Near infrared	30
	Band 5	1.57-1.78	Intermediate infrared	30
	Band 6	10.42-11.66	Thermal infrared	120
	Band 7	2.08-2.35	Mid infrared	30
MSS[c]	Band 1	0.5-0.6	Green	76
	Band 2	0.6-0.7	Red	76
	Band 3	0.7-0.8	Near infrared	76
	Band 4	0.8-1.1	Near infrared	76

[a] Launch dates: Landsat 1: 7/23/72, operation ended 1/6/78; Landsat 2: 1/22/75, operation ended 2/25/82; Landsat 3: 3/5/78, operation ended 3/31/83; Landsat 4: 7/16/82; Landsat 5: 3/1/84; Landsat 6: 10/5/1993, failed to achieve final orbit.
[b] Return beam vidicon camera.
[c] Multispectral scanner.
[d] Thematic mapper.

0.63–0.69 μm, and 0.76–0.95 μm), and the shortwave infrared (SWIR) includes four mid-infrared bands.

The IRS, which repeats every 22 days, contains a pushbroom linear imaging self-scanner (LISS) with four bands (0.45–0.52 μm, 0.52–0.62 μm, 0.65–0.69 μm, and 0.76–0.90 μm) at 72.5 m resolution (LISS-1) or 36.5 m resolutions (LISS-2).

The Radarsat which has 3 day repeat cycle includes SAR system (C- or L-band), 150 km swath, 25–30 m resolution, 4–100 km look angles. The main object-ives are for studying Arctic area, agriculture, forest, water resources management, and ocean. The Radarsat is planning to be launched in 1994.

2.2.3.2. *Weather Satellites*

Weather satellites include the NOAA- TIROS (National Oceanic and Atmospheric Administration - Television Infrared Observation Satellite) (USA), GOES (USA), Meteor (Russia), Meteosat (European Space Agency), and GMS (Geosynchron-ous Meteorological Satellite) (Japan). Most weather satellites contain both visible and thermal infrared sensors. The High Resolution Picture Transmission (HRPT) data from the Advanced Very High Resolution Radiometers (AVHRR) are primary source of TIROS satellite data. The Automatic Picture Transmission (APT) data, which have a lower spatial resolution (4 km x 4 km) than the HRPT (1 km x 1 km), also come from the AVHRR. But, the APT signal has a much lower transmission frequency (137.62 MHz) than that of HRPT (1707 MHz) and two channels of im-agery can be received simultaneously. The APT receiving system is relatively in-expensive to maintain, and can acquire usable image data (Xin and Shih, 1991). The equatorial crossing times are dependent upon the particular satellite. For in-stance, the descending equatorial crossing times are morning (7:30 a.m.) for even-numbered missions (i.e. NOAA-10 and NOAA-12), and nighttime (2:30 a.m.) for odd-numbered missions (but barely for NOAA-11, and no longer for NOAA-9). Since repeat coverage is every 12 hours, the ascending crossing times are evening (7:30 p.m.) for even-numbered missions. In other words, the NOAA-TIROS series can provide morning and afternoon coverage each day.

The GOES satellites, which are geostationary, contain two major instruments, i.e. VISSR (Visible and Infrared Spin Scan Radiometer) and VAS (VISSR At-mosphere Sounder). The VISSR has two bands (0.55–0.70 μm visible band at 2 km resolution, and 10.5–12.6 μm thermal infrared at 7 km resolution). The VAS contains 12 bands (0.55–0.70 μm visible band at 1 km resolution, and 3.9–14.7 μm thermal infrared bands at 7–14 km resolutions). This system covers Atlantic Ocean, Pacific Ocean, North America, and South America, with repeat imagery every half-hour. It was especially designed for storm-tracking, and supplies two-band weather-facsimile (WEFAX) images derived from VISSR data.

Meteor satellites have a 0.5–0.7 μm panchromatic band (resolution is 1–2 km for series 2, and 0.7–2 km for series 3) and a 8–12 μm thermal-infrared band (resol-ution is 8 km for series 2, and 3 km for series 3). Both series also carry atmospheric sensors. Coverage is repeated at least two times each day. Only APT type images are transmitted (panchromatic in day, thermal-infrared at night).

Meteosat satellites, which are geostationary, have three bands (i.e. 0.4–1.1 μm band at 2.5 km resolution, 5.7–7.1 μm band at 1 km resolution, and 10.5–12.5 μm band at 5 km resolution).

The GMS satellite, which is geostationary at 140° E, has a VISSR system with two bands. The resolutions are 1.25 km for panchromatic (0.5–0.75 μm) and 5 km for thermal-infrared. This satellite covers the western Pacific Ocean, Australia, and eastern Indian Ocean.

2.3. Image Processing System

Satellite sensed images provide data to the computer which are digitally compatible with computer software developed for interpreting physical meaning on the earth surface. A public domain image processing program used in the United States is the Earth Resources Laboratory Application Software (ELAS) (Beverley and Penton, 1989). ELAS is a software package designed and maintained by the National Aeronautical and Space Administration (NASA) to provide analysis and processing capabilities that enable the construction and manipulation of various geographic data files. The ELAS system has two major components: an operating subsystem is the main framework and supports overall operations, while the modules furnish the requirements for the accomplishment of various application purposes. A widely-used commercial domain image processing program is the Earth Resources Data Analysis System (ERDAS[1], 1991). Currently, several other image processing programs are also available in the market.

2.4. Geographic Information System

A GIS is a software system designed specifically to manage large volumes of geocoded spatial data derived from a variety of sources. The capabilities of a GIS system are to accept input data; to serve as a clearinghouse for data; to store, retrieve, manipulate, analyze, overlay and display these data based on the requirements of the user; and to create both tabular and cartographic output which reflect these requirements.

The data structure of a GIS can be based on either a raster or vector format. A raster format is a simple matrix of "x" and "y" coordinates with an appropriate relational database for each pixel. A vector format consists of lines (or arcs), points, and areas (or polygons). Lines are often used to represent roads, canals, etc., while areas are used to represent regions that are homogeneous for a single attribute. Points can be used to represent wells, sewer holes, etc. Database files are used to relate the geographic and attribute data. The attributes are artificial descriptors assigned by the user or others (i.e., soil type, land use, elevation, tax status, etc.).

---

[1] Trade names have been used to provide specific information. Their mention does not constitute endorsement by the author.

2.5. Global Positioning System

Global Positioning System (GPS) is a sophisticated, surveying system, based on reception of the signals from the GPS (U.S. Navy) satellites in conjunction with other positional differential correction data (for civilian users) to obtain location information. Under ideal conditions, GPS can provide reliable and accurate position and elevation information. August et al. (1994) indicated that there are many sources of errors that can degrade the quality of GPS-derived positional data. These include obstructions on the horizon, interference of satellite signals by forest canopy, atmospheric disturbances, selective availability, and reflection (multi-pathing) of satellite signals. Thus, users of GPS should be aware of fundamental accuracy problems and the sources of error that can degrade positional accuracy.

2.6. Interface Among Remote Sensing, GIS, and GPS

Satellite imagery can provide periodic information for a large area, and it is a very powerful tool for gathering regional information (Lillesand and Kiefer, 1987). The earth-resources satellites have been operated over 20 years, however, the public use of satellite imagery is still weak. There are many factors influencing public use and two major factors should be mentioned. The first factor is that an efficient system for storing, analyzing, and manipulating the satellite dataset was lacking until the GIS became available. The interface between remotely sensed data and GIS is still weak and many problems must be solved before it becomes widely available (Shih, 1988b). The second factor is that the satellite images are difficult to transfer accurately and rapidly to users. This problem is mainly due to the raw satellite imagery containing geometric distortions so significant that it cannot be directly overlain with base maps. Even with a geocorrection procedure, the results of rectified satellite imagery are still difficult to be located in the field. But, this problem has been overcome significantly by using the recently developed global positioning system.

The merging of remotely sensed data into GIS can be accomplished by three different methods (Curran, 1985). The first method converts all data (remotely sensed and otherwise) into polygon layers. These layers are then overlain for importation into the GIS. The second method keeps the remotely sensed data in a raster format. The data are converted into geo-referenced thematic maps and inserted with other necessary data into the geographically registered GIS database. The third method omits the conversion of remotely sensed data into thematic maps and inserts input data directly into the registered GIS. Systems that at present attempt to use one of these three methods to merge remotely sensed data and GIS include ODYSSEY, GIRAS, IBIS, and ARC/INFO-ERDAS (Lo, 1986), and (Shih, 1988b), etc.

There are several problems that exist when integrating remotely sensed data and GIS. One problem involves regions of high topographic relief. In this case, displacement of polygons from the remotely sensed data will occur and thus a digital terrain model must be employed (Goodenough, 1988). Another possible problem is distortion/displacement due to the conversion of raster satellite data to GIS vector data. Also, attributes associated with the GIS vector database present a problem. Thus, should one simply let the GIS determine the appropriate attribute assignments, or should the user manually determine these assignments through a given procedure? Another solution is to let the spatial attribute values at the center of the raster cell determine the characteristics. Barker (1988) illustrates this raster-to-vector and vector-to-raster problem.

Lam et al., (1987) analyzed the errors involved with vector-raster-vector transformations. They concluded that for their set of test areas, the maximum error increases as the raster grid cell size increases. However, this could be due to the fact the starting and ending vector area values were kept equal. One topic that needs to be studied further is raster-vector conversion error. One would suspect that the more irregular the perimeter the greater the error. Problems with raster-vector conversions include the "stair-step" look of the resulting polygon layer and small polygons which produce a "salt-and-pepper" appearance in the image (Bury, 1989). The resulting polygons do not have a "realistic" appearance. The processing required user interaction with the program to decide if any errors had occurred. The results of Bury (1989) were within National Map Accuracy Standards for 1:24,000 scale maps.

The overlaying of data sets with different scales presents a problem somewhat similar to that caused by raster-to-vector conversion. The similarity comes from the fact that distortions can occur, but the solution to this problem involves the choice of automatic or manual rectification. Most researchers have used a raster GIS with cell sizes ranging from 60 to 100 m. However, to overlay this data with vector data, a transformation must be performed.

Finally, the classification accuracy and positional accuracy of the remotely sensed data are often not comparable with those of other data in the GIS (Marble and Peuquet, 1983). Thus, for any calculations made within the GIS, the largest errors will be attributable to the remotely sensed data. However, the larger the study area, the less influence these errors will have on GIS calculations. Another concept that should encourage the merging of remote sensing with GIS is the use of multitemporal data (Xiao et al., 1989). Thus, satellite (or other remotely sensed) data from different dates would be used to increase the effectiveness of the GIS. For example, the GIS manager would then be able to overlay different "snapshots" of land use/land cover with static data (such as soil information) and dynamic data (such as cropping, zoning, and ownership).

2.7. Applications

Practical applications of the integration of remotely sensed data with GIS for hydrologic studies are exemplified in the areas of land use/land cover classification, precipitation, soil moisture, evapotranspiration, water extent, groundwater, water quality, and runoff.

2.7.1. LAND USE/LAND COVER CLASSIFICATION

Shih (1988b) used Landsat MSS data in conjunction with the unsupervised classification technique of ELAS to determine land use/land cover classifications in the Econlockhatchee (Econ) River Basin, Florida. The results showed that the scatter diagrams of band 4–band 7, and band 4– band 6 can be utilized as well as the traditional band 5–band 7 scatter diagram approach for classifying land use/land cover. Both the Zoom transfer scope and the GIS have been demonstrated to be very useful tools for land use/land cover classification of Landsat data. A 100 by 100-m cell size was used in that analysis. Another study by Tan and Shih (1988) was conducted by overlaying digitized aerial photographs and a classified Landsat image for assessing historical agricultural land use changes. Williams and Shih (1989) studied the land use change within the Florida Everglades using Landsat images captured 13 years apart. They achieved a Level I (Table 2.2) classification of the data and used the ELAS raster-based GIS to analyze the data. They discerned that agricultural development was moving southeast toward the water conservation area of the Everglades.

Jordan and Shih (1991) found, using a GIS database, that the irrigated land-use classification accuracy of Landsat TM image (0.20% error) was found have be quality superior to that of landsat MSS image (14.9% error).

The classification of SPOT imagery in conjunction with a GIS was used by Jadkowski and Ehlers (1989) to study the growth of an urban/suburban area. They used a raster-based GIS to overlay SPOT land use/land cover with zoning maps. Jampoler and Haack (1989) used aerial photography, and Landsat and SPOT imagery within a GIS framework to analyze land use change, deforestation, slope failures, and soil suitability. Their use of the remotely sensed data was particularly important due to a lack of other data sources. The study was conducted for the Kathmandu, Nepal area which lacks the historical databases available in most other countries. The ERDAS raster-based GIS with a 70-m grid cell size was used. One problem they encountered was in comparing historical land use (1972 and 1979) from aerial photography with recent land use (1987) from SPOT digital data.

Vegetation change detection using Landsat data was performed by Jakubauskas (1989). He used images from 1973 and 1982 to analyze the impact of a forest fire on the spatial extent of different tree species. The raster-based GIS used matrix manipulations to analyze the classified digital data for changes that occurred over time.

Wilkening (1989) used a Landsat/GIS merger for regional water resources management. Hydrologic sub-basins were manually determined and input into the GIS in vector format and then converted to raster format (60-m square cell size). This enabled the overlaying of the sub-basins with Landsat land cover data. Other

TABLE 2.2. Land use/land cover classification system for use with remote sensor data by the United States Geological Survey.

Level I		Level II
1. Urban or Built-Up Land	11	Residential
	12	Commercial and Services
	13	Industrial
	14	Transportation, Communications, and Utilities
	15	Industrial and Commercial Complexes
	16	Mixed Urban or Built-up Land
	17	Other Urban or Built-up Land
2. Agricultural Land	21	Cropland and Pasture
	22	Orchards, Groves, Vineyards, Nurseries, and Ornamental Horticultural Areas
	23	Confined Feeding Operations
	24	Other Agricultural Land
3. Rangeland	31	Herbaceous Rangeland
	32	Shrub and Brush Rangeland
	33	Mixed Rangeland
4. Forest Land	41	Deciduous Forest Land
	42	Evergreen Forest Land
	43	Mixed Forest Land
5. Water	51	Streams and Canals
	52	Lakes
	53	Reservoirs
	54	Bays and Estuaries
6. Wetland	61	Forested Wetland
	62	Nonforested Wetland
7. Barren Land	71	Dry Salt Flats
	72	Beaches
	73	Sandy Areas Other Than Beaches
	74	Bare Exposed Rock
	75	Strip Mines, Quarries, and Gravel Pits
	76	Transitional Areas
	77	Mixed Barren Land
8. Tundra	81	Shrub and Brush Tundra
	82	Herbaceous Tundra
	83	Bare Ground Tundra
	84	Wet Tundra
	85	Mixed Tundra
9. Perennial Snow or Ice	91	Perennial Snowfields
	92	Glaciers

data (soils, hydrologic, etc.) were also entered into the system and were used in water resource management.

One study (Johnston and Bonde, 1989) analyzed the boundaries between adjacent ecosystems using Landsat TM data (bands 2, 3, and 4). A normalized difference vegetation index (NDVI) was calculated within the GIS using brightness values from bands 2 and 4. A 3 × 3 pixel scan was used to enhance the difference among the vegetation classes. They concluded that the GIS helped in determining subtle differences between ecosystems.

Walsh et al., (1990) stressed the importance of using a digital elevation model (DEM) with Landsat TM within the GIS. This was performed to enhance the satellite data where topography is an important factor. They also used the GIS for spatial rectification of the Landsat data during image processing.

Rogers and Shih (1990) integrated the Landsat MSS with a GIS database data as a tool for an agricultural water usage permitting program. The results showed that the Landsat imagery can be used not only to assess the land-use changes and the areas for permitting the agricultural water usage, but also to update the permitted maps for future water permitted areas on a continual basis. Tan and Shih (1993b) studied the effect of citrus canopy on SPOT image responses. The results showed that the red (HRV band 2) and green (HRV band 1) bands are highly correlated with the citrus canopy. Therefore, it will be advantageous to integrate these two bands data with GIS in an agricultural land-use classification scheme, particularly in areas where citrus orchards are predominant.

2.7.2. PRECIPITATION

2.7.2.1. *Rainfall*

Over the last three decades several rainfall estimation techniques utilizing remotely sensed data have been developed. Most of these methods fall into one of six categories; i.e., cloud indexing approach, threshholding approach, life-history approach (Griffith, 1987; Scofield, 1987; Shih, 1990), pattern-classification approach (Weiss and Smith, 1987), integrating approach (Cheng and Shih, 1992), and microwave radiometry (Spencer et al., 1988). It should be noted that direct measurement of rainfall from satellite data for operational purposes has not been generally feasible. However, rainfall analysis can be improved by using both satellite and conventional ground-based data. Satellite data are most useful in providing information on the spatial distribution of potential rain producing clouds, while gage data are most useful for accurate point measurement.

Cloud-indexing approach: Using visible and thermal-infrared data to characterize the cloud type or temperature for correlation with rainfall via empirical relationships is the basic concept of the cloud-indexing approach. This approach is time independent, identifies different types of rain clouds, and estimates the rainfall from the number and duration of clouds or from their area.

Thresholding approach: This approach is based on temperature thresholding and cloud brightness to identify potential rain clouds. This approach considers all clouds with low upper-surface temperatures to be rain clouds.

Life-history approach: This approach is based on premises that significant rainfall comes mostly from convective clouds, and that convective clouds can be distinguished from other clouds in satellite images (Barrett and Martin, 1981). This approach is time-dependent and considers the rates of change in individual convective clouds or in clusters of convective clouds. Shih (1989, 1990) used GOES visible and thermal infrared data to estimate rainfall. The Scofield-Oliver method of rainfall estimation was used within the GIS environment. The estimated rainfall based on the satellite data was well correlated with raingage rainfall data ($r^2 =$ 0.81).

Pattern-classification approach: This approach uses a statistical pattern-classification technique to assign rainfall to one of several classes according to such parameters as coldest cloud-top temperature, and average cloud-top temperature difference between two consecutive images.

Integrating approach: This approach correlates a small-coverage, high-resolution sensor with a large-coverage, low-resolution sensor, and then uses the large-coverage sensor to estimate the regional rainfall in a given area which was not covered by the small-coverage sensor. The Cokriging technique has been used with a combination of ground measurements and satellite visible and thermal data to estimate spatial rainfall.

Microwave radiometry: Passive Microwave techniques offer a great potential for measuring rainfall because at some microwave frequencies clouds are essentially transparent, and the measured microwave radiation is directly related to the presence of rain drops themselves. Microwave radiometry or passive microwave techniques react to rain in two fundamental ways: (1) by emission/absorption, and (2) by scattering.

Radar: Several methods of using radar to estimate rainfall have been summarized by Doviak (1983). Over 80 articles which used different calibration parameters for the reflectivity method in radar rainfall estimation were summarized by Shih (1992). However, the Next Generation Radar (NEXRAD) system offers a great potential in remote sensing of rainfall estimation (Klazura and Imy, 1993).

Lightning flashes: The GIS database of lightning data from NLDN has been found useful for locating heavy rain regions, and for estimating rainfall rates (Shih, 1988c; Tan and Shih, 1991a).

2.7.2.2. *Snow Hydrology*
Remote sensing offers a new and valuable tool for obtaining snowfall data for predicting snow melt runoff. The reader is referred to the recent book published by Hall and Martinec (1985) for more in-depth coverage.

Gamma radiation method: This method takes advantage of the natural emission of low level gamma radiation from the soil. Aircraft passes over the same flight line before and after snow cover measure the attenuation resulting from the snow layer which is empirically related to an average snow water equivalent for that site (Carroll and Vadnais, 1980).

Visible/Near infrared method: Snow can readily be identified and mapped with the visible bands of satellite imagery because of its high reflectance in comparison to non-snow covered areas. Generally this means selecting the NOAA AVHRR visible band, Landsat MSS bands 4 and 5, SPOT HRV bands 1 and 2, or Landsat TM bands 2 and 3. The contrast between clouds and snow is greater in the near infrared and this serves as a useful discriminator between clouds and snow (Dozier, 1984).

Thermal infrared method: Thermal data can be useful for helping identify snow/non-snow boundaries and discriminating between clouds and snow with NOAA AVHRR data. Ferris and Congalton (1989) used NOAA AVHRR satellite data to estimate snowpack water volume. The estimate of water volume was based on the near infrared, thermal infrared, and aspect factor for each pixel. These data were used as layers within a raster-based GIS, which then calculated the snowmelt volumes.

2.7.3. SOIL MOISTURE

Remote Sensing in visible/near infrared, mid infrared, thermal infrared, and microwave wavelengths are briefly described as follows for soil moisture estimation.

2.7.3.1. *Visible/Near Infrared Method*
In general, wet surfaces have less reflectance values in both visible and near infrared than do dry surfaces. However, the reflected solar radiation is not only dependent upon the soil moisture condition, but also on confusing factors such as organic matter, structure, roughness, texture, mineral content, illumination geometry, angle of incidence, color, and plant cover (Engman and Gurney 1991). In other words, it is possible to develop a unique relationship between spectral reflectance and soil moisture for a specific site where the confusion factors are known.

2.7.3.2. *Mid-Infrared Method*

Mid-infrared (MIR) is inversely related to the surface soil moisture content (Bowers and Hanks, 1965). Shih and Jordan (1992) integrated the MIR data of the Landsat TM band 7 with GIS to assess regional soil moisture conditions. They found that the agricultural/irrigated and urban/clearings land-use categories had generally higher MIR response than did the forest/wetland land use, while water had the lowest MIR response. The MIR response was found to be inversely related to the qualitative soil moisture. High resolution (30 m) assessment of regional soil moisture conditions can be accomplished with Landsat TM MIR imagery. Such assessment would result in more efficient direction of ground-based investigators to areas with the most extreme (wet or dry) conditions. They also recommended that further work be conducted to improve the Landsat TM MIR based soil-moisture assessment technique. The first improvement would be to calibrate TM MIR planetary albedo to surface albedo through the use of atmospheric data and ground MIR radiometric measurements. The second improvement would be the quantification of soil-type and vegetation effects.

2.7.3.3. *Thermal Infrared Method*

Thermal-infrared (TIR) method is based on the fact that soil surface temperature is primarily dependent on the thermal inertia of the soil. The thermal inertia, in turn, is dependent upon both the thermal conductivity and volumetric heat capacity. The inertia is an indication of soil resistance to the diurnal surface temperature fluctuation. A soil with a high thermal inertia (due to a high soil moisture content) will have a lower diurnal range of surface temperature (Schmugge et al., 1980). Shih and Jordan (1993) indicated that integration of the lands surface temperature data from the Landsat TM TIR imagery (band 6) with GIS can provide useful information for periodic (8 to 16 days) monitoring of the spatial distribution of soil moisture conditions. They summarized three important issues. First, the daytime surface temperature is inversely related to the soil moisture content. Second, Landsat TM TIR data have significant potential for the detection of the soil moisture condition of various land-use categories. The results might be implemented in such ways as land-use evaluation, land-use planning, preplanting condition, soil moisture mapping, drought area assessment, and drainage zone identification. Third, further work was recommended in order to refine the Landsat TM TIR based soil moisture assessment technique. The normalization of the relation between surface temperature and soil moisture to account for different soil types and seasonal sunlight conditions, the calibration of TM radiant temperatures to surface temperatures, and the quantification of vegetation damping of maximum surface-temperature were recommended for investigation.

2.7.3.4. *Microwave Sensing Method*

Both passive (radiometric) and active (radar) microwave systems can be used to measure soil moisture (Schmugge et al., 1980). The theoretical basis of microwave sensing methods consists of the dielectric properties of a soil, which are highly correlated with the soil moisture content.

2.7.4. EVAPOTRANSPIRATION

Remote sensing techniques cannot measure evaporation or evapotranspiration (ET) directly. However, remotely sensed data offer methods for extending empirical relationships based on either vegetation mapping or climatic factors (such as temperature and solar radiation) which are used in evapotranspiration estimation. For instance, Gervins and Shih (1981) used Landsat MSS data to map the littoral zone vegetation of Lake Okeechobee, Florida. By adjusting the effective surface area according to the satellite-derived vegetation area, they were better able to account for ET and improve the total water balance. Furthermore, water surface temperature has been measured using APT thermal infrared data from the NOAA-TIROS AVHRR to estimate lake evaporation (Xin and Shih 1991). The correlation of pan evaporation with water surface temperature and solar radiation was studied. A regression model of evaporation based on satellite-derived water temperature, ground-measured water temperature, and solar radiation was developed. The results indicated that pan evaporation was highly correlated with temperature and solar radiation, and that the satellite-derived lake temperatures had a better correlation to the pan evaporation than did the ground-point measurements, especially for afternoon situations. This implies that the AVHRR APT data can be used to assess lake surface temperature and to estimate lake evaporation.

Estimates of the net radiation from geostationary satellite data were used by Heilman et al. (1977) to estimate ET. The resulting moisture flux was then used to develop a water balance model for predicting crop yields.

Estimation of regional scale ET through analysis of remotely sensed temperature data has been studied by several researchers (Soer, 1980; Price, 1982; Gurney and Camillo, 1984; Sucksdorff and Ottle, 1990; and Sandholt and Anderson, 1993). Several important parameters in ET models such as the solar radiation, surface albedo, vegetation condition, and/or surface temperature can be inferred using Landsat, SPOT, NOAA-AVHRR, and ERS-1 satellite imagery.

2.7.5. WATER EXTENT

The water extent, which includes surface water storage parameters such as water surface area, water volume, and water depth, can be determined by remote sensing techniques.

Water surface: Since land/water contrast is very strong in the near infrared band, Shih (1985) used Landsat MSS data to delineate the water surface area from the surrounding land. Shih indicated that both techniques of density slicing from band 7 (near infrared) and ELAS classification with bands 5 and 7 combination can successfully assess the water-surface area. The deviation of the surface-area assessment between two techniques was within 3%.

Water volume: Shih (1982a) developed a technique using a number of Landsat MSS data covering lakes of interest which were selected to correspond with a wide range of lake stages. After an average water surface area between two stages was measured from the Landsat data, the change in lake volume was estimated from the change in the recorded stages multiplied by the corresponding average satellite-derived water-surface area.

Water depth: Kolipinski and Higer (1969) used Landsat MSS data to differentiate water depth ranges. They compared the MSS data of an unmapped area with the spatial signatures of local control points at known depths and were able to identify depths between 0–1 m, 1–1.5 m, and 1.5- -5 m. Shih (1982b) found the Landsat MSS data to be very useful for determining the water depth in Conservation Area 3A, Florida. The multiple correlation coefficient between Landsat MSS four bands combination and water depth were all higher than 0.83 in all four days of study. Pesnell and Brown (1977) studied the littoral zone of Lake Okeechobee, Florida, and found that the major vegetation associations are linked to elevation through the hydroperiod. This concept has been implemented by Gervin and Shih (1981) to produce the littoral zone vegetation map which was derived from the landsat MSS data. The water depth is estimated based on the deviation between the recorded lake water stage and the satellite-derived ground elevation.

2.7.6. GROUNDWATER

In general, deep groundwater aquifers can not be detected directly by remote sensing techniques. However, aerial photograph interpretation has been implemented to eliminate areas of potential low-water-bearing state (Engman and Gurney, 1991), to assess abandoned flowing artesian wells (Jordan and Shih, 1988a), and to estimate artesian well flow rates (Jordan and Shih, 1990). Satellite imagery enables hydrologists to infer the aquifer location from surface features, and to find regions with a high potential for containing well sites (Jordan and Shih, 1991). Airborne radar can be used to show surface features even under dense vegetation canopies and is especially valuable for revealing topographic relief and roughness (Engman and Gurney, 1991). Ground-penetrating radar can be used to detect water-table depth (Shih et al., 1986).

2.7.6.1. *Groundwater Exploration*
Engman and Gurney (1991) indicated that satellite imagery can be most effectively used for regional exploration, and summarized the results of Landsat and SPOT image interpretations as: (1) less need for field work and slower, more expensive exploration methods; (2) identification of promising areas for more detailed study

and ground exploration; (3) new or better geologic and hydrologic information; (4) the perspective of large areal coverage available from satellite imagery, which may be unavailable from other means of exploration. Both SAR and SLAR data have a great potential for groundwater exploration, especially in arid and hyperarid regions.

2.7.6.2. *Water-Table Investigation*

Conventional water-table depth observation is reliable and provides information at a specific location. However, most investigations require groundwater inform- ation over the entire region of interest, with data indicating both depth and dis- tribution of water table. The interpretation of the regional water- table informa- tion must be inferred from a limited number of widely spaced observation wells. Errors often arise as a result of the incompleteness of sampling. Alternative ap- proaches using remote sensing techniques are available. In earthen materials, the GPR has provided continuous data of subsurface conditions from depths of less than 1 m to more than 30 m (Shih et al., 1986). Landsat MSS data bands 4, 5 and 7 have also been used in conjunction with aerial photographs to assess perched wa- ter tables (Rector, 1982). Furthermore, aerial photograph interpretation and satel- lite data analysis can provide the location of aquifers from surface features which should precede ground surveys and fieldwork (Engman and Gurney, 1991). Tem- perature difference techniques, as mentioned in the soil moisture section, can also be used to infer or identify shallow groundwater and springs or seeps.

2.7.6.3. *Abandoned Well Monitoring*

The continuous flow of saline water from thousands of deep artesian wells presents a water management problem. For instance, a 1978 estimate by the United States Geological Survey numbers these flowing wells in Florida alone at 15,000. Due to the saline water, the well casings have corroded, and leakage into shallow non- saline (via overland flow and underground seepage) aquifers has occurred. To im- plement well plugging programs as mandated by state legislation, Florida water management districts required a faster alternative to conventional ground surveys for monitoring the wells. However, there are two major problems in monitoring wells. First, location of flowing wells is difficult because many are hidden on aban- doned agricultural land, which has reverted to thick brush and subtropical forest. Second, conventional methods of identifying the location of wells and storing geo- graphic and corresponding attribute data are cumbersome and slow. Consequently, the feasibility of using remote sensing, GIS, and GPS for monitoring abandoned wells has been studied by several researchers. For instance, color infrared photo- graphy was used to analyze the spectral reflectance of land surface features that are associated with flowing wells (Shih and Jordan, 1990). Well-site soil was found to have a higher spectral reflectance than soil not associated with a well. Well water had higher spectral reflectance than natural pond and ditch water. The combina- tion of red and near-infrared spectral reflectance recorded on color infrared photography appears to be useful for distinguishing flowing well sites. In practical application, individual well-sites were first located using aerial color infrared (ACIR) photography interpretation, then verified through spot microdensitometer analysis of the ACIR film (Jordan and Shih, 1991). Sites that appear to have wells from remotely sensed data have been located on the ground using either a LORAN-C or GPS (Jordan and Shih, 1988b).

Jordan and Shih (1991) developed a well number estimation technique which was based on a well index, which represented the number of wells to be expected within an area under irrigated land-use. Statistical classification of satellite images, including Landsat MSS and TM, was successfully used to distinguish areas under irrigation within a south Florida region. The TM image was found to have an irrigated land-use classification accuracy much better than that of the MSS image. This well number estimation technique was determined to be practical for regional-scale application to flowing artesian well assessment. This implies that a method can be developed to find regions with a high potential for containing well-sites via satellite imagery.

The geographic location and attribute data of each well were successfully stored in a GIS database for monitoring and management of the abandoned wells (Tan and Shih, 1990; Myhre and Shih, 1990b). GIS has also been used in a well permitting program (Myhre and Shih, 1991), and in unused-well differentiation (Tan and Shih, 1991b). The newly developed method of using remotely sensed data, GIS, and GPS for monitoring abandoned wells is being used by Florida water management districts as a faster alternative to conventional methods.

2.7.7. WATER QUALITY

Remote sensing has an important role in water quality evaluation and management strategy. Sources of pollution from pipes or open channels are often easy to identify; non-point source pollution can perhaps be evaluated best by remote sensing.

2.7.7.1. *Determining Water Quality*

Remote sensing is mostly limited to surface measurement of turbidity, suspended sediment, chlorophyll, eutrophication, and temperature which can be used as indicators of more specific pollution problems (Shih and Engman, 1992). Remote sensing data have been used with field data to develop empirical water quality models (Shih and Gervin, 1980; Khorram, 1985). Blanchard and Leamer (1973) made spatial reflectance measurement of water from various sources containing suspended sediment and found that peak sediment reflectance generally occurred around 0.57 μm which corresponds to Landsat MSS band 5 and TM band 2, and SPOT HRV band 1. Khorram et al. (1987) refined the chlorophyll a relationship by suggesting different wavelength ratios for different concentration levels.

2.7.7.2. *Erosion and Non-Point Source Pollution*

Non-point source pollution is closely related to land use/land cover so that when coupled with soils and topography the potential water quality in runoff can be described. Several groups have used Landsat data for land cover input to their non-point source models (Schecter, 1976; Ragan and Rogers, 1978). The Ohio-Kentucky-Indiana (OKI) model estimates the erosion and sediment contribution from sub-areas. This approach is based on the evaluation of the Universal Soil-Loss Equation (Wischmeier and Smith, 1978).

Another study by Pelletier (1985) described a GIS adaptation to the Universal Soil-Loss Equation that used data from either the Landsat MSS (76 m resolution) or the Landsat TM (30 m resolution) to determine the land cover, and subsequently the cropping management factor, C. Pelletier (1985) also provided a good discussion of some of the accuracy limitations that may be encountered by automating the procedure with satellite and digital elevation data.

2.7.7.3. *Wetland Assessment*

Wetlands are of interest to water resource management as a natural vegetation filter for improving water quality. Landsat data have been used to make estimates of wetland water volumes on a monthly basis by combining depth/stage relationships with surface water area (Higer et al., 1974). Still and Shih (1991) used Landsat MSS data to classify spectrally unique wetland areas within the Econ Basin, Florida. The results showed that the scatter diagram of band 4 and band 5 can be used to identify most efficiently the wetland areas within the Econ Basin. Both the zoom transfer scope and the GIS were demonstrated to be very useful tools for wetland assessment from Landsat data. An additional aspect of wetland management includes the scope of dredging, waste lagoons, drainage, water depth control, and other man- induced changes that have an effect on the natural environment (Engman and Gurney, 1991). Remote sensing is well suited to monitor these changes and to make preliminary estimates of the environmental impact. The temporal aspect of Landsat, SPOT, NOAA-AVHRR, and ERS-1 satellite data allows changes to be observed over time and in some cases for pre-development baseline data to be obtained from early scenes.

Lillesand et al. (1989) studied multi-temporal vegetation, vegetation/soil correlation, groundwater mapping, lake level changes, and water quality parameters of different lakes. Land cover analysis, lake size, and water quality parameters were determined from Landsat TM data. ERDAS and GIS were used to establish a database.

2.7.8. RUNOFF

Runoff cannot be directly measured by remote sensing techniques. The role of remote sensing in runoff prediction is generally to provide a source of input data or as an aid to estimating equation coefficients and model parameters. There are three general areas where remote sensing data have currently been used as input data for computing runoff. First, these data are very useful for obtaining information on watershed geometry, drainage networks, and other map-type information (Haralick et al., 1985). Second, remote sensors produce input data for a class of empirical flood peak, annual runoff, or low flow equations (Allord and Scarpace, 1981). These first two applications use satellite imagery in the same way that aerial photography has been used. Third, runoff models that are based on a land use component have been modified to use land use classes from image interpretation of multispectral data (Ragan and Jackson, 1980; Rango et al., 1983; Still and Shih, 1985).

Land use/land cover information is important to the runoff process that affects infiltration, erosion, and ET. To date, most of the work on adapting remote sensing to runoff modeling has been with the Soil Conservation Service (SCS) runoff curve number (RCN) model. The RCN is an index of runoff potential and is a function of soil type, the land use condition, and the antecedent soil moisture. Recently, this method has been used as an index to assess the land-use change effect on basinwide runoff for three reasons. First, the soil type within a basin does not effectively change with time. Second, the antecedent soil moisture affected by weather conditions is assumed to be stable for long-term average conditions. Third, the basinwide runoff is a summation of subbasin runoffs and the maximum potential difference between rainfall and runoff is linearly related to the RCN. Therefore, the basinwide runoff index could be estimated from a weighted RCN which can be obtained from either the overall land-use classification or from the subbasins. Thus, the prime variable in the basinwide runoff index estimation is land-use change with time which can be obtained from the Landsat, SPOT, NOAA-AVHRR, and ERS-1 satellite data. Examples of integration of satellite data with GIS for land-use classification have been reported by Shih (1988b).

Estimation of runoff volume using Landsat images and aerial photographs was conducted by Stuebe and Johnston (1990). They used Landsat images to obtain vegetative cover and aerial photography to determine land use (both at a resolution of 100 meters). Comparisons of runoff volume calculations using the GIS and those from a manual method showed differences between 0.9% to 32.8%. Development of a watershed management model was performed by Oslin et al. (1988). Their model (called STREAMS) incorporates remote sensing and GIS. The data are used in conjunction with hydrologic, erosion, and sediment transport models. The remotely sensed data (Landsat TM and SPOT) provided land cover information. ERDAS was used for processing the satellite digital data. The models HEC-1, HEC-2, and HEC-6 were used after the data were processed in the raster-based GIS.

The calculation of RCN was the focus of a study by Berich and Smith (1985). Two Landsat scenes taken 10 years apart were used as the basis for land cover analysis. ERDAS software was used to resample the land cover data (originally 60 by 80 m) to 61 by 61 m. This was done in order to match the soil-type and zoning data which were also used. The GIS in that study was IRIS, which is a raster format system. Thus, each cell in GIS was set to correspond with the resolution of the soils/zoning data.

2.8. The Future

The Landsat 7 program is a joint Department of Defense (DOD)/NASA venture for maintaining "continuity of Landsat-type data", as well as for fulfillment of national security needs. The Landsat 7 will have an Enhanced Thematic Mapper Plus (ETM+) which is similar to the Landsat 5 TM system except that it will add a 15 m resolution panchromatic band and will change the thermal band resolution to 60 m. Landsat 7 will also have a new sensor called High Resolution Multispectral Stereo Imager (HRMSI). But, the Landsat program could suffer a drawback due to the loss of the Landsat 6 which was launched on October 5, 1993, and failed to achieve final orbit. Williams (1994) mentioned that this accident, as well as more recent FY '95 budget constraints projected for both NASA and DOD, could affect the Landsat 7 program which is proposed to be launched in March, 1998. Williams (1994) proposed four possible descope options for Landsat 7 instrumentation and some ground data processing capabilities. Option 1 suggests deletion of band 6 (thermal band), partially because of the Japanese-built ASTER instrument, which will fly on the first Earth Observing System morning equator-crossing platform (EOS AM1) in mid-1998. The EOS AM1 platform will have five 90 m thermal bands covering a 90 km swath width. Option 2 suggests the additional deletion of TM band 7 (2.08- 2.35 μm), partially because the EOS AM1 will have six 30 m bands in the short wave infrared (SWIR) region. Option 3 suggests replacement of a 15 m panchromatic band with a 15 m TM band 4 (0.78–0.91 μm). Option 4 would delete 15 m spatial resolution capability altogether, but it preserves TM bands 1, 2, 3, 4, and 5 with 185 km swath width and 16 day repeat coverage, calibration, etc.

However, current Landsat 5 TM data precision (8-bit) appears quite adequate for hydrologic studies and the 30 m resolution allows easy picking of ground-control points for geo-correction. Market price has been relatively high in recent years. EOSAT Co. customer service has been excellent for TM data. High image quality has been the hallmark of the Landsat program to date. Hopefully refinement, rather than diminution, will be implemented on future Landsat sensors.

References

Allord, G. J., and F. L. Scarpace. (1981) 'Improving streamflow estimates through use of Landsat.' In Satellite Hydrology, M. Deutsch, D. R. Wiesnet, A. Rango, eds., Am. Water Resour. Assoc., pp. 284-291.

August, P., J. Michaud, C. Labash, and C. Smith. (1994). 'GPS for environmental applications: accuracy and precision of locational data.' Photogram. Engrg. Remote Sens., 60(1):41-45.

Barker, G. R. (1988). 'Remote sensing: the unheralded component of geographic information systems.' Photogram. Engrg. Remote Sens., 54:195-199.

Barrett, E. C., and D. W. Martin. (1981). The Use of Satellite Data in Rainfall Monitoring. Academic Press, London, 340 pp.

Berich, R. H., and M. B. Smith. (1985). 'Landsat and micro-GIS for watershed modeling.' Hydraulics and Hydrology in the Small Computer Age, Am. Soc. Civil Eng. (ASCE), pp 668-673.

Beverley, A. M., and P. G. Penton. (1989). 'ELAS science and technology laboratory applications software.' Report No. 183, NASA, John C. Stennis Space Center, Sci. and Technol. Lab., Mississippi.

Blanchard, B. J., and R. W. Leamer. (1973). 'Spatial reflectance of water containing suspended sediment.' In Remote Sensing and Water Resources Management, Am. Water Resour. Assoc., pp. 339-348.

Bowers, S. A., and R. J. Hanks. (1965). 'Reflection of radiant energy from soils.' Soil Sci. 100(2):130-138.

Bury, A. S. (1989). 'Raster to vector conversion: a methodology.' Proc. GIS/LIS 1989, Am. Cong. Surv. Map., pp 9-11.

Carroll, T. R., and K. G. Vadnais. (1980). 'Operational airborne measurement of snow water equivalent using natural terrestrial gamma radiation.' Proc. 48th Annual Western Snow Conf., Laramie, WY, pp. 97- 106.

Cheng, K. S., and S. F. Shih. (1992). 'Rainfall area identification using GOES satellite data.' J. Irrig. and Drain. Engrg., ASCE, 118(1):179- 190.

Curran, P. J. (1985). Principles of Remote Sensing. Longman Inc., New York, NY.

Doviak, R. J. (1983). 'A survey of radar rain measurement techniques.' J. Climate, Appl. Meteorol., 22:832-849.

Dozier, J. (1984). 'Snow reflectance from Landsat-4 thematic mapper.' IEEE Trans. Geosci. and Remote Sens., GE-22(3):323-328.

Engman, E. T., and R. J. Gurney. (1991). Remote Sensing in Hydrology. Chapman and Hall, London, 225 pp.

ERDAS. (1991). Earth Resources Data Analysis System. Field Guide, ERDAS Inc., Atlanta, GA.

Ferris, J. S., and R. G. Congalton. (1989). 'Satellite and geographic information system estimates of Colorado River basin snowpack.' Photogram. Engrg. Remote Sens., 55(11):1629-1635.

Gervin, J. C., and S. F. Shih. (1981). 'Improvements in lake volume predictions using Landsat data.' In Satellite Hydrology, M. Deutsch, D. R. Wiesnet, and A. Rango, eds., Am. Water Resour. Assoc., pp. 479-484.

Gilman, C. S. (1964). 'Rainfall'. In Handbook of Applied Hydrology, V. T. Chow ed., McGraw-Hill Book Co., pp. (9-1)-(9-68).

Goodenough, D. G. (1988). 'Thematic mapper and SPOT integration with a geographic information system.' Photogram. Engrg. Remote Sens., 54(2):167-176.

Griffith, C. G. (1987). 'Comparisons of gauge and satellite rain estimates for the central United States during August 1979.' J. Geophys., 92(D8):9551-9566.

Gurney, R. J., and P. J. Camillo. (1984). 'Modeling daily evapotranspiration using remotely sensed data.' J. Hydrol. 69:305-324.

Hall, D. K., and J. Martinec. (1985). Remote Sensing of Ice and Snow. Chapman and Hall, London, 189 pp.

Haralick, R. M., S. Wang, L. G. Shapiro, and J. B. Campbell. (1985). 'Extraction of drainage networks by using a consistent labeling technique.' Remote Sens. Environ., 18:163-175.

Heilman, J. L., E. T. Kanemasu, J. O. Bagley, and V. P. Rasmussen. (1977). 'Evaluating soil moisture and yield of winter wheat in the Great Plains using Landsat data.' Remote Sens. Environ., 6:315-326.

Higer, A. L., A. E. Coker, E. H. Cordes. (1974). 'Water management models in Florida using ERTS-1 data.' Proc. 3rd Earth Resour. Technology Satellite -1 Sym., NASA sp-351-v1, pp. 1071-1078.

Idso, S. B., R. J. Reginato, D. C. Reicosky, and J. L. Hatfield. (1981). 'Determining soil-induced plant water potential depressions in alfalfa by means of infrared thermometry.' Agron. J., 73:826-830.

Jadkowski, M. A., and M. Ehlers. (1989). 'GIS analysis of SPOT image data.' Proc. GIS/LIS 1989, Am. Cong. Surv. Map., pp 65-74.

Jakubauskas, M. E. (1989). 'Utilizing a geographic information system for a vegetation change detection.' Proc. GIS/LIS 1989, Am. Cong. Surv. Map., pp 56-64.

Jampoler, S. M., and B. N. Haack. (1989). 'Use of GIS to identify land use change: Kathmandu, Nepal.' Proc. GIS/LIS 1989, Am. Cong. Surv. Map., pp 75-84.

Johnston, C. A., and J. Bonde. (1989). 'Quantitative analysis of ecotones using geographic information system.' Photogram. Engrg. Remote Sens., 55:1643-1647.

Jordan, J. D. and S. F. Shih. (1988a). 'Use of remote sensing in abandoned well assessment.' Trans. Am. Soc. Agric. Eng., 31(5):1416-1422.

Jordan, J. D. and S. F. Shih. (1988b). 'Ground-truthing of remotely sensed abandoned wells.' Soil Crop Sci. Soc. Fla. Proc., 47:88-91.

Jordan, J. D. and S. F. Shih. (1990). 'Photogrammetric estimation of artesian well flow rate.' Soil Crop Sci. Soc. Fla. Proc., 49:24-26.

Jordan, J. D. and S. F. Shih. (1991). 'Satellite and aerial photographic techniques for use in artesian well assessment.' Proc. Intl. Conf. on Computer Application in Water Resources, Taipei, Taiwan, R.O.C., 2:991- 998.

Jordan, J. D., and S. F. Shih (1993). 'Comparison of thermal-based soil moisture estimation techniques.' Soil Crop Sci. Soc. Fla. Proc., 52:83- 90.

Khorram, S. (1985). 'Development of water quality models applicable throughout the entire San Francisco Bay and delta.' Photogram. Engrg. Remote Sens., 51:53-62.

Khorram, S., G. P. Catts, J. E. Cloern, and A. W. Knight. (1987). 'Modeling of estuarine chlorophyll a from an airborne scanner.' IEEE Trans. Geosci. Remote Sens. GE-25:662-669.

Klazura, G. E., and D. A. Imy. (1993). 'A description of the initial set of analysis products available from the NEXRAD WSR-88D system'. Bull. Amer. Meteorol. Soc., 74(7):1293-1311.

Kolipinski, M. E., and A. L. Higer. (1969). 'Inventory of hydrobiological features using automatically processed multispectral data.' In Sym. on Remote Sens. Environ., Ann Arbor, MI, pp. 79-95.

Lam, N.S.N., P. J. Grim, and F. Jones. (1987). 'Data integration in geographic information systems: an experiment'. Proc. GIS/LIS 1987; Am. Cong. Surv. Map, pp 53-62.

Lillesand, T. M., and R. W. Kiefer. (1987). Remote Sensing and Image Interpretation. 2nd ed., John Wiley and Sons, Inc., NY.

Lillesand, T. M., M. D. MacKenzie, J.R.V. Castle, and J. J. Magnuson. (1989). 'Incorporating remote sensing and GIS technology in long-term and large-scale ecological research.' Proc. GIS/LIS 1989, Am. Cong. Surv. Map., pp 228-241.

Lo, C. P. (1986). Applied Remote Sensing. Longman Inc., New York, NY.

Marble, D. F., and D. J. Peuquet. (1983). 'Geographic information systems and remote sensing.' Manual of Remote Sensing, R. N. Colwell ed., Am. Soc. Photogram., Falls Church, VA, pp 923-958.

Miller, A., and J. C. Thompson. (1975). Elements of Meteorology. Charles E. Merrill Publishing Company, Columbus, Ohio, 362 pp.

Myhre, B. E., and S. F. Shih. (1990a). 'Using infrared thermometry to estimate soil-water content for a sandy soil.' Trans. Am. Soc. Agr. Eng., 33(5):1479-1486.

Myhre, B. E., and S. F. Shih. (1990b). 'Use of geographic information system for a well monitoring program.' Soil Crop Sci. Soc. Fla. Proc., 49:18-20.

Myhre, B. E., and S. F. Shih. (1991). 'Using geographic information system in well permitting program.' Soil Crop Sci. Soc. Fla. Proc., 50:102- 105.

Orville, R. E., R. W. Henderson, and L. F. Bosart. (1983). 'An east coast lightning detection network.' Bull. Am. Meteorol. Soc., 64:1029-1037.

Orville, R. E. (1991). 'Lightning ground flash density in the contiguous United States-1989 annual summary.' Bull. Am. Meteorol. Soc., 119:573- 577.

Oslin, A. J., R. A. Westsmith, and D. S. Morgan. (1988). 'STREAMS: a basin and soil erosion model using CADD remote sensing and GIS to facilitate watershed management.' In Modeling Agricultural, Forest, and Rangeland Hydrol., K. L. Campbell ed., Am. Soc. Agric. Eng., pp. 470-477.

Pelletier, R. E. (1985). 'Evaluating non-point pollution using remotely sensed data in soil erosion models.' J. Soil Water Conserv., 40:332- 335.

Pesnell, G. L., and Brown, R. T. (1977). 'The major plant communities of Lake Okeechobee, Fla., and their associated inundation characteristics as determined by gradient analysis.' South Fla. Water Manag. Dist., West Palm Beach, Fla., Technical Publication 77-1:68 pp.

Piepgrass, M. V., E. P. Kridder, and C. B. Moore. (1982). 'Lightning and surface rainfall during Florida thunderstorms.' J. Geophy. Res., 87(C13), 11193-11201.

Price, J. C. (1982). 'Estimation of regional scale evapotranspiration through analysis of satellite thermal-infrared data.' IEEE Trans. Geosci. Remote Sens., GE-20(3):287-292.

Ragan, R. M., and Rogers, R. H. (1978). 'Use of Landsat satellite remote sensing for regional environmental planning and management.' XV Convention Pan Am. Federal Engrg. Soc., Santiago, Chile.

Ragan, R. M., and T. J. Jackson. (1980). 'Runoff synthesis using Landsat and SCS model.' J. Hydraul. Div., ASCE, 106(HY5):667-678.

Rango, A., A. Feldman, T. S. George, III, and R. M. Ragan. (1983). 'Effective use of Landsat data in hydrologic models.' Water Resour. Bull., 19:165-174.

Rector, M. R. (1982). 'Remote sensing and water resources management: a California example.' In Remote Sensing for Resource Manag., C. J. Johannsen and J. L. Sanders, eds., Soil Conserv. Soc. Am., pp. 255-263.

Rogers, J. W., and S. F. Shih. (1990). 'Land use classification in agricultural water usage permitting program via Landsat data.' Applied Engrg. in Agric., 6(1):54-58. Sandholt, I. and H. S. Anderson. (1993). 'Derivation of actual evapotranspiration in the Senegalese Sahel using NOAA-AVHRR data during the 1987 growing season.' Remote Sens. Environ., 46:164-172.

Schecter, R. N. (1976). 'Resource inventory using landsat data for area wide water quality planning.' Proc. Sym. on Machine Processing of Remotely Sensed Data, Lab. for Applications of Remote Sens., Purdue Univ., West Lafayette, IN.

Schmugge, T. J., T. J. Jackson, and H. L. McKim. (1980). 'Survey of methods for soil moisture determination.' Water Resour. Res., 16(6):961-979.

Scofield, R. A. (1987). 'The NESDIS operational convective precipitation estimation technique.' Monthly Weather Review, 115:1773-1792.

Shih, S. F. and J. C. Gervin. (1980). 'Ridge regression technique applied to Landsat data investigation of water quality in Lake Okeechobee.' Water Resour. Bull., 16(5):790-796.

Shih, S. F. (1982a). 'Using landsat data to estimate reservoir storage'. Proc. 8th Intl. Sym. on Machine Processing of Remotely Sensed Data. Lab. for Applications of Remote Sens., Purdue Univ., West Lafayette, IN, pp. 321-326.

Shih, S. F. (1982b). 'Use of Landsat data to improve water storage information in Conservation Area, Florida.' In Applied Modeling in Catchment Hydrology, V. P. Singh ed., Water Resour. Publication, Littleton, CO, pp. 511-518.

Shih, S. F., and E. Y. Chen. (1984). 'On the use of GOES thermal data to study effects of land uses on diurnal temperature fluctuation'. J. Climate and Appl. Meteorol., 23(3):426-433.

Shih, S. F. (1985). 'Comparison of ELAS classifications and density slicing Landsat data for water-surface area assessment'. In Hydrologic Applications of Space Technology (Publication No. 160), A. I. Johnson, ed., Intl. Assoc. Hydrological Sci., pp. 91-97.

Shih, S. F., J. A. Doolittle, D. L. Myhre, and G. W. Schellentrager. (1986). 'Using radar for ground-water investigation'. J. Irrig. and Drain. Engrg. ASCE, 112(2):110-118.

Shih, S. F. (1988a). 'Using lightning for rainfall estimation in Florida'. Trans. Am. Soc. Agric. Engrg., 31(3):750-755.

Shih, S. F. (1988b). 'Satellite data and geographical information system for land-use classification'. J. Irrig. and Drain. Engrg., ASCE, 114(3):505-519.

Shih, S. F. (1989). 'Geographical information system and GOES image enhancement in rainfall estimation'. Proc. GIS/LIS 1989, Am. Cong. Surv. Map., pp 468-477.

Shih, S. F. (1990). 'Satellite data and geographic information system for rainfall estimation'. J. Irrig. and Drain. Engrg., ASCE, 116(3):319-331.

Shih, S. F., D. S. Harrison, A. G. Smajstrla, and F. S. Zazueta. (1991). 'Infrared thermometry to estimate soil-water content in pasture areas'. Soil Crop Sci. Soc. Fla. Proc., 50:158-162.

Shih, S. F. (1992). 'Calibration parameters for radar rainfall estimation'. Soil Crop Sci. Soc. Fla. Proc., 51:23-29.

Shih, S. F., and E. T. Engman. (1992). 'Applications of remote sensing to hydrology'. Proc. of Water forum's 92, ASCE, pp. 535-540.

Shih, S. F., and J. D. Jordan. (1992). 'Landsat mid-infrared data and GIS in regional surface soil moisture assessment'. Water Resour. Bull., 28(4):713-719.

Shih, S. F., and J. D. Jordan. (1993). 'Use of Landsat thermal infrared data and GIS in soil moisture assessment'. J. Irrig. and Drain. Engrg., ASCE, 119(5):868-879.

Shih, S. F., and D. L. Myhre. (1994). 'Ground-penetrating radar for salt- affected soil assessment'. J. Irrig. and Drain. Engrg., ASCE, 120(2):322-333.

Soer, G.J.R. (1980). 'Estimation of regional evapotranspiration and soil moisture conditions using remotely sensed crop surface temperature'. Remote Sens. Environ., 9:27-45.

Spencer, R. W., H. M. Goodman, and R. E. Wood. (1988). 'Precipitation retrieval over land and ocean with SSM/I, Part 1: identification and characteristics of the scattering signal'. J. Atmos. Oceanic

Tech., 2:254-263.

Still, D. A., and S. F. Shih. (1985). 'Using Landsat data to classify land use for assessing the basin-wide runoff index'. Water Resour. Bull., 21:931-940.

Still, D. A., and S. F. Shih. (1991). 'Using Landsat data and geographic information system for wetland assessment in water quality management'. Soil Crop Sci. Soc. Fla. Proc., 50:98-102.

Stuebe, M. M., and D. M. Johnston. (1990). 'Runoff volume estimation using GIS techniques.' Water Resour. Bull., 26(4):611-620.

Sucksdorff, Y., and C. Ottle. (1990). 'Applications of satellite remote sensing to estimate areal evapotranspiration over a watershed'. J. Hydrol., 121:321-333.

Tan, C. H., and S. F. Shih. (1991a). 'The Potential for application of lightning data to rainfall estimation'. Proc. Intl. Conf. on Computer Application in Water Resources, Taipei, Taiwan, R.O.C., 2:999-1005.

Tan, C. H., and S. F. Shih. (1993a). 'Use of geographic information system to manage lightning data in Florida'. Soil Crop Sci. Soc. Fla. Proc., 52:65-69.

Tan, Y. R. and S. F. Shih. (1988). 'A geographic information system for study of agricultural land-use changes in St. Lucie County, Florida'. Soil Crop Sci. Soc. Fla. Proc., 47:102-105.

Tan, Y. R. and S. F. Shih. (1990). 'Current agricultural landuse for abandoned well assessment.' Soil Crop Sci. Soc. Fla. Proc., 49:20-24.

Tan, Y. R. and S. F. Shih. (1991b). 'Geographic Information System for differentiating unused wells'. Soil Crop Sci. Soc. Fla. Proc., 50:110- 116.

Tan, Y. R., and S. F. Shih. (1993b). 'Effect of citrus canopy on the spectral characteristics of SPOT satellite images'. Soil Crop Sci. Soc. Fla. Proc., 52:61-64.

Walsh, S. J., J. W. Cooper, I. E. VonEssen, and K. R. Gallager. (1990). 'Image enhancement of Landsat thematic mapper data and GIS data integration for evaluation of resource characteristics'. Photogram. Engrg. Remote Sens., 56(8):1135-1141.

Weiss, M., and E. A. Smith. (1987). 'Precipitation discrimination from satellite infrared temperatures over the CCOPE mesonet region'. J. Climate and Appl. Meteorol., 26(6):687-697.

Wilkening III, H. A. (1989). 'Landsat data processing and GIS for regional water resources management in Northeast Florida'. Proc. GIS/LIS 1989, Am. Cong. Surv. Map., pp 110-119.

Williams, D. L., C. L. Walthall, and S. N. Groward. (1984). 'Collection of in sites forecast canopy spectra using a helicopter: A discussion of methodology and preliminary results'. Proc. Sym. on Machine Processing of Remotely Sensed Data, Lab. for Application of Remote Sens. Purdue Univ., West Lafayette, IN, pp. 94-105.

Williams, D. L. (1994). 'Letter explaining NASA's position on Landsat 7 program'. NASA Goddard Space Flight Center, Greenbelt, Maryland.

Williams, L. E. and S. F. Shih. (1989). 'Landsat data and GIS for land use change assessment in the Florida Everglades'. Proc. GIS/LIS 1989, Am. Cong. Surv. Map., 1:305-313.

Wischmeier, W. H., and D. D. Smith. (1978). 'Predicting rainfall erosion losses'. Agricultural Handbook No. 537, U.S. Dept. Agri. Sci. and Education Administration, Washington, DC.

Xiao, Q., H. Raafat, and D. Gauthier. (1989). 'A temporal database structure for remotely sensed image data management within GIS.' Proc. GIS/LIS 1989, Am. Cong. Surv. Map., pp 116-122.

Xin, J. N., and S. F. Shih. (1991). 'NOAA polar-orbiting satellite APT data in lake evaporation estimation.' J. Irrig. and Drain. Engrg., ASCE, 117(4):547-557.

CHAPTER 3

Hydrologic Data Development

M.L. Wolfe

Abstract.   Hydrologic applications of geographic information systems require
the development of appropriate databases. The process includes determining what
data are needed, identifying sources of that data, converting the data into appropriate format for the GIS software, and maintaining the database. Quality control
measures are important throughout the development process.

3.1. Introduction

Many potential applications for geographic information systems (GIS) exist in hydrology. In this chapter, the term GIS is used to include both the software for data
analysis, storage, management, display, etc. and the database itself. This chapter
focuses on developing a database for applications in hydrology.

A GIS database is comprised of a number of data layers, a few or many, depending on the application. Data layers often used in hydrologic applications include soils, land use, stream network, watershed boundaries, well locations, and
topography. The two main methods of database development are obtaining existing digital data and converting existing analog (paper) maps to digital form through
either manual digitizing or automated scanning.

A number of questions must be answered before developing a GIS database for
the first time. Major questions include:

- What types of data are included in a GIS database?
- What are sources of data?
- How are the data entered into the GIS?
- What is the quality of the data?
- How can data from different sources be integrated?
- What is the cost of developing a database?

43

V. P. Singh and M. Fiorentino (eds.), Geographical Information Systems in Hydrology, 43–63.
© 1996 *Kluwer Academic Publishers. Printed in the Netherlands.*

This chapter addresses the questions posed above, discusses database maintenance and update, and concludes with some recommendations regarding database
development and use.

3.2. A GIS Database

A GIS database includes two general types of data: spatial (or locational or graphic) and attribute (or nongraphic). The spatial component of the GIS database is
often described as a series of *layers*, or *coverages*, each of which contains map
features that are related functionally. For example, one layer could be soil classes,
while another could be land use. Each layer is registered positionally to the other
database layers through a common coordinate system.

Spatial data include six types of graphic elements to depict map features (Antenucci, 1991):

point   a zero-dimensional object that specifies a geometric location through a set
          of coordinates. A *node* is a special type of point, also a zero-dimensional ob-
          ject, that is a topological junction or end point and may specify a geometric
          location.

line   a one-dimensional object. A line segment is a direct line between two points.
          Special forms of lines include the following: string, a series of line segments;
          arc, a locus of points forming a curve defined by a mathematical function; and
          chain, a directed sequence of nonintersecting line segments or arcs with nodes
          at each end.

area   a bounded, continuous two-dimensional object that may or may not include
          its boundary; individual areas are represented as polygons.

grid cell   a two-dimensional object that represents a single element of a continuous
          surface.

pixel   a two-dimensional picture element that is the smallest indivisible element of
          an image.

symbol graphic element that represents features at points on a map.

The format of a database can be either *raster* or *vector* or contain layers in both
formats. In raster format, graphic images are stored as values for uniform grid cells
or pixels. In vector format, graphic images are stored as strings of coordinates,
which represent features as points, lines, and polygons.

Nongraphic data describe the characteristics of the graphic layers and phenomena that occur at specific geographic locations. For example, the graphic layer may
include locations of weather stations. The nongraphic data may include the number of times lightning has struck each station and the weather variables that are
measured at each. Antenucci et al. (1991) define four classes of nongraphic data in
a GIS database:

> *Nongraphic Attributes* provide descriptive information, both qualitative and
> quantitative, about the characteristics of map features, represented by points,
> lines, polygons, or networks in the database. Attributes are usually stored sep-

arately from graphic data files, but are linked to the graphic elements through common identifiers, often called geocodes, that are stored in both spatial and nongraphic records.

Geographically Referenced Data describe incidents or phenomena that occur at a specific location. Unlike attributes, they do not describe the map feature itself. This type of data describes items or actions (such as sampling and analysis protocols) that can be related to specific geographic locations (such as water quality monitoring stations). Geographically referenced data typically are stored and managed in separate files and systems that are not directly associated with the graphic features in a GIS database, but do contain elements that identify the location of the item or phenomena.

Geographic Indexes help locate map features and data based on their geographic identifiers.

Spatial Relationships are descriptions of the proximity, adjacency, and connectivity of map features.

The GIS maintains the relationship between spatial and attribute data throughout the analytical, display, and other GIS functions.

3.3. Data Sources

Potential sources of data for a GIS database include primary sources and secondary sources. Primary sources include direct measurement of data, e.g., field sampling or remote sensing. The density of sampling determines the resolution of the data. For example, water samples taken every day will capture day-to-day variation, but miss shorter-term variation. Soil samples taken every 100 m will miss any variation at resolutions less than 100 m.

Secondary sources include existing analog data, such as paper and mylar maps and aerial photographs, as well as existing digital databases. Maps are available from a number of agencies and organizations. Antenucci et al. (1991) describe the mapping efforts of a number of U.S. government agencies. Various state and local agencies also have mapping responsibilities for different resources and features. Some of these agencies are using or are converting to computer-aided mapping, sometimes in conjunction with a GIS.

3.3.1. MAPS

The most important source of analog spatial data is the map. Many problems can occur in developing digital geographical databases from maps. Users of digital databases are often unaware of the limitations of conventional maps and, consequently, may make unreasonable or inappropriate assumptions about the information from those maps (Fisher, 1991).

The majority of digital data in use today is based on maps. A GIS database user should be familiar with the characteristics and limitations of maps to facilitate appropriate use of data taken from maps.

3.3.1.1. *Scale*

Map scale is the ratio of units of measurement on the map to units of measurement on the earth. Scale is often stated in the format of 1:X, where one unit of measurement on the map is equal to X identical units of measurement on the earth. Maps are described as being large scale or small scale. Large scale, e.g. 1:10 000, maps show great detail and small features. In contrast, a small scale, e.g. 1:250 000, map shows only large features.

Scale determines the smallest area that can be drawn and recognized on a paper map (Fisher, 1991). On a 1:50 000 scale topographic map, it is not possible to represent accurately any object of dimensions less than one line width, or less than about 25 m across. However, small features can be important, so cartographers have devised methods for selecting and symbolizing small but significant features, even though their physical dimensions on the ground may be less than one line width (Robinson et al., 1984; Tobler, 1988; Muller, 1991). Thus, many roads and rivers that are less than 25 m across are shown on 1:50 000 maps.

Scale influences the material included on thematic maps as well as on general topographic maps (Fisher, 1991). Two thematic maps of interest for many hydrologic applications are vegetation and soils. On a vegetation map, a minimum area, or mapping unit, is established based on the map scale. Areas smaller than this size are merged with neighboring areas, despite having distinctly different vegetation. Thus, the class of vegetation assigned to an area is merely the dominant class in the area. It is accepted that there may be significant inclusions of other classes present. If the area was mapped at another scale, and thus a different minimum mapping area, the original area might either be broken into more specific units, or combined with others to yield a more general unit, depending on the direction of the scale change. Similarly, in the case of soil maps, different scales are characterized by different mapping units which reflect different degrees of map unit homogeneity.

3.3.1.2. *Projection*

Map projection is the transformation of positions on the curved surface of the earth onto a flat map surface (Antenucci et al., 1991). Some distortion is inevitable in this transformation. Distortion is least for maps of small areas, and greatest when a map attempts to show the entire surface of the earth. Numerous projections have been invented, and arguments continue about which is best for which purposes (Goodchild and Kemp, 1990). Maling (1991) describes methods for selecting a suitable projection for a particular geographic area, based on the location, size, and shape of the area.

Some map projections are very common (Antenucci et al., 1991). Each state in the U.S. has a standard State Plane Coordinate (SPC) system based on one or more projections. Usually one of two projections - Transverse Mercator (for states with a north-south orientation) or Lambert Conformal (for states with an east-west orientation) - defines specific zones for each state. The SPC system is tied to the National Geodetic Reference System (NGRS) and provides the accuracy and compatibility of that geodetic framework. The SPC system is easy to use for areas within a single zone. The coordinate system relates to a national structure that can be transformed to other projections with relative ease.

The Universal Transverse Mercator (UTM) is an internationally accepted projection that covers the entire globe as one continuous system. This worldwide system is composed of zones 6 degrees wide that are based on the Transverse Mercator projection. UTM values are generally recorded in meters. The projection provides sufficient accuracy for surveying and other detailed purposes. The latitude and longitude coordinate system, which records angular measurements relative to the equator and prime meridian, is also commonly used.

3.3.1.3. *Temporal Considerations*

A map is a representation of features in space as they existed at the time they were surveyed. The real world of geographical information changes continuously, but many maps remain static. Thus, maps become increasingly inaccurate in their representation of the world over time.

Vegetation and land use maps, used often in hydrologic analyses, require constant revision. Although soils and geology are less subject to change, even these classes of maps must be updated regularly to accommodate new field work and general improvements in the level of human understanding of soils and geology (Fisher, 1991).

In generating a digital database, it may not be possible to find a truly current map of an area, but before digitizing every effort should be made to acquire the most up-to-date information. Care should be taken to ensure that the information used was derived by survey (Fisher, 1991). For example, the location of a dam should be entered from as-built surveys rather than from the construction plans, since modifications sometimes occur during construction.

3.3.1.4. *Extent of Existing Map Coverage*

Scales and completeness of map coverage of different geographical areas are highly variable. In the U.S., the most detailed complete coverage scale of topographic maps is 1:24 000, while in Great Britain, it is a combination of 1:1250, 1:2500, and 1:10 000 (Starr and Anderson, 1991; Sowton, 1991). Thematic map coverages are even more patchy. Although there are some global scale map coverages of some environmental themes, such as soil and geology (Clark et al., 1991), the extent is much more irregular at larger scales (Fisher, 1991). For example, in the U.S., soil mapping at the county level scales of 1: 15 840 or 1:31 680 is nearly completed, although much remains unpublished. In the United Kingdom, although there is now full coverage at 1:250 000 as a result of an EEC initiative, larger scale soil maps are relatively uncommon, covering a small fraction of the country. By contrast, geological mapping of the UK at 1:63 360 is complete and being extensively revised, while less than 50 percent of the U.S. has been mapped at scales of 1:250 000 or larger (Thompson, 1988). In contrast to these two developed nations, many countries in the world have neither topographic nor thematic maps at any detailed scale (Brandenberger and Ghosh, 1985).

3.3.1.5. *Map Accuracy*

The most commonly used measure for the quality of maps is positional accuracy, a measurement of the variance of the position of a map feature from the true position of the entity (Antenucci et al., 1991). Relative accuracy is a measure of the accuracy of individual features on a map when compared to other features on the same map. Absolute accuracy is a measure of the location of features on a map compared to their true position on the face of the earth.

Many agencies have established accuracy standards for geographical data. Accuracy standards generally are stated in terms of an acceptable tolerance that must be achieved and the proportion of measured features that must meet the criteria (Antenucci et al., 1991). Location accuracy standards are commonly decided from the scale of source materials. For example, the U.S. National Map Accuracy Standards, summarized in Table 3.1, differentiate between scales larger and smaller than 1:20 000. The American Society of Photogrammetry and Remote Sensing specifies acceptable root-mean-square error terms for horizontal locations on various maps (Merchant, 1987).

Scale is often misused as a measure of accuracy (Fisher, 1991). As described in Section 3.3.1.1, scale determines the smallest area that can be drawn and recognized on a map. It indicates nothing about accuracy.

Names and attribute information on maps are not generally subject to a specific standard (Fisher, 1991). However, standards have been set for mapping land attributes as part of the Land Use and Land Cover mapping program of the U.S. Geological Survey (USGS) (Table 3.2). The U.S. Department of Agriculture-Soil Conservation Service (USDA-SCS) (now Natural Resources Conservation Service, NRCS) has accuracy specifications for soil maps (Table 3.3).

The existence of standards does not ensure adherence to them. Fisher (1991) discusses the lack of implementation of standards and subsequent testing. He points out that the main factors preventing implementation of accuracy standards are time and cost. All mapping programs are subject to deadlines and cost limits, and coverage and timeliness will always be given more weight than accuracy in allocating scarce resources.

3.3.1.6. *Map Sheets and Series*

Some map series are designed, drafted, and published as a collection of individual map sheets. Each separate map sheet or quadrangle is intended to stand alone as

TABLE 3.1. Summary of important parts of the U. S. National Map Accuracy Standard U. S. Bureau of the Budget (Thompson, 1988)

1. On scales smaller than 1:20 000, not more than 10 percent of points tested should be more than 1/50 inch in horizontal error, where points refer only to points which can be well defined on the ground.

2. On maps with scales larger than 1:20 000, the corresponding error term is 1/30 inch.

3. At no more than 10 percent of the elevations tested will contours be in error by more than one half the contour interval.

4. Accuracy should be tested by comparison of actual map data with survey data of higher accuracy (not necessarily with ground truth).

5. If maps have been tested and do meet these standards, a statement should be made to that effect in the legend.

6. Maps that have been tested but fail to meet the requirements should omit all mention of the standards on the legend.

TABLE 3.2. Accuracy specifications for USGS Land Use and Land Cover maps (Anderson et al., 1976)

1. 85 percent is the minimum level of accuracy in identifying land use and land cover categories.

2. The several categories shown should have about the same accuracy.

3. Accuracy should be maintained between interpreters and times of sensing.

a single entity. However, there is no guarantee of conformity across the seam of the maps. Edge-matching between map sheets can be a major problem. Two map series, USDA-SCS soil survey maps and USGS topographic quadrangles, used often in hydrologic applications, fit this category. In fact, Fisher (1991) cites county soil reports by USDA-SCS as an example of a particularly problematic situation, i.e., maps prepared on poorly rectified orthophotomaps.

TABLE 3.3. Accuracy specifications for USDA-SCS soil maps (SCS, 1984 in Fisher, 1991)

1. Up to 25 percent of pedons may be of other soil types than that named if they do not present a major hindrance to land management.

2. Up to only 10 percent of pedons may be of other soil types than that named if they do present a major hindrance to land management.

3. No single included soil type may occupy more than 10 percent of the area of the map unit.

3.3.2. EXISTING DIGITAL SPATIAL DATA

Many public and private organizations around the world offer digital data products. There are numerous ongoing data conversion projects being conducted, making it impossible to produce a complete, current list of available data. The following sections describe some sources of existing digital data.

United States Government Sources.   Many U.S. government agencies have digital databases. The National Cartographic Information Center (NCIC) collects, sorts, and describes all types of cartographic information from Federal, State, and local government agencies and, where possible, from private companies in the mapping business. The NCIC national headquarters is in Reston, Virginia. Regional offices are located in Missouri, Mississippi, Colorado, California, and Alaska. The NCIC has also established affiliated offices with many State governments. Addresses of affiliated offices can be obtained by contacting the nearest regional NCIC office.

The USGS produces and distributes, through the NCIC, a variety of digital cartographic and geographic data under the name US GeoData. Some current products include: digital line graphs (DLGs) from 1:24 000 scale maps; digital line graphs from 1:100 000 scale maps; digital line graphs from 1:2 000 000 scale maps; land use and land cover from 1:250 000 and 1:100 000 scale maps; digital elevation models (DEMs); and geographic names information system.

The USDA-SCS has established three soil databases representing different scales of mapping (Reybold and TeSelle, 1989). The three databases are: the Soil Survey Geographic Data Base (SSURGO) at scales from 1:15 840 to 1:31 680; the State Soil Geographic Data Base (STATSGO) at a scale of 1:250 000, and the National Soil Geographic Data Base (NATSGO) at a scale of 1:7 500 000. Each database links digitized soil map unit delineations with computerized data for each map unit, giving the proportionate extent of the component soils and their properties. Reybold and TeSelle (1989) describe development of the databases, including the digitizing standards that were followed.

State and Local Government Sources (U.S.). The use of geographic information technology is widespread within state governments. At least forty-five states are using a GIS or a related spatial system in at least one state agency, and at least thirty-four states have more than one GIS. States' uses of GIS include natural resources management, transportation, planning, public lands management, and property assessment applications (Antenucci et al., 1991). Obviously, a variety of digital databases have been developed to support these applications.

The development of GIS databases has followed different patterns in different U.S. states. Millette (1990) states that the Vermont State Geographic Information System (Vermont GIS) is the first state-wide GIS to be legislatively mandated and funded in the U.S. He contrasts this to a number of other states where the GIS programs are nested within a state planning or natural resource agency where the GIS applications are restricted to a narrow focus.

For example, the Virginia Geographic Information System (VirGIS) (Shanholtz et al., 1987) was developed through a contract between a state agency, the Department of Conservation and Recreation-Division of Soil and Water Conservation, and the Information Support Systems Laboratory in the Department of Agricultural Engineering (now Biological Systems Engineering) at Virginia Tech. The VirGIS was initiated as a part of the Chesapeake Bay environment protection program, with the primary objective of establishing procedures for identifying and prioritizing agricultural land areas with serious nonpoint source pollution problems.

International Government Sources. Development of digital databases and applications of GIS have occurred around the world. The existence and types of databases vary. Several examples are cited in this section.

In a review of GIS applications in Europe, Scholten and van der Vlugt (1990) cite a number of projects that included database development. Examples of national databases on land use that have resulted from database development for land information systems include: Finland (Ahonen and Rainio, 1988), Norway (Engebretsen, 1987), and Sweden (Falk, 1987; Piscator, 1986). Sweden also has two national elevation databases (Bogaerts, 1987). Other citations include: a community-wide environmental database by the Commission of the European Community (Briggs, 1988); the MEDASE project by several nations bordering the Mediterranean to improve the information base for environmental management (Montanari, 1989); and the RIA project (Scholten and Meijer, 1989), an overall spatial information system in the Netherlands.

O'Callaghan and Garner (1991) discuss database development in Australia, including specific examples. The National Resources Information Center was established in Australia to provide rapid identification, access to, and integration of the available data to support government decision making concerning natural resources management. An Environmental Resources Information Network was also established to provide geographically related environmental information to federal agencies.

Applications of GIS in developing countries are discussed by Taylor (1991). Taylor describes GIS and database development activities in India and China, as well as mentioning activities in other developing countries. He argues that, although GIS has potential to be of use in the struggle for development, that potential has not been realized and there are many problems to be overcome.

Commercial Sources.   The 1993 International GIS Sourcebook (GIS World, 1992) includes a directory of spatial data suppliers. The directory includes 141 listings from 20 countries. Thirty states are represented from within the U.S. The sourcebook also includes directories of organizations that provide services related to database development, i.e., digitizing (table and scan) (214 listings); remote sensing, aerial photography, and photo interpretation (122 listings); surveying services (128 listings); and mapping and field data acquisition (234 listings).

Additional Sources.   Descriptions of numerous GIS applications have been published (e.g. Antenucci et al., 1991), including applications by federal government, state, and local agencies; electric, gas, water, wastewater, and telephone utilities; railroad companies; and private industry, including oil and gas, timber, transportation, marketing and sales, and real estate companies. Digital databases were developed for many of these applications; some used existing databases. Some of these databases may be useful for hydrologic applications.

3.4. Data Input

Regardless of the source of the data, the data must be entered into the GIS to be used in hydrologic (or any other) applications. Entering data into the GIS database can be accomplished in a number of ways. The method of data entry depends on factors such as the data source and format and the available hardware, software, and personnel.

3.4.1. PRIMARY SOURCE DATA

Direct field measurements can be entered into a GIS database in several ways. Indirect methods include recording the values in the field manually on paper and then entering them into the GIS using a keyboard. An intermediate step could be taken; the values could be entered into a separate digital database and then converted into the GIS format. Direct field measurements could also be recorded through a data logger, resulting in a digital data record that could be transferred to the GIS, with any needed format conversions.

Many hydrologic applications require topographic data and locations of objects such as monitoring stations and wells. Traditionally, survey data for topographic maps and locations of objects were recorded in notebooks in the field and then processed manually. Automated surveying equipment now can directly determine the horizontal and vertical positions of objects. The total station captures distance and direction data in digital form. The data are downloaded to a host computer at the end of each session for direct input to GIS and other programs (Goodchild and Kemp, 1990).

The Global Positioning System (GPS) is the newest tool for determining accurate positions on the surface of the earth. The GPS computes positions from signals received from a series of satellites (NAVSTAR). A radio receiver with appropriate electronics is connected to a small antenna, and depending on the method used, in one hour to less than 1 second, the system is able to determine the location of the receiver in 3-D space. Developed and operated by the U.S. Armed Forces (Getting, 1993), GPS access is generally available and civilian interest is high. The GPS is particularly valuable for establishing accurate positional control in remote areas (Goodchild and Kemp, 1990).

3.4.2. SECONDARY SOURCE DATA

Data from other digital sources must be converted to the required format for the GIS. Digital data can be obtained through electronic media, such as tapes, diskettes, CD-ROM, or file transfer over a computer network. The data can be entered into the GIS using appropriate software and hardware. The software may be internal or external to the GIS.

Some of the data layers commonly used in hydrologic applications include soils, land use, topography, and stream network. As described earlier, some of these databases are available from commercial or government sources. However, for many hydrologic applications, suitable digital data do not exist. Coverage of the specific geographic area of interest may not be available or the scale of the available data may not be appropriate for the application. For many hydrologic applications, existing digital databases were derived from maps of too small a scale for many hydrologic applications. As a result, for many hydrologic applications, digital data must be created by conversion of analog maps.

Conversion of existing paper, mylar, or other manually produced maps to digital form can be accomplished through manual digitizing or scanning. Techniques used to convert other analog materials include photogrammetric digitizing, coordinate geometry, and key entry (Antenucci et al., 1991). The following sections describe the listed data conversion techniques.

3.4.2.1. *Manual Digitizing*

Manual digitizing is accomplished using a digitizing table or tablet. The operator moves a cursor over the surface of the digitizing tablet. The position of the cursor is detected by the computer and interpreted as pairs of x,y coordinates. Frequently, there are control buttons on the cursor which permit control of the system without having to turn attention from the digitizing tablet to a computer terminal.

A number of references, including Antenucci et al. (1991); Goodchild and Kemp (1990); and Lai (1988), describe the digitizing operation in detail and provide advice regarding the operation. The following descriptions of the steps in the digitizing operation encompass information from the cited references as well as from the author's personal experience.

Step 1   Planning.   Digitizing is neither a fast nor an inexpensive process. Lai (1988) cautions that haphazard digitizing can result in digital data with a short life or require a great deal of remedial effort to be converted into a useable form. Careful planning and an understanding of how the digital data are to be used will not only ensure the timely delivery of useful databases but will help to eliminate unnecessary waste of personnel and computer resources.

Step 2   Prepare Map for Digitizing.   Most maps were not drafted for the purpose of digitizing. Ambiguities or feature representations may need clarification. For example, boundaries may be difficult to define (Antenucci et al., 1991). It may be unclear whether a boundary follows a street centerline or an edge, yet the difference may be important. Background research may be necessary to prepare materials for digitizing. Usually it is necessary to prepare the source materials by reviewing and marking them to enable rapid, high-volume entry of data by persons not familiar with the sources.

Discrepancies across map sheet boundaries can cause discrepancies in the total digital database. Two sources of maps for hydrologic studies in which this can be a problem are soil surveys and USGS topographic maps. Very often, the area of interest will lie on two or more sheets which will require edge mapping. Sometimes, roads, streams or contour lines do not meet exactly when two map sheets are placed next to each other. These discrepancies should be resolved before digitizing, e.g., through surveying, if possible. If not possible, they will have to be corrected after digitizing, which is likely to be more difficult.

Step 3   Affix Map to Digitizing Table.   Place the map so that all areas to be digitized can be reached comfortably by the operator. Extensive reaching could cause errors in the digitizing. Be sure the map is flat, with no ridges or fold lines that will disrupt the smooth movement of the cursor. If a map has been folded, it may be necessary to prepare a traced copy of the map to have a smooth sheet. Photocopying may be an alternative, but care must be taken because photocopying tends to distort the image that is copied. Depending on the scale of the map, the distortion may be significant. In addition, the size of the map may preclude photocopying.

Step 4   Digitize Control Points.   Control points are easily identified points, such as intersections of major streets. The coordinates of the control points will be known in the coordinate system to be used in the final database. The control points are used by the GIS to calculate the necessary mathematical transformations to convert all coordinates to the final coordinate system. The larger the number of control points, the more accurate the digital map will be.

Step 5   Digitize Map.   Two modes of digitizing can be used: point mode and stream mode. Point mode can be used to enter discrete sites, such as well locations and water quality monitoring sites. Point mode can also be used to enter lines or polygons. For lines, the operator identifies the beginning and ending points of the line and indicates that they should be joined by a line. The method of indicating this depends on the digitizing hardware and software. A polygon can be entered as a series of line segments, with the beginning and ending point of each segment entered. One disadvantage to point mode is that since the operator selects points subjectively, two operators will not code a line in the same way.

In stream mode, points are captured at set time intervals or on movement of the cursor by a fixed amount. Stream mode generates large numbers of points, many of which may be redundant. Stream mode is more demanding on the user while point mode requires some judgement about how to represent the line. Goodchild and Kemp (1990) stated that most digitizing is currently done in point mode.

Step 6   Validate and Edit Data.   Quality control is essential in digitizing. Many errors can occur in digitizing. Errors caused by the operator include overshoots, undershoots (gaps), and spikes at intersections of lines. Operator fatigue and boredom lead to digitizing errors. It is necessary to edit the spatial data using interactive graphics functions in the GIS or other digitizing software.

The digitized data should be plotted to compare against source maps and corrected until all point and line features fall within a specified tolerance of the original map. The tolerance will vary with application. For example, Lai (1988) specified a tolerance of 1 mm. Digital non-spatial data should also be compared to source documents to assure accuracy.

3.4.2.2.   *Scanning*

Manual digitizing is labor-intensive, time-consuming, and costly. Scanning technology has been emerging for many years as a solution to these constraints. Scanning involves systematic sampling of the source document, by either transmitted or reflected light. Most scanners produce digital data in raster format, recording a value of dark (e.g., representing a line or symbol) or light (no line or symbol) for each grid cell or pixel of the scan. Compared to manual digitizing, scanning is very quick and less costly.

Source documents for scanning must be clean (no smudges or extra markings). Lines should be at least 0.1 mm wide. Complex line work provides greater chance of error in scanning. Relatively simple maps are better. Text may be accidently scanned as line features. Contour lines cannot be broken with text. Automatic feature recognition is not easy (two contour lines vs. road symbols). Special symbols (e.g. marsh symbols) must be recognized and dealt with. If good source documents are available, scanning can be an efficient time saving mode of data input (Goodchild and Kemp, 1990).

Scanning produces raster data layers and, thus, is more appropriate for applications for which raster rather than vector data are suitable. Algorithms have been developed to convert raster data to vector format and to recognize line, symbol, and annotation patterns that supply the necessary intelligence to the scanned data (Antenucci et al., 1991; Goodchild and Kemp, 1990).

Successful vectorization requires a clean line to be scanned. To create a sufficiently clean line, it is often necessary to redraft input documents. Since the scanner can be color sensitive, vectorizing may be aided by the use of special inks for certain features. Although scanning is much less labor intensive, problems with vectorization lead to costs which are often as high as manual digitizing. Two stages of error correction may be necessary: edit the raster image prior to vectorization and then edit the vectorized features.

3.4.2.3. *Photogrammetric Digitizing*

Photogrammetric data capture is based on the stereoscopic interpretation of aerial photographs or satellite imagery, using suitable photogrammetric equipment, i.e. manual or analytical stereoplotters (Weibel and Heller, 1991). Photogrammetric digitizing most often is used to record very precise, accurate digital planimetric features and elevation data from stereophotographs. Elevation data can be recorded either as continuous-contour lines of a specific interval or as point data (Antenucci et al., 1991).

3.4.2.4. *Coordinate Geometry*

In the coordinate geometry technique, geometric descriptions of map features are keyed into the computer. Mathematical algorithms compute the resulting coordinates. This approach requires the definition of a point of origin through digitizing or entry of coordinate values. Bearings and distances or other geometric descriptions of the features to be mapped are entered on a keyboard. The coordinate geometry technique is often used to enter survey data. As mentioned earlier, recent developments support the direct entry of digital data collected in the field from modern electronic survey instruments such as total stations and electronic distance measurement (EDM) devices. The coordinate geometry approach can produce very accurate cartographic data, more accurate than is practical with conventional manual digitizing of existing maps (Antenucci et al., 1991).

3.4.2.5. *Key Entry*

Attribute data and map annotation commonly are entered through a key-entry procedure. The data may be keyed directly into a GIS workstation or into other less expensive terminals. Data entered separately can be bulk-loaded later into the GIS database. Annotation generally is key entered and then positioned on the map image through the interactive graphic capabilities of the GIS.

3.4.3. CRITERIA FOR CHOOSING MODES OF INPUT

Several factors should be considered in choosing the mode of data input. If the data source is an image, scanning would be appropriate, while maps can be scanned or digitized. If the database model of the GIS is raster, then scanning is appropriate; digitizing is better for a vector format. For some data, entry in vector form is more efficient, followed by conversion to raster. For example, a county boundary could be digitized in vector form by capturing the locations of points along the boundary, assuming that the points are connected by straight line segments. The vector representation of the boundary could be converted to a raster format through vector-raster conversion, a function that most GISs have. The computer determines which county each cell is in using the vector representation of the boundary and outputs a raster data layer. Digitizing the boundary is much less work than cell by cell entry. Other examples of data for which vector digitizing would be appropriate include watersheds, Thiessen polygons, and land use, particularly where large areas have the same attributes.

The resolution of raster input depends on the size of the grid cell, so it is important to select an optimal grid size to suit the requirements. The larger the size, the less precise the information; the smaller the grid size, the larger the database.

3.5. Quality of Digital Data

The quality of any GIS analysis depends on the quality of the GIS database. A great quantity of data is available for developing GIS databases for hydrologic applications, however, more important than the quantity is the quality of the data. Regardless of the technique used to develop the database, it is the responsibility of the user to confirm the quality of the database used in GIS analyses.

The goal in describing data quality is to describe fitness for some use (Chrisman, 1991). Criteria to be verified include positional accuracy of the map features, completeness of the data entered, proper definition of all elements (such as assignment to the proper layer or feature type), topological integrity of spatial relationships, and logical consistency of data values (Antenucci et al., 1991). Chrisman (1991) discusses most of these criteria in detail.

The following discussion of errors in digital data is taken largely from Goodchild and Kemp (1990). There is a nearly universal tendency to lose sight of errors once data are in digital form. However, there is the potential for numerous errors of various types to be in a database. Any errors in the original sources (source errors) are transferred into the database. If higher accuracy is required than the source materials, then better source materials are needed. The user must determine if the added cost is justified by the objectives of the study. Accuracy standards should be determined by considering both the value of information and the cost of collection.

Accuracy of analog maps was discussed in Section 3.3.1.5. Existing digital data may be distributed with a statement as to its quality. However, it is the user's responsibility to determine the level of accuracy of the data and whether that level of accuracy is appropriate for the application of interest. When a user of a spatial database does not know how the data were captured and processed prior to input, it can often lead to misinterpretation and false expectations about accuracy. Thus, it is important to obtain information in addition to the data, e.g., information on the procedures used to collect and compile the data, information on coding schemes, quality control measures used, and accuracy of instruments. Unfortunately, such information is often not available.

In many cases, existing digital data were derived from analog maps that are available to the user. One accuracy check the user should make is comparing the digital coverage to the analog map. If the accuracy is not acceptable, the user may be able to correct the errors through manual computer-assisted digitizing in-house. In some cases, the existing digital data may not be useable.

Source errors are extremely common in non-mapped source data, such as location of wells. Source errors can be caused by misinterpreting aerial photography and other remotely sensed data. Source errors often occur because base maps are relied on too heavily. Goodchild and Kemp (1990) cite an attempt in Minnesota to overlay Department of Transportation bridge locations on USGS transportation data that resulted in bridges lying neither beneath roads, nor over water, and roads lying apparently under rivers. Until the two data sets were compared in this way, it was assumed that each was locationally acceptable. The ability of GIS to overlay may expose previously unsuspected errors.

Data capture errors occur in manual digitizing because eye-hand coordination varies from operator to operator and from time to time. Digitizing is a tedious task. It is difficult to maintain quality over long periods of time. The accuracy of the digital data can be determined through comparison with the map. Comparison with the map includes overlaying the digital coverage with the map and determining if lines and points are within some tolerance level. The tolerance may vary with the application. For example, Lai (1988) specified a tolerance level of 0.1 cm (all points and lines were within 0.1 cm of those on source documents) in an application with a 1:24 000 USGS quadrangle.

Classification errors can be identified by displaying map layers. Simple typing errors may be invisible until presented graphically. Floodplain soils may appear on hilltops. Pastureland may appear to be misinterpreted as marsh. More complex classification errors may be due to the sampling strategies that produced the original data. Timber appraisal is commonly done using a few, randomly selected points to describe large stands. Information may exist that documents the error of the sampling technique, however, such information is seldom included in the GIS database.

The quality of GIS data should be measured periodically as the database is developed and updated. Quality descriptions can be entered in the database as attributes of the map features and related data. This aspect of the GIS database too often is overlooked, but its importance is gaining appreciation. The presence of quality characteristics allows users to make valid decisions about the usefulness of data for specific applications (Antenucci et al., 1991).

3.6. Integrating Data from Different Sources

Development of a GIS database usually involves integration of data from multiple sources with varying accuracies, scales, geometric structures, spatial resolutions, and other characteristics. The differences in these data layer characteristics must be considered in the integration of the data to ensure an acceptable quality database.

Digital data derived from maps by scanning or digitizing retain the map's projection. With data from different sources, a GIS database often contains information in more than one projection. The data must be converted to a common projection to enable integration or comparison. A good GIS can convert data from one projection to another and to latitude/longitude (Goodchild and Kemp, 1990). Conversion of data from one map projection to another generally can be accomplished without significant loss of accuracy (Antenucci et al., 1991). Maling (1991) describes some of the methods of transformation that can be used in a GIS.

Raster data from different sources may use different pixel sizes, orientations, positions, and projections. Resampling is the process of interpolating information from one set of pixels to another. Resampling to larger pixels is comparatively safe; resampling to smaller pixels is very dangerous (Goodchild and Kemp, 1990).

Scale is an important factor in integration of data from various sources. Maps of the same area at different scales will often show the same features, e.g., features are generalized at smaller scales, enhanced in detail at larger scales. Variation in scales can be a major problem in integrating data. For example, suppose the scale of topography, soils, and land cover maps is 1:250,000, but the only geological mapping available is 1:7,000,000. If integrated with the other layers, the user may believe the geological layer is equally accurate.

3.7. Cost of Building a Database

Developing the database is by far the most expensive cost in a typical GIS implementation. The cost of developing a GIS database can vary considerably, depending on the number of features included, the accuracy to be achieved, the condition of source materials, the availability of existing digital files, and whether new data must be acquired. It is not uncommon for the cost of data collection to exceed the cost of hardware and software by a factor of two (Maguire, 1991). Antenucci et al. (1991) estimate that the hardware and software may represent as little as 5 percent, and seldom more than 20 percent, of a project's total cost.

3.8. Database Administration and Update

After a database is developed, it is essential to provide for its administration, maintenance, and update. Chorley (1988) relates the following lesson from experience: "The custodian of a database should have long-term interest in the information; only in this way will the custodian have confidence in the data and keep them up to date."

A data directory is essential to effective management of a GIS database (Antenucci et al., 1991). The directory defines the entities, their attributes, and associated domain values and conveys those meanings to the database administrator and GIS users. A directory allows system users to identify map features and attribute data available for specific applications, as well as the definitions, quality, and other characteristics of those data. Grimshaw (1988) recommends including methods of updating, amending, and deleting data.

Database administration maintains backup copies of the database and system software to allow restoration should a mechanical, electrical, or human failure occur. A specific "backup" cycle should be established to duplicate parts of and the whole database at regular intervals. It is a good idea to store copies of the backup data at a separate site in case of a natural disaster, vandalism, or other unforeseen circumstance at the system facility.

Database updating involves an important question (Hunter, 1988): should noncurrent data be archived for future use or purged from the system as a cost-control measure? There are many applications for which it is necessary to recover previous versions of the information to determine changes which have occurred over time. For example, a hydrologic analysis of the effect of land use change on water quality over time requires land use data over time. Traditionally, the need for historical information has been satisfied by retaining copies of previous versions of hard copy maps. The destruction of paper records involves a much more physical effort than the deletion of machine-readable data, consequently, paper records are not as easily destroyed as digital data.

3.9. Summary and Recommendations

Developing a database for GIS applications in hydrology requires thorough planning, identification of sources of data, procurement of the data, and entry of data into the GIS, followed by maintenance and updating of the database. Consistent quality control measures must be implemented throughout the effort.

Many people have cautioned against letting GIS activities become activities in support of the technology rather than in support of the user's agenda. Similarly, available data should not drive the particular hydrologic analyses conducted. The planning process is important. First, the required analyses should be identified. Then, required data, including the specific data layers, for the analyses should be determined. It is important to determine the specific data layers before digitizing data because it is much easier to combine data layers in the GIS than to disaggregate layers.

Next, sources of data should be identified. Sources include both analog materials, such as maps and aerial photographs, and existing digital databases. Data can be obtained from government agencies at various levels and from private organizations. The procedure and cost for procuring data vary depending on the source.

Entry of the data into the GIS is usually the most time-consuming and expensive part of a GIS application. Many hydrologic applications require data derived from maps at larger scales than those from which many existing digital databases were derived. This means that digitizing is required for many hydrologic applications. Manual digitizing is a tedious operation that requires training and practice to do well.

Regardless of the source of the data and the method of data entry into the GIS, quality control is of utmost importance. The quality of any analyses conducted using the data depends on the quality of the data. It is the responsibility of the analyst to determine the quality of the data.

A database requires maintenance and update. Quality control must be implemented in this phase as well to maintain the integrity of the database. Caution must be used in updating databases over time. There are many cases, including many instances in hydrology, when it is necessary or desirable to have data at different points in time, sometimes over years. Experience has shown that archiving digital data is preferable to deleting or overwriting it.

The use of GIS databases and analyses tools has the potential to contribute significantly to addressing hydrologic problems and issues. However, caution must be exercised to prevent thinking that the data are more accurate than they actually are. Also, combining data to "create" new data should be approached with caution. It is easy to believe that what the computer outputs is accurate. Responsible development and use of GIS databases is essential if GIS technology is to be applied successfully in hydrology.

References

Ahonen, P. and A. Rainio. 1988. Developing a query system for joint use of spatial data in Finland. In: J.C. Muller (ed), Environmental Applications of Digital Mapping, Proceedings of Eurocarto Seven, ITC, Enschede, The Netherlands, pp. 77-86.

Anderson, J.R., E.E. Hardy, J.T. Roach, and R.E. Witmer. 1976. A land use and land cover classification system for use with remote sensor data. Professional Paper 964. USGS, Reston, Virginia.

Antenucci, J.C., K. Brown, P.L. Croswell, M.J. Kevany, and H. Archer. 1991. Geographic information systems: a guide to the technology. Van Nostrand, New York. 301p.

Bogaerts, M.J.M. 1987. Grootschalige ruimtelijke informatiestemen in Nederland - nieuwe ontwikkelingen in de periode 1984-1987 (Large-scale spatial information systems in the Netherlands - new developments in the period 1984-1987). Kartografisch Tijdschrift 13:38-42.

Brandenberger, A.J. and S.K. Ghosh. 1985. The world's topographic and cadastral mapping operation. Photogrammetric Engineering and Remote Sensing 51(4):437-444.

Briggs, D.J. 1988. The use of broad-scale GIS for modelling the environmental impacts of land use change in Europe. Paper presented to the Polish Academy of Sciences, Radzikow, Poland.

Chorley, R. 1988. Some reflections on the handling of geographical information. Int. J. Geographical Information Systems 2(1):3-9.

Chrisman, N.R. 1991. The error component in spatial data. In: Maguire, D.J., M.F. Goodchild, D.W. Rhind (eds). Geographical Information Systems: Principles and Applications. Longman, London, pp. 165-174, Vol 1.

Clark, D.M., D.A. Hastings, and J.J. Kineman. 1991. Global databases and their implications for GIS. In: Maguire, D.J., M.F. Goodchild, D.W. Rhind (eds.). Geographical Information Systems: Principles and Applications. Longman, London, pp. 217-231, Vol 2.

Engebretsen, O. 1987. The Norwegian land accounting system. In: ECE-Seminar on New Techniques to Collect and Process Land-use Data, Volume II: Response Papers, Gvle, Sweden.

Falk, T. 1987. Land-use information for physical planning and land consolidation in Sweden. In: Proceedings of the Seventh Annual ESRI User Conference, Redlands, California.

Fisher, P.F. 1991. Spatial data sources and data problems. In: Maguire, D.J., M.F. Goodchild, D.W. Rhind (eds.). Geographical Information Systems: Principles and Applications. Longman, London, pp. 175-189, Vol 1.

Getting, I.A. 1993. The global positioning system. IEEE Spectrum 30(12):36-38 and 43-47.

GIS World. 1992. The 1993 International GIS Sourcebook, 4th edition, October 1992. Fort Collins, CO: GIS World, Inc.

Goodchild, M.F. and K.K. Kemp. 1990. Introduction to GIS: NCGIA Core Curriculum. National Center for Geographic Information and Analysis, University of California, Santa Barbara.

Grimshaw, D.J. 1988. The use of land and property information systems. Int. J. Geographical Information Systems 2(1):57-65.

Hunter, G.J. 1988. Non-current data and geographical information systems-a case for retention. Int. J. Geographical Information Systems 2(3):281-286.

Lai, P. 1988. A case study of the Patuxent basin geographical information system database. Int. J. Geographical Information Systems 2(4):329-345.

Maguire, D.J. 1991. An overview and definition of GIS. In: Maguire, D.J., M.F. Goodchild, D.W. Rhind (eds.). Geographical Information Systems: Principles and Applications. Longman, London, pp. 9-20, Vol 1.

Maling, D.H. 1991. Coordinate systems and map projections for GIS. In: Maguire, D.J., M.F. Goodchild, D.W. Rhind (eds.). Geographical Information Systems: Principles and Applications. Longman, London, pp. 135-146, Vol 1.

Merchant, D.C. 1987. Spatial accuracy specification for large scale topographic maps. Photogrammetric Engineering and Remote Sensing 53(7):958-961.

Millette, T. 1990. The Vermont GIS: a model for using Regional Planning Commissions to deliver GIS in support of growth management. In Geographic Information Systems: Developments & Applications, L. Worrall, ed. London and New York: Belhaven Press. pp. 65-86.

Montanari, A. 1989. Strategic project the Mediterranean: environment and economic development. National Research Council, Naples.

Muller, J-C. 1991. Generalization of spatial databases. In: Maguire, D.J., M.F. Goodchild, D.W. Rhind (eds). Geographical Information Systems: Principles and Applications. Longman, London, pp. 457-475, Vol 1.

O'Callaghan and B.J. Garner. 1991. Land and geographical information systems in Australia. In: Maguire, D.J., M.F. Goodchild, D.W. Rhind (eds.). Geographical Information Systems: Principles and Applications. Longman, London, pp. 57-70, Vol 2.

Piscator, I. 1986. The Swedish land data bank system and its use by local authorities. In: B. Rystedt (ed), Land-Use Information in Sweden. Applications of New Technology in Urban and Regional Planning and in the Management of Natural Resources, Swedish Council for Building Research, Stockholm, pp. 59-68.

Reybold, W.U. and G.W. TeSelle. 1989. Soil geographic data bases. Journal of Soil and Water Conservation 44(1):28-29.

Robinson, A., R. Sale, J. Morrison, and P.C. Muehrcke. 1984. Elements of Cartography, 5th edn. Wiley, New York.

SCS. 1984. Technical Specifications for Line Segment Digitizing of Detailed Soil Survey Maps. Government Printing Office, Washington, DC.

Scholten, H.J. and E. Meijer. 1988. From GIS to RIA. Paper presented at the URSA- NET Conference, Patras, Greece.

Scholten, H. and M. van der Vlugt. 1990. A review of geographic information systems in Europe. In Geographic Information Systems: Developments & Applications, L. Worrall, ed. London and New York: Belhaven Press. pp. 13-40.

Shanholtz, V.O., P.A. Hellmund, R.K. Byler, S. Mostaghimi, and T.A. Dillaha. 1987. Agricultural pollution potential database - phase I. Virginia Geographic Information Systems Laboratory, Agricultural Engineering Department and Landscape Architecture Program. Virginia Polytechnic Institute and State University, Blacksburg, Virginia. 133p.

Sowton, M. 1991. Development of GIS-related activities at the Ordnance Survey. In: Maguire, D.J., M.F. Goodchild, D.W. Rhind (eds.). Geographical Information Systems: Principles and Applications. Longman, London, pp. 23-38, Vol 2.

Starr, L.E. an K.E. Anderson. 1991. A USGS perspective on GIS. In: Maguire, D.J., M.F. Goodchild, D.W. Rhind (eds.). Geographical Information Systems: Principles and Applications. Longman, London, pp. 11-22, Vol 2.

Taylor, D.R.F. 1991. GIS and developing nations. In: Maguire, D.J., M.F. Goodchild, D.W. Rhind (eds.). Geographical Information Systems: Principles and Applications. Longman, London, pp. 71-84, Vol 2.

Thompson, M.M. 1988. Maps for America, 3rd edition. US Geological Survey, Reston, Virginia.

Tobler, W.R. 1988. Resolution, resampling and all that. In: Mounsey, H.M. (ed). Building Databases for Global Science. Taylor and Francis, London, pp. 129-137.

Weibel, R. and M. Heller. 1991. Digital terrain modeling. In: Maguire, D.J., M.F. Goodchild, D.W. Rhind (eds.). Geographical Information Systems: Principles and Applications. Longman, London, pp. 269-297, Vol 1.

CHAPTER 4

Spatial Data Characteristics

C.A. Quiroga, V.P. Singh and S.S. Iyengar

4.1. Introduction

The use of GIS technology in engineering, in general, and in hydrology, in particular, has become feasible due primarily to the possibility of handling and integrating enormous amounts of geographic data efficiently. As shown in Figure 4.1, a GIS

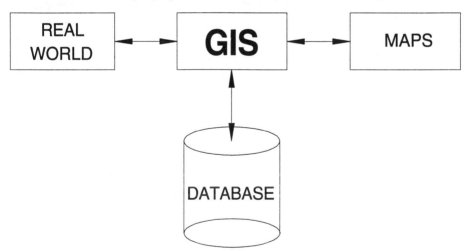

Figure 4.1. General Configuration of a GIS

can be thought of as a powerful tool capable of creating and manipulating maps, with the necessary links to retrieve, manipulate, and update data from a database. Both maps and attribute data constitute abstractions or simplifications of the real world.

Most geographic databases contain both spatial and nonspatial attributes and are mounted on relational database management systems (DBMS). However, spatial data has unique characteristics which limit the use of standard query languages.

V. P. Singh and M. Fiorentino (eds.), Geographical Information Systems in Hydrology, 65–89.
© 1996 *Kluwer Academic Publishers. Printed in the Netherlands.*

As a result, geographic databases usually contain two sets of candidate keys: one for linking the graphical or spatial features with the database, and the other one as primary key for query purposes in the DBMS. This duality is further complicated by the fact that spatial data is multidimensional (X + Y + Z + T + other attributes), even though for computer implementations, only three dimensions are usually considered: X, Y, and Z. The third dimension, Z, is used to represent both elevation or other attributes. Time is usually either ignored altogether or accounted for in different layers of information.

Spatial data, particularly geographic data, tends to follow patterns, i.e., nearby locations tend to influence each other in what is normally known as spatial dependence. How to conceptualize this dependence, though, is heavily influenced by human perceptions and needs. So strong is this influence that, to date, a complete and rigorous framework for spatial data modeling has not been developed. As a result, there has been a proliferation of data models, although for practical purposes, most of them can be classified either as raster or vector models.

This chapter deals with basic concepts related to spatial data characteristics. Nonspatial data management and GIS software issues are described in Chapter 6. The presentation is divided as follows: Section 4.2 presents approaches for spatial data modeling encountered in the literature. Section 4.3 examines some of the most common data structures used for handling spatial information. Section 4.4 covers issues related to conceptualization and exchange of geographic data. Finally, section 4.5 addresses the problem of time in GIS. An effort was made to include aspects of specific interest in hydrology, even though most concepts described in the chapter are general.

4.2. Spatial Models

Two kinds of data are usually associated with geographic features: spatial and nonspatial data. Spatial data refers to the shape, size and location of the feature. Nonspatial data refers to other attributes associated with the feature such as name, length, area, volume, population, soil type, etc.. Each type of data has specific characteristics and, as a result, different conceptual models and data structures have been developed for their implementation. Spatial data can be conceptualized using a category approach, an object approach, or a combination of both. Nonspatial data is normally conceptualized using a relational data model, because of the popularity of relational databases. In this chapter, spatial modeling approaches and data structures are discussed.

4.2.1. CATEGORICAL APPROACH

Spatial data can be thought of as a template composed of cells and/or points and lines, to which specific information is linked. As more than one kind of information is usually handled at the same time, e.g., drainage network, soil associations, land use, and so on, several layers, each containing the spatial structure associated with a single category in the map, can be defined. In general, categories are implemented separately and, as a result, the attributes that describe specific objects in the map are scattered among several, disconnected layers of information (Figure 4.2). If one

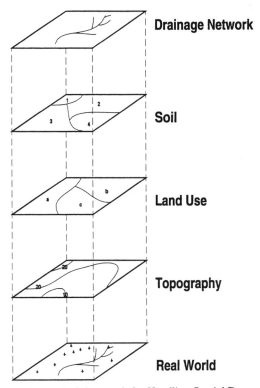

Figure 4.2. Categorical Approach for Handling Spatial Data

of the layers contains the boundaries of the basin, spatial operators can be used to define boundaries in the remaining layers, as well as to define specific relationships among them.

This approach is conceptually simple, and is based on the assumption that events in the real world can be represented as linear superpositions of information categories. In many cases, this assumption is valid. In fact, most GIS are based on it (Aronoff, 1989; Burrough, 1986). Indiscriminate aggregation, however, may lead to erroneous or, at least, meaningless results. For example, superimposing soil associations, land use types, and ground slopes in rural areas might lead to useful conclusions as far as defining types of crops best suited to specific soil and drainage conditions. Such an exercise in urban settings might be meaningless, at least on a short-term basis, because of the permanent character of buildings and infraestructure. In addition to this, in most cities, man-made drainage networks usually play a more important role in defining drainage conditions than ground slope.

Because most GIS are based on the categorical approach, practically all geographic databases in existence are built around the concept of data layering. Since building geographic databases usually require a considerable investment both in terms of time and money, an obvious conclusion of this situation is that the categorical approach will most likely remain in use, despite its shortcomings.

4.2.2. OBJECT APPROACH

One way to avoid the problem of indiscriminate aggregation, typical of the categorical approach, is by assuming that spatial data is composed of objects. With this approach, each object has a clearly defined identity (for example, Mississippi River, Amite River Basin), defined in terms of attributes. Strictly speaking, spatial and nonspatial information could be handled simultaneously. The object identity is kept all the time and, as a result, the correlation between the real world event and its representation is not lost, solving, therefore, the integrity problem that characterizes the categorical approach. Objects can be integrated with other objects to form more complex objects. However, since objects are treated as self-contained units of information, associations with other objects are not automatically defined and, therefore, interobject references must be made (Elmasri and Navathe, 1989).

Only a few, mainly experimental GIS, are based on this object- oriented approach (David et.al., 1993; Healey, 1991). Two examples of conceptual models encountered in the literature are described below. It may be worth mentioning, however, that efforts have been made towards coupling the category and object-oriented approaches to make relationships among objects in a map better grasped and visualized, and to provide more meaningful correlations among layers (Kim et.al., 1993; Healey, 1991).

4.2.3. DEDUCTIVE OBJECT-ORIENTED MODEL

As mentioned before, most GIS handle spatial and nonspatial data separately. A problem with this approach is that the model used for conceptualizing the nonspatial component, usually the relational data model, is not well suited for handling hierarchies and complex relationships that are normally associated with spatial data. Abdelmoty et.al. (1993) attempted to solve this situation by integrating both spatial and nonspatial data into a single data model, based on an object-oriented approach. They use deductive mechanisms to derive data and to express queries and integrity constraints.

As shown in Figure 4.3, a deductive object-oriented geographic database can be defined using two disjointed sets of relationships: (1) a set of base or extensional database (EDB) relationships, and (2) a set of derived or intensional database (IDB) relationships. The EDB component, in turn, contains two levels of data representation: a primitive level and an object level. The primitive level is composed of the spatial data structures used to represent geographic objects, and can either be in vector or raster form, as described in Section 4.3. The object level includes the spatial model used to represent geographic phenomena in terms of semantic relationships such as aggregation, specialization, and association. The IDB

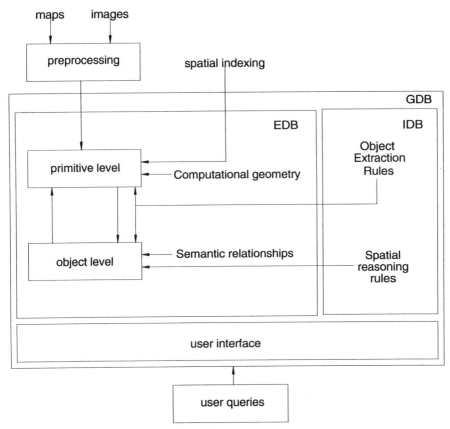

Figure 4.3. Components of a deductive object-oriented geographic database (after Abdelmoty, Williams and Paton, 1993)

component contains the set of rules that are used for spatial reasoning over the geographic database, and for the feature extraction needed to infer object level entities from the corresponding primitive level objects. Here, spatial reasoning is defined as the method to deduce spatial information which has not been recorded explicitly.

Geographic objects are assumed to be instances of specific object classes. For implementation purposes, therefore, they can be defined as views of other geographic objects and/or through specific relationships with those objects. Deductive rules based on open-set and closed-set topology can be applied over entities on the object level to define these relationships.

Object classes can be generalized into themes. Instances of themes are called map layers. The data set associated with each map can be analyzed separately, or can be combined with other data sets to derive meaningful information. Deductive rules can be used to represent those object classes involved in the combination process.

The use of deductive mechanisms can also be used to automatically extract object level concepts from primitive level representations. Two cases are possible: attribute-based object recognition, and structure-based object recognition. Attribute-based object recognition is suitable in relatively simple cases in which objects can be distinguished by a set of unique identifiers. This assumes that each pattern of object class is considered in isolation with other patterns. No search strategy is needed. However, possible relationships between objects are ignored. Structure-based object recognition, on the other hand, considers both spatial attributes of objects and spatial relationships between objects. This approach requires the selection of a strategy for selecting objects and for defining how the interpretation process is carried out.

4.2.4. SPATIAL DATA TRANSFER STANDARD (SDTS) MODEL

A problem with current GIS systems is their lack of consistency as far as data types, structures and storage, which makes the transfer of data between dissimilar computer systems a difficult task. In an effort to facilitate and systematize such a transfer, the National Institute of Standards and Technology developed the Spatial Data Transfer Standard (NIST, 1992). In SDTS, standard procedures for mapping from the conceptual level to the low or physical encoding level are defined. Part of SDTS is a conceptual, object-oriented model for spatial data.

The SDTS spatial model is divided into three parts: a model of spatial phenomena; a model of spatial objects; and a model of spatial features. These components are briefly explained below.

Spatial Phenomena Model.   Spatial phenomena are facts or occurrences that are defined in time and space in the real world (Amite River, Interstate 10, etc.). Spatial phenomena that share the same types of attributes (length, slope, area) are said to belong to a class (Amite River belongs to River). Classes that contain characteristics that make them distinguishable from other classes are called entity types. Individual phenomena are called entity instances. Each entity instance is uniquely identified by a specific attribute value or set of attribute values. The set of attributes used for this purpose is called key attribute. For example, for an entity type called 'Basin', with key attribute <u>Name</u> and <u>Outlet</u>, and non-key attribute Area, a valid entity instance might be "Amite" at "Darlington", with an area of 1542 km^2.

Three processes that affect spatial phenomena and classes are defined. **Generalization** is a process by means of which classes are grouped into themes or categories. **Aggregation** is a process that allows the construction of spatial phenomena out of simpler, component, phenomena. **Association** is the assignment of phenomena to sets, using criteria different from those used for classification. **Relationships** are special cases of associations that exist between classes.

Entity types can be generalized into themes according to attribute characteristics. Themes can also have their own attributes. For example, the theme 'Water

Body' is a generalization of the entity types 'River' and 'Lake'. 'Utility' is a generalization of the entity types 'Sewer' and 'Water Pipe'.

Entity instances can be aggregated to form instances of a different entity type. For example, a river and its tributaries can be aggregated into a drainage basin system.

Entity instances can be associated with other entity instances using domains different from those used to define an entity type. Two commonly used association domains are space and time. An example of association by space is given by the distance between several entity instances and a common point (Zachary, Denham Springs and Baker are all located within a 30 km radius from downtown Baton Rouge). An example of association by time is given by age (Mississippi River floodplains before and after the construction of flood control works).

Associations among entity types are called relationships. A relationship type defines the relationship characteristics. A particular relationship value defines a relationship instance between entity instances.

Spatial Objects Model. Spatial objects are graphical elements used to represent spatial phenomena in a map. Such a representation can be assumed to be digital. For implementation purposes, a spatial object can result from the aggregation of other spatial objects, not all of which represent entity instances. For this reason, SDTS makes a distinction between spatial objects and entity objects. An **entity** object is a spatial object that represents all of a single entity instance.

Like spatial phenomena, entity objects have attributes (spatial and nonspatial) and topology (relationships). Spatial objects can have attributes and topology even if they are not entity objects. Entity objects that share the same types of attributes are said to belong to an entity object class. As with entity types, it is possible to define generalizations and associations among entity object classes. The representation of an entity theme is an object theme. The representation of an entity association is an object association. The correspondence between spatial phenomena and their digital representation is summarized in Table 4.1.

TABLE 4.1. Correspondence Between Spatial Phenomena and their Digital Representation in the SDTS Model (Adapted from NIST, 1992)

Level	Phenomena	Digital Representation
Class	Entity Type	Entity Object Class
Instance	Entity Instance	Entity Object
Generalization	Entity Theme	Object Theme
Association	Entity Association	Object Association

Spatial objects can be aggregated into more complex spatial objects. For standardization purposes, all spatial objects are assumed to be constructed using simple spatial objects. Depending on whether the spatial object has topology or not, two types of simple spatial objects can be defined: (a) for geometry only; and (b) for geometry and topology. For certain operations, like network analysis for example, only topology is strictly needed; in this case, the coordinates associated with each node can eventually be removed. A summary of simple spatial objects is presented in Figure 4.4.

Spatial Features Model. A feature is defined as the combination of the phenomenon in the real world and its representation in the map. A feature type consists of a set of feature instances, each resulting from the combination of an entity instance and the entity object used to represent it.

4.3. Spatial Data Structures

Spatial models are implemented using spatial data structures. Using the three-schema architecture found in most data base systems as an analogy (Elmasri and Navathe, 1989), the spatial model would be the high-level data model used for the conceptual schema, whereas the spatial data structure would be the low-level/implementation data model used for the internal schema. This analogy is useful for notation purposes, as the use of the terms 'spatial model' and 'spatial data structure' in the literature has not always been consistent (Egenhofer and Herring, 1991; Aronoff, 1989; Burrough, 1986). In some cases, the term 'data model' has been used indiscriminately, with no qualifier, to denote both the conceptual spatial model and the spatial data structure. In some others, the term 'data structure', which is widely used in traditional implementation database models such as the relational, network and hierarchical data models, has also been used to denote both the conceptual spatial model and the spatial data structure. For consistency, therefore, 'spatial model' is used to denote the high-level data model, whereas 'spatial data structure' is used to denote the low-level/implementation data model.

Spatial data structures are classified depending on which form, either raster or vector, spatial objects are defined. In either case, however, since spatial data structures deal with the low- level/implementation model used for defining internal schemas, they are designed to minimize computer storage requirements, and to optimize spatial operations. This makes spatial data structures system dependent (Egenhofer and Herring, 1991).

In the following sections, a short description of the most commonly used spatial data structures is given. The discussion includes both raster- and vector-based spatial data structures.

Number of Dimensions	Geometry	Geometry and Topology	Graphical Representation	Purpose
Zero	Point		•	Geometric location of point features
		Node		Topological junction
One	Line segment			Direct line between two points
	String			Connected nonbranching sequence of line segments
	Arc			Curve defined by a mathematical expression
		Link		Topological connection between nodes
		Chain		Sequence of nonintersecting lines and/or arcs bounded by nodes
	G-ring			Sequence of nonintersecting strings and/or arcs, with closure
		GT-ring		Ring created with chains
Two	Interior area			Area not including its boundary
	G-polygon		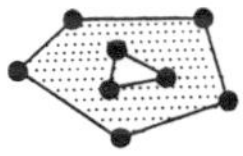	Area bounded by one outer G-ring
		GT-polygon	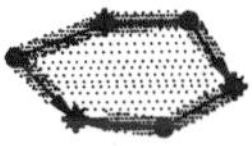	Area bounded by one or more GT-rings
	pixel			Smallest nondivisible element of a digital image
	grid cell			Smallest nondivisible element of a grid

Figure 4.4. Summary of Simple Spatial Objects in the SDTS Model (NIST, 1992)

4.3.1. RASTER SPATIAL DATA STRUCTURES

A spatial object is in raster form if it is based on the aggregation of grid cells (or pixels, in the case of a digital image). Most implementations are built around square or rectangular grid cells, but other shapes, such as triangular and hexagonal shapes, are also possible. In the simplest case, grid cells are uniform and can, therefore, be referenced by their row and column numbers. Each grid cell may be assigned only one value, which represents the value of the attribute being mapped. Different types of attributes are handled by creating data layers or overlays, as shown in Figure 4.5.

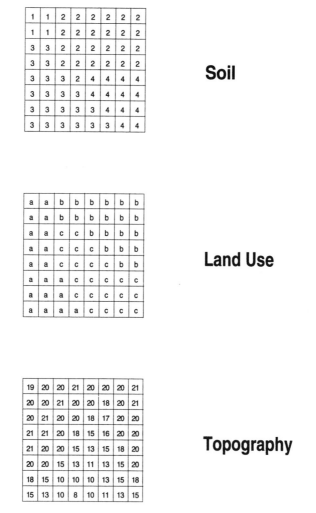

Figure 4.5. Raster Representation of the Soil, Land Use, and Topography Categories shown in Figure 4.2

Raster spatial data structures are simple to conceptualize and to use for overlay analysis. Further, they can be used efficiently for modeling high spatial variability, which makes them appealing for the manipulation of remotely sensed data and digital images. On the negative side, data files tend to be huge, even though data compression techniques can reduce this burden.

Strictly speaking, raster spatial data structures are two- dimensional arrays. This means that the area that each grid cell represents can be used both to define map resolution and the number of grid cells needed to describe the spatial distribution of the attribute under study. The finer the resolution, i.e. the smaller the area each grid cell represents, the larger the number of grid cells needed. The number of grid cells needed is inversely proportional to the square of the resolution and, consequently, the need to develop methods to minimize storage requirements is crucial. Most data file compression methods are based on the fact that adjacent grid cells often have the same attribute value and, as a result, it makes sense to group them under the same category. Two of such methods, run-length encoding and quadtrees, are described below.

Run-Length Encoding.   With this technique, grid cells having the same attribute value are organized into blocks or runs. A table can be formed to describe, for each row, the number of consecutive grid cells that are associated with the same attribute value. Table 4.2 shows, as an example, the run-length encoding that would correspond to category 'soils' in Figure 4.5.

TABLE 4.2. Run-Length Encoding of the Raster Spatial Data Structure of Figure 4.5

Row	Value	Length	Row	Value	Length	Row	Value	Length
1	1	2	4	3	2	6	4	4
1	2	6	4	2	6	7	3	5
2	1	2	5	3	3	7	4	3
2	2	6	5	2	1	8	3	6
3	3	2	5	4	4	8	4	2
3	2	6	6	3	4			

Quadtrees.   Quadtrees are perhaps the most common structure used to represent raster data. In general, the technique is based on a successive division of the map or image into quadrants until every subdivision can be assumed to be spatially homogeneous, i.e., one single attribute value can be associated with it. In practical terms, grid cells can adopt variable sizes, depending on the number and spatial distribution of the original grid cells having the same attribute value.

Figure 4.6 shows a sample region and its quadtree representation. For easy visualization, a node with children is represented by a circle; a node without children and no attribute values is represented by a blank square; and a node with an

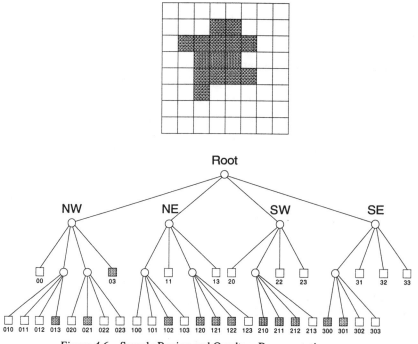

Figure 4.6. Sample Region and Quadtree Representation

attribute value is represented by a hashed square. NW, NE, SW and SE are used to designate quadrants, even though for computer implementation, a numerical coding sequence, such as 0, 1, 2 and 3, is more convenient. If Mortons numbering sequence is used, the quadtree encoding adopts the form shown in Figure 4.7. Many algorithms to accomplish this goal are available in the literature. Reviews on the subject can be found in Samet (1984, 1990), and Ooi (1990).

Alternative approaches include the use of forest of quadtrees (Jones and Iyengar, 1984; Iyengar and Miller, 1986). In the forest-of- quadtrees representation, a quadtree is decomposed into a collection of subtrees, each of which corresponds to a maximal square, as shown in Figure 4.8. L and K correspond to the coordinates of each node, assuming a 0-1-2-3 numerical sequence. The root has coordinates (1,0). If the coordinates of a node are denoted by (L,K), those of its children are given by (L+1,4K+d), where d is the digit representing the direction of the child. L is the level of P in the tree structure, and K is a code that represents the number of sequential choices from root to node P.

4.3.2. VECTOR SPATIAL DATA STRUCTURES

A spatial object is in vector form if it is based on the aggregation of topologic objects. Vector spatial data structures are designed to represent object geometry as accurately as possible. In theory, this implies that the map area must be continu-

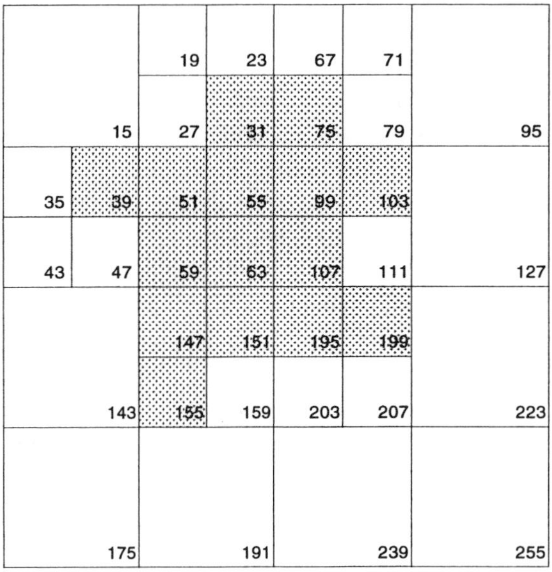

Morton #	Attribute
15	W
19	W
23	W
27	W
31	B
35	W
39	B
43	W
47	W
51	B
55	B
59	B
63	B
67	W
71	W
75	B
79	W
95	W
99	B
103	B
107	B
111	W
127	W
143	W
147	B
151	B
155	B
159	W
175	W
191	W
195	B
199	B
203	W
207	W
223	W
239	W
255	W

Figure 4.7. Sample Region and Encoding using Morton's Numbering Scheme

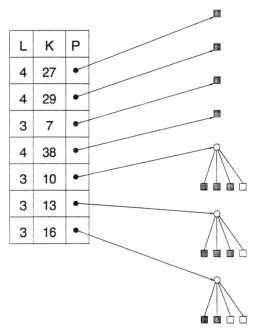

L	K	P
4	27	•
4	29	•
3	7	•
4	38	•
3	10	•
3	13	•
3	16	•

Figure 4.8. Forest-of-Quadtrees Representation of Sample Region of Figure 4.6

ous, so that exact coordinates can be defined. In practice, accuracy is limited by the number of bits used to define the numerical values associated with these coordinates.

Vector objects can be assigned several attributes, usually associated with an attribute table. As with raster spatial data structures, different types of attributes are handled by creating data layers or overlays. Vector spatial data structures provide a more compact, but also more complex, way to conceptualize geographic objects than raster spatial data structures. As a result, data files are usually much smaller than those making use of raster structures. They are also more efficient for modeling topology. On the negative side, spatial variability is more difficult to handle, and overlay operations require special care to avoid the occurrence of spurious polygons.

As opposed to raster spatial data, resolution for vector data is a relative term that depends both on the spatial variability of the geographic object and the number of spatial objects (nodes, links and chains) needed for its definition. For this reason, the term map scale, used in a similar context as in traditional cartography, is usually preferred.

A variety of spatial data structures have been developed to handle vector data. In general, differences among them are related to the way they handle topological relationships. Three of them, spaguetti data structures, topologic data structures and TINs, are described below.

Spaguetti Data Structure. With this technique, geometry is defined in terms of strings, basically imitating the digitizing process. Strings are not interconnected, even if they appear to be so in the map. This means that if polygons are formed, data for adjacent boundary segments must be entered twice. An example of application of the spaguetti data structure is shown in Figure 4.9. Spaguetti data structures are very inefficient for handling topology. However, they are well suited for scanning purposes in which spatial connectivity is not required. Like most vector-based data structures, they do not constitute the best solution for modeling surfaces, and perform poorly at representing transition zones.

Topologic Data Structure. With this technique, spatial relationships among entities, mainly connectivity, are explicitly recorded. In the general case, polygons may be formed using several connected chains or arcs, as shown in the example of Figure 4.10. Adjacency is defined based on the polygons that surround every chain. As with other data structures, tables can be constructed to describe coordinates, connectivity and adjacency characteristics. In Figure 4.10, these tables are represented as summarized lists. The polygon-arc list includes the arcs associated with each polygon, even those that are common to more than one polygon. This list can be related to the arc-coordinate list in order to determine the coordinates of all points that constitute each polygon. The arc topology list defines the adjacency characteristics of all arcs.

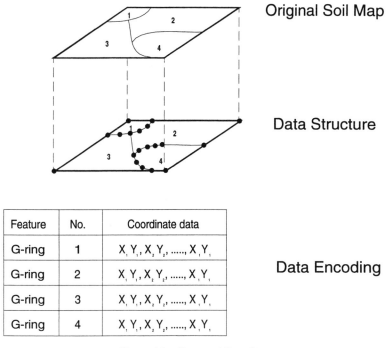

Feature	No.	Coordinate data
G-ring	1	$X_1 Y_1, X_2 Y_2, ..., X_1 Y_1$
G-ring	2	$X_1 Y_1, X_2 Y_2, ..., X_1 Y_1$
G-ring	3	$X_1 Y_1, X_2 Y_2, ..., X_1 Y_1$
G-ring	4	$X_1 Y_1, X_2 Y_2, ..., X_1 Y_1$

Figure 4.9. Spaguetti Data Structure

Topologic data structures are more sophisticated than spaguetti data structures and are included in practically all vector-based GIS. Their ability to handle topology makes them well suited for contiguity and connectivity analysis (Aronoff, 1989), both of which have ample applications in hydrology. Contiguity capabilities, for example, can be applied to determine flood impact zones, correlations between soil types and land use, and others. Connectivity capabilities can be applied to drainage network and transportation analysis, emergency evacuation routes, and so on. Like spaguetti data structures, however, topologic data structures do not constitute the best solution for representing surfaces and transition zones.

TINs. Triangular Irregular Networks (TINs) constitute a special case of topologic data structures in which nodes are interconnected using single links (Figure 4.11), resulting in a set of triangular facets that covers the area of interest. The location of each node is given by its X, Y, and Z coordinates, which provides TINs with a powerful tool to simulate highly variable surfaces. TIN topology is based on triangle adjacency and the definition of the nodes that compose each triangle.

The number of possibilities for defining and locating nodes is nearly endless. As a result, many algorithms and techniques have been devised to optimize the process of TIN generation. Many GIS contain built-in routines that do the work, but customized procedures are continuously being developed for particular applica-

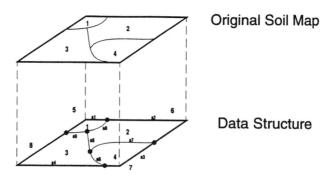

Original Soil Map

Data Structure

Polygon-Arc List

Polygon	Arcs
1	a1, a8, a9
2	a2, a7, a5, a8
3	a5, a6, a4, a9
4	a7, a3, a6

Arc Topology List

Arc	Left pol.	Right pol.
a1	5	1
a2	6	2
a3	7	4
a4	8	3
a5	2	3
a6	4	3
a7	2	4
a8	1	2
a9	1	3

Arc Coordinate List

Arc	X, Y pairs
a1	0,4 0,6 1,6
a2	1,6 6,6 6,3
a3	6,3 6,0 ...
a4	5,0 0,0 ...
a5	1,5 3,2
a6	3,2 4,1...
a7	3,2 3,3 ...
a8	1,5 2,5 ...
a9	0,4 1,4 1,5

Figure 4.10. Topologic Data Structure

tions, some of which are directly related to hydrology (see, for example, Kuniansky and Lowther (1993); and Tachikawa et.al., (1993)). In general, algorithms begin with a regular mesh or uniform grid of triangular elements upon which a series of point and/or linear well-defined features are drawn. Point features include peaks and wells, among others, and are maintained in the TIN as triangle nodes. Linear features include channel networks, lake shorelines, etc, and are maintained in the TIN as additional triangle edges. The algorithm then takes the location of these additional nodes and edges, and subdivides the original grid into finer triangles, causing, therefore, a more dense spatial resolution. The shape and spatial resolution of the original mesh may actually be the result of an interpolation process based on contour data or a digital elevation model. Chapter 7 provides a comprehensive presentation on this subject.

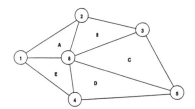

Node Coordinates	
Node	Coordinates
1	X_1, Y_1
2	X_2, Y_2
3	X_3, Y_3
4	X_4, Y_4
5	X_5, Y_5

Triangle Adjacency	
Triangle	Adjacent Triangle
A	B, E
B	A, C
C	B, D
D	C, E
E	A, D

Triangle Nodes	
Triangle	Nodes
A	1, 2, 6
B	2, 3, 6
C	3, 5, 6
D	4, 5, 6
E	1, 4, 6

Z Coordinates	
Node	Value
1	Z_1
2	Z_2
3	Z_3
4	Z_4
5	Z_5

Figure 4.11. Triangulated Irregular Network

TINs are well suited for representing surface spatial variability. Raster data structures can also be used for this purpose, although their applicability may not necessarily be overlapping. TINs usually provide a more compact representation for the same level of accuracy. On the other hand, raster data structures are more suited for the evaluation of parameters such as slope and aspect, which work best when systematic sampling is used.

4.4. Geographic Conceptualization and Standardization Issues

In Section 4.2.4, the issue of data exchange between dissimilar computer systems was briefly discussed, and a spatial model, based on standard procedures for mapping from the conceptual level to the low or physical encoding level, was presented. Inherent to the use of such standard procedures is the assumption that geographic entities can be uniquely defined and characterized in all situations, regardless of human interpretation. However, geographic conceptualization is heavily dependent on cultural ties and personal perception. As a result, spatial categorization and feature definition are likely to vary on a personal, cultural, and regional basis.

One way to solve this problem is by providing multiple translations to the same concept. For example, in the case of the standards for digital data exchange developed by the Digital Geographic Information Working Group (DGIWG) within NATO (Mark, 1993), every geographic entity is characterized by a definitional phrase in English, followed by terms in six different languages, plus an additional version in British English. Table 4.3 shows an application of this approach for the geographic entities lake, pond, and lagoon. In column Code, B refers to "Hydrography", one of 10 top-level themes. H refers to "Inland Water", one of 10 subthemes. The number corresponds to one of 28 kinds of inland water features.

TABLE 4.3. Equivalent Terms for Lake, Pond, and Lagoon in Six Languages, as included in the DGIWG Standard (Adapted from Mark, 1993)

Code	Definition	Language	Terms
BH080	A body of water surrounded by land	US English	Lake/Pond
		French	Lac/Etang
		German	See/Teich
		Italian	Lago/Stagno
		Dutch	Meer/Plas/Vijver
		Spanish	Lago/Laguna
		UK English	Lake/Pond
BH190	Open body of water separated from the sea by sand bank of coral reef	US English	Lagoon/Reef Pool
		French	Lagon/Lagune
		German	Lagune
		Italian	Laguna
		Dutch	Lagune/Strandmeer
		Spanish	Albufera
		UK English	Lagoon/Reef Pool

In the real world, however, the situation is more complicated. In English, 'lake' is used to denote most inland standing water bodies, large and small. In fact, the SDTS definition of 'lake' is 'any standard body of inland water' (NIST, 1992). 'Pond' usually denotes very small water bodies, although there are cases in which small water bodies are called 'lake', and cases in which large water bodies are called 'pond'. SDTS does not even consider 'pond' as part of the standard entity type set, as opposed to 'lake'. 'Lagoon' is defined by SDTS as a sheet of salt water separated from the open sea by sand or shingle banks. In French, a large inland water body is usually called 'lac'. 'Etang' is used to denote ponds, although in France 'étang' may include large inland water bodies and even lagoons. 'Lagune' is used to represent lagoons. In Spanish, 'lago' is normally used to represent lakes. 'Laguna' is used to denote small lakes, although it is also applied sometimes to describe wetlands. 'Ciénaga' is also used sometimes for these purposes. In Spain, 'albufera' is used to represent lagoons.

The problem with this approach is that equivalence among various languages is given at the attribute level only. However, the same would still be true if equivalence was to be made at the geographic entity level because there is no guarantee that they will be equally interpreted by people from different cultures. Since supposedly equivalent words can actually represent nonequivalent geographic entities and vice versa, the risk of misinterpreting maps and attribute information after data exchange has taken place is present and must be taken into account. Unfortunately, current GIS do not include any procedure to translate both graphical features and attribute data from one language to another. As a result, equivalence must be carried out manually.

Data exchange, of course, is not limited to cross-language type of translations. Even different groups within the same organization are faced with similar tasks. A typical problem is that of linking linear features and their attributes, which were developed by one group, with point features and their attributes, which were developed by another group. In the case of the Louisiana Department of Transportation and Development (LaDOTD), for example, one group is in charge of highway design, and another group is in charge of traffic engineering (Bullock and Quiroga, 1995). The roadway design group typically reference their plans by "Control Section-Log Mile". With this referencing scheme, a unique identifier is assigned to all sections of roads maintained by the state. Items or segments within a particular section are located using longitudinal distances (measured approximately along the center line) from a predefined reference point. By contrast, the traffic engineering group reference the location of their traffic signal equipment by numerical codes corresponding to parish, town, and intersecting roads. The problem arises when one of the groups requires data from the other one, or when a third group or state agency requires data from both groups. Because automated data sharing was considered essential, a procedure for linking linear and point features had to be developed. Such a linking involved the development of a unique geographic referencing model by means of which signalized intersections were placed along the highway network map, and the development of database modules or groups of data, each of which was related to a set of fundamental tables. These fundamental tables contained the basic linkage between graphical features and the remaining attribute tables.

A similar application in hydrology would be the linkage between a watershed drainage network map, which is composed of linear features, and rain gages and streamflow stations, which are point features. The location of the drainage network in the map may be given in geographic coordinates such as latitude and longitude, but this does not necessarily mean that the coordinates given reflect the actual location of streams and rivers. At best, they indicate the location of the center line at a specific point in time. The location of rain gages and streamflow stations is also given in latitude and longitude, but this does not imply registration between the station map and the drainage network map. In addition to this, station location may still be given in degrees and minutes, which may imply uncertainties on the order of hundreds of meters with respect to the actual location of the station.

4.5. Time in Geographic Information Systems

Most geographic databases are built assuming static geographic feature conditions. However, it is a known fact that geographic features evolve through time. Many of these features change as a result of natural processes, but in many other cases, they change as a result of human intervention. Land cover, land use, land slope, and stream network geometry including alignment, slope and cross section, are just some examples of landscape characteristics that may change over time and that may have a direct impact on watershed hydrologic modeling.

Unfortunately, current GIS are normally atemporal in that they describe only one data state. Because most GIS in existence are based on the categorical approach described in Section 4.2.1, they are not capable of providing explicit linkages for objects through time. However, a reasonable goal for GIS would be to be able to respond to queries such as what, where, when, how fast, and how often changes have taken place. Many applications in engineering, in general, and in hydrology, in particular, would surely benefit from having a tool capable of solving such types of queries.

One of the reasons that explain the lack of temporal capabilities of current GIS is that a general model for spatiotemporal data does not exist. While the issue of time in GIS has attracted the attention of researchers over the years, knowledge acquired so far is not sufficient for handling functions which would be typical of a temporal GIS, such as updates, quality control, and scheduling. Nonetheless, a brief introduction to the subject is convenient because of the increasing awareness that changes in geographic databases over time should not be ignored, and that there should be some means available to account for such changes. At the very least, geographic databases should be designed and populated with time explicitly specified at the attribute level because, as temporal GIS become available, the conversion from the current static geographic databases to the new temporal geographic databases will likely be more straightforward.

Emphasis is given here to basic concepts dealing with time in GIS, rather than to complicated implementation issues. The brief summary that follows draws heavily from the work by Langran (1992) because of the conciseness and completeness of her presentation.

4.5.1. CARTOGRAPHIC TIME

As mentioned before, geographic data is comprised of three basic components: location, non-spatial attributes, and time. From a cartographic perspective, location is usually represented in two dimensions, because maps have traditionally been developed on sheets of paper. A third dimension is added to represent attributes through the use of symbols and tones. When either location or attributes vary over time, a fourth dimension, called cartographic time, is added to the picture. Strictly speaking, cartographic time becomes the third dimension, and attributes become the fourth dimension.

For all practical purposes, cartographic time and each of the axes used to represent cartographic space (X and Y) can be considered to be independent of each other. However, many descriptors used in cartographic space are equivalent or, at least, similar to descriptors used in cartographic time. For example, a **map** in cartographic space is equivalent to a **state** in cartographic time. Maps are separated by **sheet lines**, whereas states are separated by **events**. A map is composed of **objects**, which are limited by **boundaries**, just like a state is composed of **versions**, which are limited by **mutations**. Objects in a map are measured by **length** or **area**, whereas state versions are measured by **duration**. Objects in a map can have other **adjacent** objects as neighbors, just like state versions can have **previous** and **next versions** as neighbors.

4.5.2. MODELS OF SPATIOTEMPORALITY

Sequential snapshots. The usual approach for conceptualizing spatiotemporality is by considering separate map states, i.e., by assuming a snapshot sequence of time slices (Figure 4.12). Each snapshot would capture the map state associated

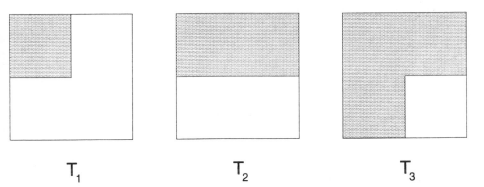

T_1 T_2 T_3

Figure 4.12. Sequential Snapshot Approach for Evaluating Change through Time

with a particular point in time. Unfortunately, snapshots are a crude means of representing change over time. In many ways, snapshots are the temporal equivalent of the spatial spaguetti data structure previously described because no topology is built into the model. Therefore, not only is it difficult to detect actual changes but also there is no systematic procedure included to trap inconsistencies. In addition to this, data redundancy is prevalent.

Base state with amendments.   One way to solve the problems of data redundancy is by assuming a base map with amendments, as shown in Figure 4.13. In this

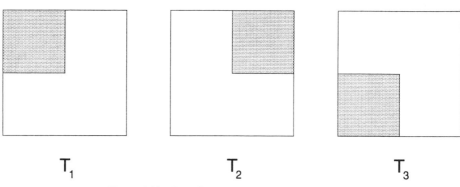

T_1 T_2 T_3

Figure 4.13.   Base State with Amendments Example

case, each snapshot contains only those changes that have taken place since the last map state. Map inconsistencies can be trapped, but still, connectivity is not obvious. Note that the process could actually be executed backwards, with the base map composed of the most recent geographic features. This approach is ideal for cases in which only the latest information is usually of interest, and for which event history is retrieved on an occasional basis.

Space-time composite.   Connectivity through time can be made explicit by providing lifespans to every feature in the map. This way, the base map with amendments model can actually be reduced from a 3D model to a 2D model because it is always possible to retrieve those features which are active at any point in time (Figure 4.14). For this approach to work, it becomes necessary to allow for the possibility of having null attributes for those time intervals in which the feature is not supposed to exist.

In any case, nonspatial attributes must also include time explicitly. This can be accomplished in several ways, depending on the role played by each of the components of geographic information, location, attributes, and time in a particular application. In general, several fields containing dates representing milestones in the lifespan of a feature may be required. Another solution may be to define events by special codes which refer to other codes in such a way that the history of events for every feature can always be retrieved. For actual implementation, both world time and database time must be explicitly recorded (Hiland et.al., 1992; Elmasri and Navathe, 1989). Database time is defined here as the actual moment of event recording into the database.

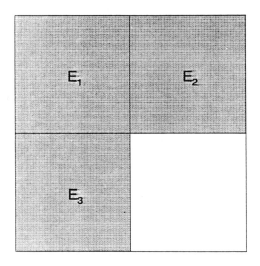

	History of Events		
	T_1	T_2	T_3
E_1	pattern	pattern	pattern
E_2	blank	pattern	pattern
E_3	blank	blank	pattern
	blank	blank	blank

Figure 4.14. Space-Time Composite Example

4.6. Summary

Spatial data can be modeled using a categorical approach, an object approach, or a combination of both. In the categorical approach, spatial data is assumed to be composed of disconnected layers of spatial features. Most GIS are based on such an assumption. In the object approach, spatial data is assumed to be composed of objects having a clearly defined identity, solving, therefore, the integrity problem that characterizes the categorical approach. Only a few, mainly experimental GIS, are based on the object approach.

Spatial models are implemented either by using raster data structures or vector data structures. Raster spatial data structures are simple to conceptualize and to use for overlay analysis. They can be used efficiently for modeling high spatial variability, which makes them suitable for the manipulation of remotely sensed data and digital images. Unfortunately, data files tend to be huge. Vector spatial data structures provide a more compact way to conceptualize spatial data, and can be used more efficiently for modeling topology. As a result, data files are usually much smaller than those making use of raster structures. On the negative side, spatial variability is more difficult to handle.

In recent years, a trend towards standardization in the modeling of geographic information has taken place. Such a trend is based on the assumption that geographic entity definition can be made regardless of human interpretation. Problems are likely to occur, however, because people from different cultures, or even from different groups within the same organization, conceptualize geographic entities differently. This means that solving the problem of data sharing is actually more complex than just providing equivalent definitions for spatial objects.

Current GIS are normally atemporal in that they describe only one data state. However, a reasonable goal for GIS would be to be able to respond to queries such as what, where, when, how fast, and how often changes have taken place. This implies taking into consideration cartographic time as an inherent dimension of geographic data. A general model of spatiotemporality does not exist, which means that the problem of providing temporal linkages between spatial objects is not fully solved yet. However, time can be explicitly included in a geographic database at the attribute level, by providing world and database date control fields.

References

Abdelmoty, A., Williams, M., and Paton, N., 1993. Deduction and Deductive Databases for Geographic Data Handling, Advances in Spatial Databases, Proceedings of the Third International Symposium, SSD '93, Singapore, June 1993, Abel, D., and Ooi, B. (eds.), Lecture Notes in Computer Science 692, Springer- Verlag, pp. 443-464.

Aronoff, S., 1989. Geographic Information Systems: A Management Perspective, WDL Publications, Ottawa, Ontario, Canada, 294 p.

Bullock, D., and Quiroga, C., 1995. A Data Storage and Retrieval Model for Louisiana Traffic Operations Data, Final Report, Louisiana Department of Transportation and Development, Louisiana Transportation Research Center, Baton Rouge, 60 p.

Burrough, P., 1986. Principles of Geographical Information Systems for Land Resources Management, Oxford University Press, New York, 193 p.

David, B., Raynal, L., Schorter, G., and Mansart, V., 1993. GeO2: Why Objects in a Geographical DBMS?, Advances in Spatial Databases, Proceedings of the Third International Symposium, SSD '93, Singapore, June 1993, Abel, D., and Ooi, B. (eds.), Lecture Notes in Computer Science 692, Springer-Verlag, pp. 264-276.

Egenhofer, M., and Herring, J., 1991. High-Level Spatial Data Structures for GIS, in Geographical Information Systems, Volume 1: Principles, Maguire, D., Goodchild, M. and Rhind, D. (eds.), Longman Scientific and Technical, Essex, England, pp. 227-237.

Elmasri, R., and Navathe, S., 1989. Fundamentals of Database Systems, Benjamin/Cummings Publishing Company, Redwood City, California, 802 p.

Healey, R., 1991. Database Management Systems, in Geographical Information Systems, Volume 1: Principles, Maguire, D., Goodchild, M. and Rhind D., (eds.), Longman Scientific Technical, Essex, England, pp. 251-267.

Hiland, M., Wayne, L., and Streiffer, H., 1992. Louisiana Coastal GIS Network: Relational Database Design for a Spatially Indexed Cataloging System, GIS/LIS 92 Proceedings, ACSM, ASPRS,

AAG, URISA, AM/FM International, pp. 322-338.

Iyengar, S., and Miller, S., 1986. Efficient Algorithm for Polygon Overlay for Dense Map Image Data Sets, Journal of Image Vision and Computing, Vol. 4, No. 3, pp. 167-174.

Jones, L., and Iyengar, S., 1984. Space and Time Efficient Virtual Quadtrees, IEEE Transactions on Pattern Analysis and Machine Intelligence, Vol. PAMI-6, No. 2, March, pp. 244-247.

Kim, W., Garza, J., and Keskin, A., 1993. Spatial Data Management in Database Systems: Research Directions, Advances in Spatial Databases, Proceedings of the Third International Symposium, SSD '93, Singapore, June 1993, Abel, D., and Ooi, B. (eds.), Lecture Notes in Computer Science 692, Springer-Verlag, pp. 1-13.

Kuniansky, E., and Lowther, R. Finite-Element Mesh Generation Using Geographic Information Systems, Proceedings of the Symposium on Geographic Information Systems and Water Resources, AWRA, Bethesda, Maryland, pp.387-396.

Langran, G., 1992. Time in Geographic Information Systems, Taylor Francis, New York, 189 p.

Mark, D., 1993. Toward a Theoretical Framework for Geographic Entity Types, Spatial Information Theory, Proceedings of the European Conference, COSIT '93, Marciana Marina, Italy, September 1993, Frank, A. and Campari, I. (eds.), Lecture Notes in Computer Science 716, Springer-Verlag, pp. 270-283.

NIST, 1992. Spatial Data Transfer Standard (SDTS), FIPS PUB 173, US Department of Commerce, Computer Systems Laboratory, Gaithersburg, Maryland, SuDocs C13.52:173.

Ooi, B., 1990. Efficient Query Processing in Geographic Information Systems, Notes in Computer Science 471, Springer-Verlag, 208 p.

Samet, H., 1984. The Quadtree and Related Hierarchical Data Structures, ACM Computing Surveys, Vol. 16, July, pp. 187-260.

Samet, H., 1990. Applications of Spatial Data Structures: Computer Graphics, Image Processing, ang GIS, Addison-Wesley, 507 p.

Tachikawa, Y., Shiiba, M., and Takasao, T., 1993. Development of a Basin Geomorphic Information System Using a TIN-DEM Data Structure. Proceedings of the Symposium on Geographic Information Systems and Water Resources, AWRA, Bethesda, Maryland, pp.163-172.

CHAPTER 5

Methods For Spatial Analysis

E.B. Moser and R.E. Macchiavelli

5.1. Introduction

Creating, manipulating, and storing the GIS data base are rarely the final end-products of GIS in the sciences. Here, we are interested in understanding and modeling the dynamic processes that generated and shaped the information stored in the data base. Very often the researcher will want to make predictions concerning the process to unobserved locations. Where sampling is expensive, much cheaper auxiliary information sampled more intensely than the quantity of interest may be used to provide estimates of the desired entity. Further, we may be interested in testing various ideas or hypotheses concerning the process. In practice we are unable to entirely sample our region of interest and so our GIS is built using data containing some degree of uncertainty. Thus, it is often convenient to model our spatial process as a stochastic spatial process, permitting our models of the process to accommodate the uncertainty in our measurements and to help explain the observed process. It is commonly accepted that measurements made near in time or in space are much more likely to be alike than would be measurements widely separated in time or space. It is upon this assumption that spatial data analysis methods have been developed. In this chapter, we will address several of the common problems encountered in spatial data analysis as it relates to GIS and will illustrate methods for their analysis.

Let s and t be the spatial locations of 2 points of interest of the spatial process $Z(\bullet)$, where for a 2-dimensional process $\mathbf{s} = (x, y)$, and $Z(\mathbf{s})$ is the value of the response at location s. Cressie (1991) provides a useful decomposition of $Z(\mathbf{s})$ that permits the use of standard statistical methodology to model $Z(\mathbf{s})$. Let our spatial model be of the form

$$Z(\mathbf{s}) = \mu(\mathbf{s}) + \kappa(\mathbf{s}) + \eta(\mathbf{s}) + \varepsilon(\mathbf{s}) \tag{5.1}$$

V. P. Singh and M. Fiorentino (eds.), Geographical Information Systems in Hydrology, 91–113.
© 1996 Kluwer Academic Publishers. Printed in the Netherlands.

where $\mu(s)$ is the non-random average or mean value of the spatial process at location s, $\kappa(s)$ is a zero-mean, intrinsically stationary stochastic process which spatially links our observations at distances for which we have observed information, $\eta(s)$ is also a zero-mean, intrinsically stationary stochastic process, but this process operates at distances smaller than those actually measured in our data base, and $\varepsilon(s)$ is a zero-mean, white-noise process. Cressie (1991) refers to $\mu(s)$ as large-scale variation, $\kappa(s)$ as smooth small-scale variation, $\eta(s)$ as micro-scale variation, and $\varepsilon(s)$ as measurement error. Here, we will refer to $\mu(s)$ as the "trend" in the process. Notice that the spatial dependency among our observations is contained in the components $\kappa(s)$ and $\eta(s)$. However, we will only be able to model the process $\kappa(s)$ since we will not have information for estimating $\eta(s)$. Each of the random components $\kappa(s), \eta(s)$ and $\varepsilon(s)$ are assumed to be mutually independent, which means that we can specify models for each component separately. However, estimation will generally not proceed independently. Thus, model (5.1) can be viewed as a mixed-model (see Stroup 1989) where $\mu(s)$ represents the fixed effects and $\kappa(s), \eta(s)$ and $\varepsilon(s)$ are the random effects. Let

$$Z(s) = \mu(s) + \delta(s) \qquad\qquad (5.2)$$

be a traditional mixed-model, where $\delta(s) = \kappa(s) + \eta(s) + \varepsilon(s)$ is the correlated random process. This formulation can make it easier to incorporate experimental treatments or additional sampling errors into a model of the process using standard mixed-models techniques.

To study the hydrology of an agricultural field, we may develop a GIS that includes data layers of soil types, soil pH, soil water content, sand, silt and clay composition, elevation, and other variables. These data may actually have been collected through sampling at specific locations within the field, but the processes themselves are actually continuous with respect to location. We may need in our GIS these data stored in terms of the estimated underlying continuous process rather than as point location data. For example, we might like to have isoclines (contours of constant value) of water content stored as a layer in the GIS. This layer could then be used in the study of the physical hydrological processes of the field, the study of the relationships of water content with other variables, and for predictions of water content to locations that were not sampled. Our job as a modeler is to determine how best to model the water content, $Z(\bullet)$, by specifying structures for $\mu(s)$, the global trend over the field, and for $\delta(s)$, the dependency among measurements made in the field and any measurement errors associated with them.

The standard approaches to modeling $Z(s)$ typically begin by placing restrictions upon various components of the spatial model. The trend surface model assumes that all of the structure in the spatial variation can be explained through the non-random mean component $\mu(s)$ and that the random process $\delta(s)$ is uncorrelated white noise. The ordinary kriging model assumes that $\mu(s)$ is a constant and that $\kappa(s)$, and hence $\delta(s)$, has a structure that can be modeled using only a few parameters. Universal kriging permits both a trend and a spatial dependency among the observations. And, as stated earlier, we can use mixed-models technology to model both the trend and the spatial dependency among observations and to assess model fit and adequacy. To summarize the variability in spatial data we most commonly use models of the variogram.

5.2. The Variogram

The variogram is the model for the dependencies among the spatially referenced observations. The variogram (see Matheron 1963) is based upon our assumption that observations made near in space are more alike than those made farther apart. Let $h = s - t$ be the vector connecting 2 locations s and t. Then the variogram is defined as

$$\text{Var}(Z(s) - Z(t)) = 2\gamma(s - t) = 2\gamma(h) \tag{5.3}$$

where $\gamma(h)$ is called the semivariogram and is a function of the distance and direction between the locations s and t. Also note that since $\text{Var}(Z(s) - Z(t)) = \text{Var}(Z(t) - Z(s))$, then $\gamma(h) = \gamma(-h)$. In addition, $\text{Var}(0) = 0$, but in practice the variogram may not approach 0 as $h \to 0$, but rather may approach a constant C_0 that is usually referred to as the nugget effect. Cressie (1991:59) describes the nugget effect as composed of variation due to measurement error and to microscale processes that occur at distances shorter than those observed in the data. If the semivariogram is a positive constant for all $h \neq 0$, then the measurements are uncorrelated. More generally, the semivariogram is a monotonically non-decreasing function that is assumed to reach a plateau or asymptote C_s at a distance A, where no spatial association occurs. The theoretical asymptote C_s is called the sill, and A is called the range of the semivariogram.

An alternative way of describing the spatial association among observations is through the covariogram, which would be more commonly used in describing time-series or mixed-models in other statistical applications. Denote the covariogram as

$$\text{Cov}(Z(s), Z(t)) = C(s - t) = C(h) \tag{5.4}$$

and note that $C(0) = \text{Var}(Z(s))$. The variogram and covariogram are directly related since

$$\text{Var}(Z(s) - Z(t)) = \text{Var}(Z(s)) + \text{Var}(Z(t)) - 2\text{Cov}(Z(s), Z(t)) \tag{5.5}$$

For a second-order stationary process,

$$\text{Var}(Z(s)) = \text{Var}(Z(t)) = C(0) \tag{5.6}$$

so that

$$\text{Var}(Z(s) - Z(t)) = 2\gamma(h) = 2C(0) - 2C(h) \tag{5.7}$$

which states that

$$\gamma(h) = C(0) - C(h) \tag{5.8}$$

where $C(\mathbf{0})$ is the same as the sill of the semivariogram, C_s.

If the semivariogram (or covariogram) is a function only of the distance, $\|\mathbf{h}\|$, between the locations, then the spatial process is further said to be isotropic. Otherwise it is called anisotropic.

5.2.1. VARIOGRAM MODELS

A number of models have been proposed to explain the spatial dependencies among observations and the selection of a good variogram model is a very important part of the spatial analysis process. Most models are constructed such that the variogram is a non-decreasing function with increasing distance, but which typically is asymptotic to the constant C_s, which is the sill of the semivariogram. Other models can be postulated but these may not guarantee that, for example, variances are positive. Three common isotropic models that are used are the spherical, exponential, and Gaussian models, which can be defined in terms of the nugget, C_0, sill, C_s, and range, A. Spherical semivariogram:

$$\gamma(\mathbf{h}) = \begin{cases} 0 & \|\mathbf{h}\| = 0 \\ C_0 + C_s[(3/2)(\|\mathbf{h}\|/A) - (1/2)(\|\mathbf{h}\|/A)^3] & 0 \le \|\mathbf{h}\| \le A \quad (5.9) \\ C_0 + C_s & \|\mathbf{h}\| > A \end{cases}$$

Exponential semivariogram:

$$\gamma(\mathbf{h}) = \begin{cases} 0 & \|\mathbf{h}\| = 0 \\ C_0 + C_s[1 - \exp(-\|\mathbf{h}\|/A)] & \|\mathbf{h}\| > 0 \end{cases} \quad (5.10)$$

Gaussian semivariogram:

$$\gamma(\mathbf{h}) = \begin{cases} 0 & \|\mathbf{h}\| = 0 \\ C_0 + C_s[1 - \exp(-\|\mathbf{h}\|^2/A^2)] & \|\mathbf{h}\| > 0 \end{cases} \quad (5.11)$$

Note that the restrictions on the parameters are C_0, C_s and $A \ge 0$. Other variogram models that are used include the linear semivariogram and the nugget semivariogram. Linear semivariogram:

$$\gamma(\mathbf{h}) = \begin{cases} 0 & \|\mathbf{h}\| = 0 \\ C_0 + C_s\|\mathbf{h}\| & \|\mathbf{h}\| > 0 \end{cases} \quad (5.12)$$

Nugget semivariogram:

$$\gamma(\mathbf{h}) = \begin{cases} 0 & \|\mathbf{h}\| = 0 \\ C_0 + C_s & \|\mathbf{h}\| > 0 \end{cases} \quad (5.13)$$

In Figure 5.1 are found examples of the above 5 semivariogram models. Cressie (1991: 58-64) and Haining (1990: 90-97) discuss other parametric semivariogram models and discuss conditions under which a model is a valid semivariogram.

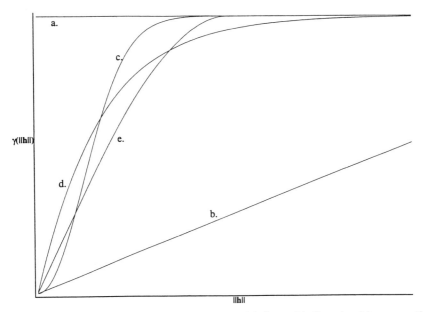

Figure 5.1. Example semivariogram models: nugget (a), linear (b), Gaussian (c), exponential (d), and spherical (e).

5.2.2. SEMIVARIOGRAM ESTIMATION

Cressie (1991:68) has stated that the variogram is more reliably estimated than the covariogram and several regression methods have been used to estimate the semivariogram. Alternative methods such as maximum likelihood (ML) and restricted maximum likelihood (REML), however, are designed to work with the covariogram. Thus, we will discuss both approaches. Zimmerman and Zimmerman (1991) summarize and compare several semivariogram estimators. The semivariogram is most often estimated using regression methods. In order to use the regression methods, the empirical semivariogram is constructed first. Matheron's (1963) isotropic semivariogram estimator is given by

$$\hat{\gamma}(\|\mathbf{h}\|) = \frac{1}{n(\|\mathbf{h}\|)} \sum_{n(\|\mathbf{h}\|)} (Z(\mathbf{s}_i) - Z(\mathbf{s}_j))^2 \qquad (5.14)$$

where the summation is over all pairs of points i and j such that $\|\mathbf{s}_i - \mathbf{s}_j\| = \|\mathbf{h}\|$, and $n(\|\mathbf{h}\|)$ is the number of pairs of locations that are $\|\mathbf{h}\|$ units apart. In practice, the data may need to be binned so that $n(\|\mathbf{h}\|)$ is not too small for all values of $\|\mathbf{h}\|$. Cressie and Hawkins (1980) proposed a more robust estimator of the semivariogram as

$$\tilde{\gamma}(\|\mathbf{h}\|) = \frac{\left[\frac{1}{n(\|\mathbf{h}\|)} \sum_{n(\|\mathbf{h}\|)} \sqrt{|Z(\mathbf{s}_i) - Z(\mathbf{s}_j)|} \right]^4}{0.914 + [0.988/n(\|\mathbf{h}\|)]} \qquad (5.15)$$

which is more resistant to extreme deviations.

Given $\hat{\gamma}(\|\mathbf{h}\|)$ or $\tilde{\gamma}(\|\mathbf{h}\|)$, a semivariogram model may be fit using least-squares regression methods. Ordinary least-squares regression (OLS) minimizes the residual sum of squares RSS,

$$\text{RSS} = \sum_{i=1}^{k}[\hat{\gamma}(\|\mathbf{h}\|_i) - \gamma(\|\mathbf{h}\|_i; \theta)]^2 \tag{5.16}$$

where k is the number of distance groups or bins of $\|\mathbf{h}\|$, $\gamma(\|\mathbf{h}\|_i; \theta)$ is the fitted semivariogram model, and θ is the vector of parameters for a specific semivariogram model. One difficulty with the OLS semivariogram estimator is that the responses making up each distance group or bin are correlated (since the points contribute to more than a single distance group) and typically the semivariances are heteroscedastic. Weighted least-squares estimators (WLS) have been developed to better account for this dependency (Cressie 1985). Zimmerman and Zimmerman (1991) summarize the WLS criterion as

$$\text{WRSS} = \sum_{i=1}^{k}\{n(\|\mathbf{h}\|_i)/[\gamma(\|\mathbf{h}\|_i; \theta)]^2\}[\hat{\gamma}(\|\mathbf{h}\|_i) - \gamma(\|\mathbf{h}\|_i; \theta)]^2 \tag{5.17}$$

In either the OLS or WLS method, $\tilde{\gamma}(\|\mathbf{h}\|)$ may be substituted for $\hat{\gamma}(\|\mathbf{h}\|)$ to obtain the robust-resistant semivariogram estimator.

Let $\mathbf{Z}$ be a vector of responses for the spatial data set and let Σ be the variance-covariance matrix of these responses. If the responses can be assumed to be distributed as the multivariate normal distribution with mean vector μ and covariance matrix Σ, then the log-likelihood for the data can be written as

$$\log L(\mu, \Sigma) = c - \frac{1}{2}n\log|\Sigma| - \frac{1}{2}\sum_{i=1}^{n}[(Z_i - \mu)'\Sigma^{-1}(Z_i - \mu)] \tag{5.18}$$

where $c = -\frac{1}{2}nd\log 2\pi$, n is the number of replications of the spatial process, d is the number of points, and $|\Sigma|$ is the determinant of Σ. In most kriging problems $n = 1$ while the case of $n > 1$ typically occurs in the analysis of experiments with spatially correlated data (e.g., Moser et al. 1994). Selection of values for the parameters of the semivariogram or covariogram model then specifies an estimator for Σ. The maximum likelihood estimator for Σ, and hence of the semivariogram or covariogram, is that choice of values for which $\log L(\mu, \Sigma)$ is largest. Restricted maximum likelihood (REML) estimation is similar to the above but operates on "error contrasts" and provides unbiased parameter estimates for certain linear models (see Kitanidis 1983, Zimmerman and Zimmerman 1991). Maximum likelihood and REML estimation will be considered further in the next section.

5.3. Trend Surface Models

The usual trend surface model assumes that the random process $\delta(\mathbf{s})$ from (5.2) is pure white noise and, therefore, nothing about the trend or magnitude of adjacent values can be learned by modeling the random process. Rather, it is hoped that the mean structure $\mu(s)$ is sufficiently smooth that a surface model, such as a regression polynomial model can be used to predict the response at unobserved locations.

5.3.1. ORDINARY LEAST-SQUARES ESTIMATION

For the specific case of a white noise random process we may be willing to make the assumptions that

$$E(Z(s)) = \mu(s) \quad \text{and} \quad \text{Cov}(Z(s), Z(t)) = \begin{cases} \sigma^2 & \text{for } s = t \\ 0 & \text{for } s \neq t \end{cases} \tag{5.19}$$

and then propose a model for $\mu(s)$. It is very common to model $\mu(s)$ as a linear function of the coordinates in the form of a polynomial trend surface. In regression analysis, we would define a design matrix $\mathbf{X}$ whose rows correspond with each measurement and whose first column is a column of 1's for the intercept, and additional columns correspond with the polynomial terms of the coordinates (x, y). For a quadratic trend surface, $\mathbf{X}$ would have a column of 1's, and columns corresponding to x, x^2, y, y^2 and xy. This trend surface model could then be written as

$$\mathbf{Z} = \mathbf{X}\beta + \varepsilon \tag{5.20}$$

with best linear unbiased estimator (BLUE) of β,

$$\hat{\beta} = (\mathbf{X}'\mathbf{X})^{-1}\mathbf{X}'\mathbf{Z} \tag{5.21}$$

when the covariance among the observations is as given in (5.19). Let $s_0 = (x_0, y_0)$ be a location for which a prediction, $\hat{Z}_0$, is desired. Then

$$\hat{Z}_0 = \mathbf{X}_0\hat{\beta} \tag{5.22}$$

is   the   best   linear   unbiased predictor (BLUP) of Z_0, where $\mathbf{X}_0 = \{1, x_0, y_0, x_0^2, y_0^2, x_0y_0\}$. The variance of a predicted value is given by

$$\text{Var}(\hat{Z}_0) = (1 + \mathbf{X}_0(\mathbf{X}'\mathbf{X})^{-1}\mathbf{X}_0')\sigma^2 \tag{5.23}$$

This type of linear model easily incorporates experimental factors and covariates by adding additional columns in $\mathbf{X}$ for these additional factors (see Neter et al. 1990). Standard statistical computing packages containing modules for multiple regression and general linear models can be used to fit these trend surfaces.

5.3.2. MAXIMUM-LIKELIHOOD AND RESTRICTED MAXIMUM-LIKELIHOOD ESTIMATION

The assumption of an uncorrelated white noise random process may not be reasonable for many applications. However, the trend $\mu(s)$ may still be the dominant part of the spatial process. In these instances, the OLS estimator will still be unbiased, but more efficient unbiased predictors are possible. Since a goal of the analysis should be efficient predictions, a more efficient predictor should be used. Maximum likelihood is an estimation technique that can simultaneously model the trend and the correlated error structure. In these instances, we would specify a covariance structure that corresponds to one of the semivariogram models. For the Gaussian isotropic semivariogram model, the corresponding covariance structure would be

$$\text{Cov}(Z_s, Z_t) = [\exp(-3\|\mathbf{h}\|^2/A^2)]\sigma^2, \tag{5.24}$$

where $\|\mathbf{h}\|$ is the distance between locations s and t as before.

Thus, in addition to estimating the trend parameters β, the parameters A and σ^2 must also be estimated. Maximum likelihood makes the assumption that the response variable, $Z(\bullet)$, is normally distributed. The likelihood of the data is formed under the normal distribution assumption and is a function of the parameters of the mean structure and of the covariance structure. Estimates of the parameters are selected by maximizing the likelihood with respect to the parameters (see Haining 1990:147-151). A practical problem with this method is that the likelihood surface may be complex and multimodal, and therefore, standard numerical procedures for maximization may not converge or converge properly (see Ripley 1988). It is recommended that a grid of starting values be evaluated, which could be supplemented with initial fits of the empirical variogram using the least-squares procedures, and that plots of the likelihood surface be constructed and studied whenever possible. The standard likelihood approach is not robust to outliers although robust M-estimators can be used (Haining 1990: 241-245).

Alternative methods to maximum likelihood and least-squares estimation of covariance parameters are available. Restricted maximum likelihood or REML is a procedure widely used in the analysis of mixed linear models and essentially transforms the data to remove the fixed effects or trend from the data and then maximizes the profile likelihood with respect to the covariance parameters (see Haining 1990: 150-151, Kitanidis 1983, SAS Institute 1992: 323-325, Zimmerman and Zimmerman 1991). This approach removes the problem of simultaneously estimating the trend and covariance parameters, but it is still sensitive to model misspecification of the trend. In addition to maximum likelihood and REML estimation, Haining (1990: 153-157) discusses other estimators that are used for mixed linear models analysis that can be directly applied to estimation of spatial covariance parameters.

A feature of maximum likelihood and REML estimation procedures is that tests of hypotheses concerning trend and covariance parameters are possible, and they are derived from well-known standard statistical principles of likelihoods. These procedures also permit the comparison of several competing models, either by likelihood ratio tests for nested models, or using various information criteria developed for likelihoods. Akaike's information criterion (AIC) and Schwarz's Bayesian criterion (BIC) are frequently used for this purpose and have been used for comparing covariance structure models (Macchiavelli 1992, Moser et al. 1994).

These criteria can be written in the form

$$\text{IC}(k) = \log L - \frac{d_k}{2} c(n) \tag{5.25}$$

where k indexes the model under consideration, d_k is the number of covariance parameters in the kth model, $\log L$ is the log-likelihood evaluated at the maximum likelihood estimates, and $c(n) = 2$ for AIC and $c(n) = \log n$ for BIC, where n is the number of observations.

5.4. Ordinary Kriging

In this section we will assume that the mean structure of the model given in (5.2), $\mu(\mathbf{s})$, is a constant, i.e., $\mu(\mathbf{s}) = \mu$ for all locations s, but that the random effects, $\delta(\mathbf{s})$, impose a spatial correlation among the observations. For the moment we will assume that $\delta(\mathbf{s})$ contains no measurement error.

5.4.1. OPTIMAL INTERPOLATION

As in the OLS case, we might be interested in an unbiased linear predictor of Z_0 at s_0. Further, we might also desire that any prediction at an observed location return the sampled value. This would produce an interpolated surface prediction where the surface would go through each of the observed values. Providing that there is no measurement error, then one interpolator that has this property is the ordinary kriging predictor. The ordinary kriging predictor is a linear combination of the observed values,

$$\hat{Z}_0 = \sum_{i=1}^{n} \lambda_i Z(\mathbf{s}_i), \tag{5.26}$$

where the λ_i are the unknown coefficients to be estimated, and n is the number of locations to be used in the prediction. In order to assure uniform unbiasedness, we require that

$$\sum_{i=1}^{n} \lambda_i = 1 \tag{5.27}$$

since the random process is stationary with mean 0 and

$$E(\hat{Z}_0) = \sum_{i=1}^{n} \lambda_i E(Z(\mathbf{s}_i)) = \mu \sum_{i=1}^{n} \lambda_i. \tag{5.28}$$

5.4.2. THE KRIGING EQUATIONS

If the variogram for the spatial process is given by

$$2\gamma(\mathbf{h}) = \text{Var}(Z(\mathbf{s} + \mathbf{h}) - Z(\mathbf{s})), \tag{5.29}$$

then the minimum mean-square error of prediction estimator of the λ_i for predicting Z_0 would be

$$\lambda_0 = \Gamma_0^{-1}\gamma_0, \tag{5.30}$$

where

$$\lambda_0 = (\lambda_1, \lambda_2, \ldots, \lambda_n, m)',$$

$$\gamma_0 = (\gamma(s_0 - s_1), \gamma(s_0 - s_2), \ldots, \gamma(s_0 - s_n), 1)'$$

$$\Gamma_0 = \begin{cases} \gamma(s_i - s_j) & i = 1, 2, \ldots, n; j = 1, 2, \ldots, n \\ 1 & i = n + 1; j = 1, 2, \ldots, n \\ 0 & i = n + 1; j = n + 1 \end{cases} \tag{5.31}$$

and Γ_0 is symmetric. Let $\lambda' = (\lambda_1, \lambda_2, \ldots, \lambda_n)$, then

$$\lambda' = (\gamma + 1\frac{(1 - 1'\Gamma^{-1}\gamma)}{1'\Gamma^{-1}1})\Gamma^{-1} \tag{5.32}$$

and

$$m = -(1 - 1'\Gamma^{-1}\gamma)/(1'\Gamma^{-1}1) \tag{5.33}$$

where $\gamma_0' = [\gamma'\ 1]$ and $\Gamma_0 = \begin{bmatrix} \Gamma & 1 \\ 1' & 0 \end{bmatrix}$. The kriging prediction variance is given by

$$\text{Var}(\hat{Z}_0) = \lambda_0'\gamma_0 = \sum_{i=1}^{n} \lambda_i\gamma(s_0 - s_i) + m = \gamma'\Gamma^{-1}\gamma - (1'\Gamma^{-1}\gamma - 1)^2/(1'\Gamma^{-1}1). \tag{5.34}$$

If multiple measurements are made at the same location (and at the same time), then a measurement error or sampling error model would be needed to account for the additional source of variability. This implies that $\varepsilon(s)$ in model (5.1) is not negligible. Since the measurement error does not provide spatial information, the spatial predictions should be independent of the measurement errors. This requires then that the predictions, even to observed locations, should be a "smoothed" version of the process and that exact interpolation is no longer important. The ordinary kriging equations given above can be modified by replacing $\gamma(0) = 0$ with $\gamma(0) = \sigma_\varepsilon^2$ to get new equations that accommodate the measurement or sampling error. The new kriging prediction variance is given by

$$\text{Var}(\hat{Z}_0) = \sum_{i=1}^{n} \lambda_i\gamma(s_0 - s_i) + m - \sigma_\varepsilon^2 \tag{5.35}$$

Cressie (1991:128-129) points out that the nugget effect contains both variation due to the microscale process $\eta(s)$ and to the measurement error process $\varepsilon(s)$, such that

$$C_0 = \sigma_\eta^2 + \sigma_\varepsilon^2 \tag{5.36}$$

where σ_η^2 is the sill for the microscale process, and therefore, one should not simply use C_0 for σ_ϵ^2 to account for measurement error. Rather, one must make multiple measurements at the locations.

5.4.3. DE-TRENDING AND MEDIAN POLISH

Note that ordinary kriging requires that there be no trend in the spatial process. For some applications it might be reasonable to first de-trend the data and secondly to estimate the spatial variogram. One approach might be to estimate a trend surface for the data and then use the residuals from this analysis to estimate the variogram. Unfortunately, the trend surface model assumes that there are no spatial dependencies among the observations and the functional form of the trend surface may not be known, and thus must be determined via model selection procedures. Therefore the trend may not be effectively removed or the spatial associations may be altered. Cressie (1991) recommends that a resistant technique such as Median-Polish (see Emerson and Hoaglin 1983) be used to remove the trend. This technique essentially bins the data according to a 2-dimensional grid and then adjusts the data values according to the grid row and column medians using an iterative procedure. The residuals from this technique are then used in the estimation of the variogram. Ordinary kriging may then be used to predict a "residual" for a new location. The trend is then added to the "residual" using, say, a planar interpolant (see Cressie 1991:185, 188-190) to produce the final prediction.

5.5. Universal Kriging

Universal kriging permits both a trend and a covariance structure among the spatial observations. Thus, ordinary kriging and the trend surface analysis could be considered as special cases of universal kriging. This results in the model

$$Z(\mathbf{s}) = \mathbf{X}\beta + \delta(\mathbf{s}). \tag{5.37}$$

The universal kriging equations for the prediction

$$\hat{Z}_0 = \sum_{i=1}^{n} \lambda_i Z(\mathbf{s}_i) \tag{5.38}$$

are given by

$$\lambda_u = \Gamma_u^{-1}\gamma_u \tag{5.39}$$

where

$$\lambda_u = (\lambda_1, \lambda_2, \ldots, \lambda_n, m_0, m_1, \ldots, m_p)',$$

$$\gamma_u = (\gamma(\mathbf{s}_0 - \mathbf{s}_1), \gamma(\mathbf{s}_0 - \mathbf{s}_2), \ldots, \gamma(\mathbf{s}_0 - \mathbf{s}_n), 1, f_1(\mathbf{s}_0), f_2(\mathbf{s}_0), \ldots, f_p(\mathbf{s}_0))',$$

$$\Gamma_u = \begin{cases} \gamma(\mathbf{s}_i - \mathbf{s}_j) & i = 1, 2, \ldots, n; j = 1, 2, \ldots, n \\ f_{j-1-n}(\mathbf{s}_i) & i = 1, 2, \ldots, n; j = n+1, \ldots, n+p+1 \\ 0 & i = n+1, \ldots, n+p+1; j = n+1, \ldots, n+p+1 \end{cases}$$
(5.40)

and $f_0(\mathbf{s}_0) \equiv 1$. The $f_i(\mathbf{s}_0)$ are functions for modeling the trend, such as the functions for a second order polynomial trend surface. Solving for the unknown quantities in λ_u yields

$$\lambda' = \{\gamma + X(X'\Gamma^{-1}X)^{-1}(\mathbf{x} - X'\Gamma^{-1}\gamma)\}'\Gamma^{-1}$$
(5.41)

and

$$\mathbf{m}' = -(\mathbf{x} - X'\Gamma^{-1}\gamma)'(X'\Gamma^{-1}X)^{-1}$$
(5.42)

where $\gamma = (\gamma(\mathbf{s}_0 - \mathbf{s}_1), \gamma(\mathbf{s}_0 - \mathbf{s}_2), \ldots, \gamma(\mathbf{s}_0 - \mathbf{s}_n))'$ and Γ is $n \times n$ with i, j elements $\gamma(\mathbf{s}_i - \mathbf{s}_j)$. The kriging (prediction) variance can then be computed as (Cressie 1991:153-154)

$$\text{Var}(\hat{Z}_0) = \lambda_u'\gamma_u = \sum_{i=1}^{n} \lambda_i\gamma(\mathbf{s}_0 - \mathbf{s}_i) + \sum_{j=1}^{p+1} m_{j-1}f_{j-1}(\mathbf{s}_0)$$
(5.43)

$$= \gamma'\Gamma^{-1}\gamma - (\mathbf{x} - X'\Gamma^{-1}\gamma)'(X'\Gamma^{-1}X)^{-1}(\mathbf{x} - X'\Gamma^{-1}\gamma)$$
(5.44)

or it may be computed as

$$\text{Var}(\hat{Z}_0) = 2\sum_{i=1}^{n} \lambda_i\gamma(\mathbf{s}_0 - \mathbf{s}_i) - \sum_{i=1}^{n}\sum_{j=1}^{n} \lambda_i\lambda_j\gamma(\mathbf{s}_0 - \mathbf{s}_i).$$
(5.45)

If the process $Z(\mathbf{s})$ is second-order stationary, then the covariogram can be used. Let Σ be the variance-covariance matrix derived from the covariances among the observed locations $\mathbf{s}_1, \mathbf{s}_2, \ldots, \mathbf{s}_n$. Then

$$\lambda' = \{\mathbf{c} + X(X'\Sigma^{-1}X)^{-1}(\mathbf{x} - X'\Sigma^{-1}\mathbf{c})\}'\Sigma^{-1}$$
(5.46)

and

$$\mathbf{m}' = (\mathbf{x} - X'\Sigma^{-1}\mathbf{c})'(X'\Sigma^{-1}X)^{-1}$$
(5.47)

where $\mathbf{c} \equiv (C(\mathbf{s}_0 - \mathbf{s}_1), \ldots, C(\mathbf{s}_0 - \mathbf{s}_n))$ are the covariances between the point to be predicted and the observed data locations. The kriging variance can then be written as

$$\text{Var}(\hat{Z}_0) = C(0) - 2\sum_{i=1}^{n} \lambda_iC(\mathbf{s}_0 - \mathbf{s}_i) + \sum_{i=1}^{n}\sum_{j=1}^{n} \lambda_i\lambda_jC(\mathbf{s}_i - \mathbf{s}_j).$$
(5.48)

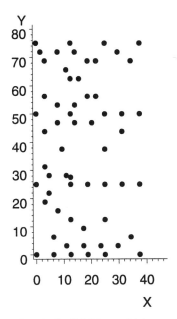

Figure 5.2. Locations in the field from which samples were taken

5.6. Examples

Water (mm^3/cm^3) content was measured at 59 locations in a rectangular field of 50 m by 100 m and the data were mapped for spatial analysis (Clarke and Dane 1991). A simple X-Y plot of the sampling sites is given in Figure 5.2. A GIS could be used to store this information along with other attributes of the field, or predictions could first be performed, with predictions or contours stored in the GIS, or both types of information recorded. We will focus on the development of a spatial model for the prediction of water content at locations not sampled in the study. This would permit construction of isoclines of constant water content which could then be imported as a layer into a GIS for overlay with other features of the study site.

The observed water content is plotted in a 3-dimensional display where linear interpolation is used to interpolate the surface over the irregularly spaced sample locations (Figure 5.3). There appears to be an increasing trend from small to large values of Y, the northing, with a surface that is concave downward. This curvature also seems to be interacting with the values of X, the easting, to orient the surface along a NW to SE axis. Two general types of models are suggested by the surface and are (1) a constant mean with stationary spatial process which extends over a much larger area than that sampled, and (2) a quadratic trend surface that may also have a spatial process for the residuals. We will examine each of these models. Note that other models may also be appropriate and could also be examined.

The empirical semivariogram for the constant mean model was constructed using Matheron's estimator (5.14) (Figure 5.4) and also using Cressie and Hawkins'

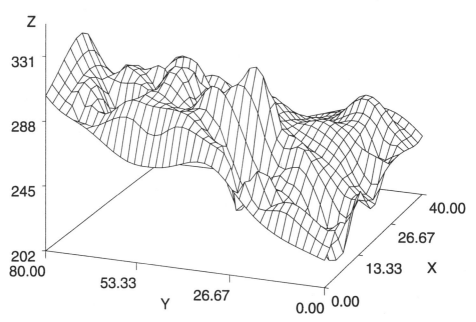

Figure 5.3. Linear interpolation of the percent water content presented as a surface over the sampled area

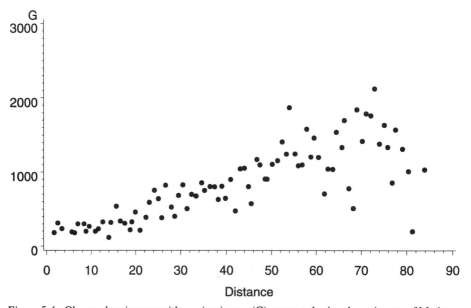

Figure 5.4. Observed variogram with semivariances (G) computed using the estimator of Matheron

robust estimator (5.15) (Figure 5.5). Both estimators suggest spatial dependency

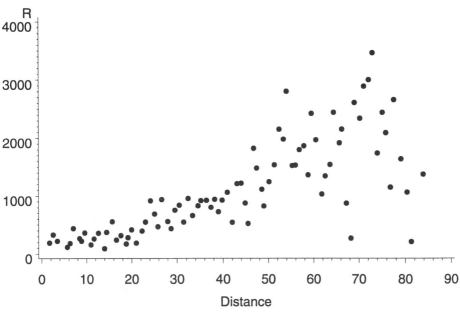

Figure 5.5. Empirical variogram with semivariances (R) estimated using therobust estimator of Cressie and Hawkins (1980)

as a function of distance. The empirical semivariogram suggests that a Gaussian semivariogram model could explain the spatial process, especially for small distances, as the Gaussian model is concave upward for small distances. The linear variogram is also fit to the data for comparison (Figure 5.6) but shows some lack-of-fit relative to the Gaussian model. Further, convergence of the maximum likelihood estimation procedure was not attained for the linear model as the parameter estimates were driven to a boundary where the variance-covariance matrix was not positive-definite (some variances could have been negative). The Gaussian semivariogram appears to fit the data well (Figure 5.7) and has the largest information criteria values compared with the model of independence and with the exponential and spherical spatial models (Table 5.1). The parameters of the Gaussian model were fit using the maximum likelihood approach (Table 5.2) and will be used subsequently to predict the water content surface using ordinary kriging. Note that the estimate of the sill for the Gaussian model has a very large standard error and is deemed not significantly different from zero using a Wald test (p=0.1664). Referring to the plot of the empirical semivariogram (Figure 5.7) confirms that the leveling off of the semivariogram is not well defined and that a wide range of choices is likely. A plot of the log-likelihood of the Gaussian model as a function of the sill and range with the nugget held constant near the maximum likelihood estimate of the nugget shows that the sill cannot be well defined given the data on hand (Figure 5.8), as the maximum occurs along a ridge of potential

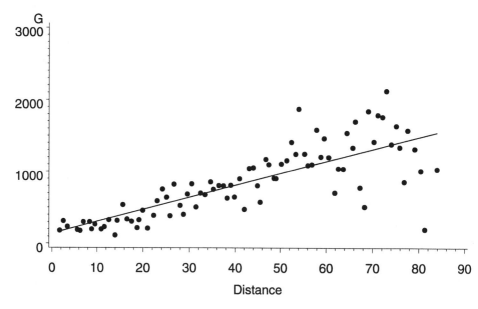

Figure 5.6. Linear variogram model with a nugget fit to the empirical semivariogram

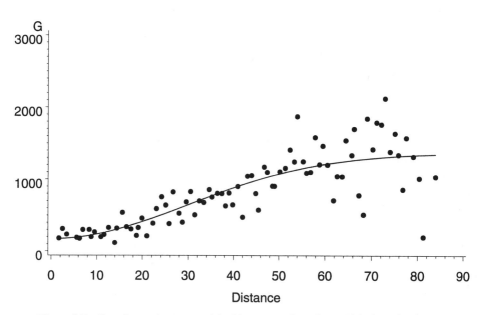

Figure 5.7. Gaussian variogram model with a nugget fit to the empirical semivariogram

candidates. This might suggest that the spatial process is not stationary and that a trend is present and must be accounted for. The empirical semivariogram was also constructed using the residuals from a quadratic trend surface model of the water

TABLE 5.1. Model fitting results for the soil water content data. The quadratic trend surface information criteria do not include a penalty for the fixed-effects terms. d_k is the number of parameters fit in the covariance structure. The information criteria are Akaike's information criterion (AIC) and Schwarz's Bayesian criterion (BIC).

Model	Log-likelihood	d_k	AIC	BIC
Simple	-278.168	1	-279.168	-280.207
Gaussian	-247.138	3	-250.138	-253.255
Exponential	-250.657	3	-253.657	-256.773
Spherical	-249.656	3	-252.656	-255.773
Quadratic Trend Surface	-236.521	1	-237.521	-238.560

TABLE 5.2. Maximum likelihood parameter estimation results for fitting the Gaussian spatial covariance structure to a constant mean model for the water content data.

Parameter	Ratio	Estimate	Standard Error	Z	Pr>$\mid Z \mid$
Sill	4.72819033	767.54577327	554.69382068	1.38	0.1664
Range	0.20349377	33.03394604	8.05560415	4.10	0.0000
Nugget	1.00000000	162.33394160	33.61846636	4.83	0.0000

content data (Figure 5.9). The spread of the semivariances and lack of trend indicate that a model of independence is plausible for the covariance structure of this trend surface model. All of the trend parameters for the trend surface model were significantly different from zero (Table 5.3).

TABLE 5.3. Trend surface parameters for a quadratic trend surface for the water content data assuming an independence covariance structure. The error degrees of freedom used in the model are 53.

Parameter	Estimate	Standard Error	T	Pr>$\mid T \mid$
INTERCEPT	206.19704660	7.08940377	29.09	0.0000
X	2.73904770	0.60982771	4.49	0.0000
Y	2.21667130	0.28312738	7.83	0.0000
X^2	-0.05651513	0.01454435	-3.89	0.0003
Y^2	-0.01325794	0.00338141	-3.92	0.0003
XY	-0.02820611	0.00605532	-4.66	0.0000

The parameter estimates (Table 5.2) for the Gaussian semivariogram were used in a kriging model to predict water content at the intersections of a 21 by 21 grid

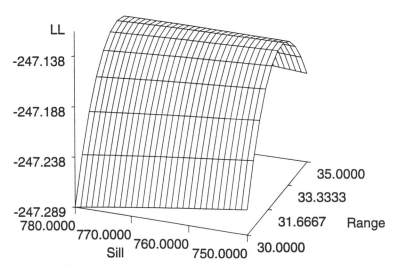

Figure 5.8. Likelihood surface of the constant mean model with Gaussian semivariogram plotted as a function of the Sill and Range parameters with the Nugget parameter held constant at 162. The maximum likelihood estimates of the parameters are Sill=767.8, Range=33.0 and Nugget=162.3. The software used to construct and maximize the likelihood uses a parameterization which differs slightly from that presented in the text. Multiply the Range by 3 to get the parameterization presented in the text

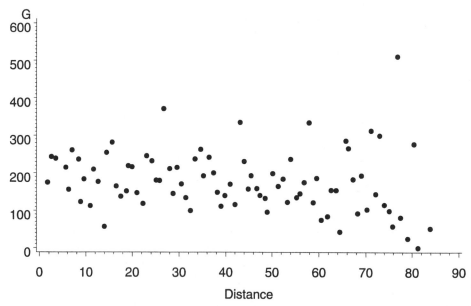

Figure 5.9. Empirical semivariogram fit to the residuals from a quadratic trend model estimated using REML

placed over the area. The estimator produced a smooth surface (Figure 5.10) that

Point Kriging of Soil Water Content

Predicted Values

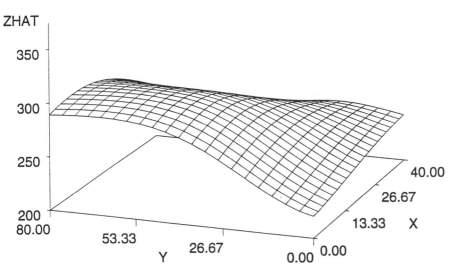

Figure 5.10. Estimated water content using a kriging estimator with a Gaussian semivariogram

approximated the general form of the original surface. Standard errors of the predicted values are greatest, as expected, around the borders of the field (Figure 5.11) with coefficients of variation of around 5 to 6 percent, which suggests good prediction ability. Cross-validation is often used to get a better idea of the predictive ability of a model. In cross-validation each of the data points is removed from the data set in turn and its water content value predicted at the sampled location. Cross-validation errors are then computed as the difference between the observed and predicted water content values at each of the sampled locations. Predicted residual sums of squares (PRESS) can also be computed for the data set by squaring and summing these errors. The cross-validation errors (Figure 5.12) are much larger than expected based upon the standard errors alone. This may imply, for example, that (1) the Gaussian semivariogram is not appropriate or is poorly estimated, (2) a constant mean large scale process is not appropriate and the process is non-stationary, and/or (3) a measurement error model is appropriate and that exact interpolation should not be used.

The quadratic trend surface model (Figure 5.13) provides a fit that is very similar to the kriging surface model. However, it is important to note that the kriging estimator will force exact interpolation to the sampled data points while the trend surface model will not. The information criteria indicate that the trend surface model provides a much better explanation of the data than do the variogram

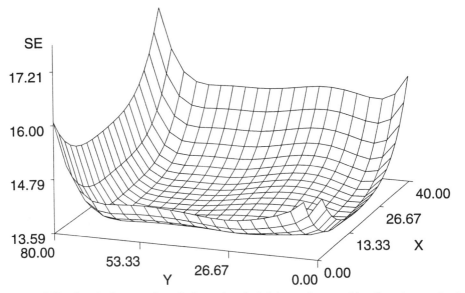

Figure 5.11. Standard errors of prediction using the kriging estimator with a Gaussian semivariogram

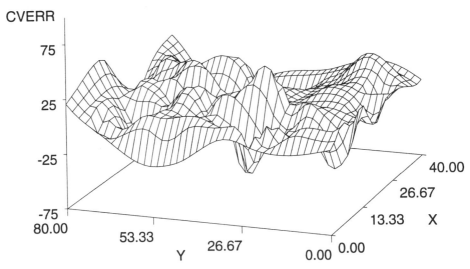

Figure 5.12. Cross-validation prediction errors (CVERR) for the kriging model using a Gaussian semivariogram

models (Table 5.1). A comparison of the surfaces by taking differences between the model predictions at the grid intersections does indicate that the models are capturing different structures within the data (Figure 5.14). The final choice of model should be based upon the investigators' knowledge of the hydrology of the physical process and the eventual use of the model.

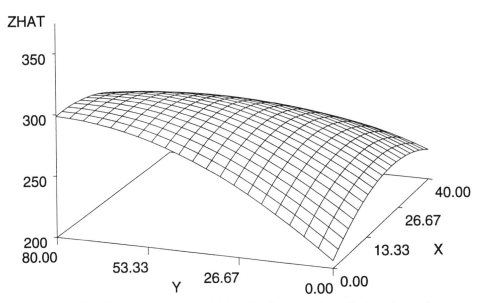

Figure 5.13. Quadratic trend surface model fit under the assumption of independence of errors

Point Kriging of Soil Water Content

Differences Between Model Predictions

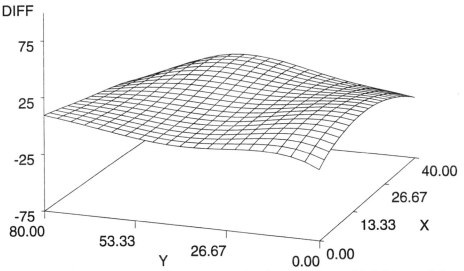

Figure 5.14. Differences between the quadratic trend surface predictions and the kriging predictions that used a Gaussian semivariogram

5.7. Integration Into GIS Software

One of the most important features that GIS software must possess is the ability to export and import the data in a form suitable for statistical analysis with other software packages. The main reason for this is that spatial data analysis covers a very broad spectrum of statistical methods, and to do a comprehensive statistical analysis requires more than simply the fitting of a model. The model must be diagnosed and alternative models tried. If kriging methods are to be used, then empirical semivariograms should be constructed, probably using different binning techniques and different variogram estimators. Trend removal may not involve a simple polynomial model, but may require more complicated surfaces. If likelihood approaches are used, then the likelihoods should be plotted and other likelihood diagnostics computed. Further, spatial analysis methods are being developed and refined constantly, and so access to current methods is typically much more available through statistical computing packages. Thus, the GIS is unlikely to provide many of the tools that will be needed for the statistical analysis.

On the other hand, many statistical computing packages do not have easy ways of plotting some of the diagnostics in such a way that other layers of the GIS data base can be simultaneously viewed. Thus, it is important that diagnostic information, model fitting results, and other relevant results be imported back into the GIS data base and viewed along with the other layers. This can help spot outliers or help explain them. It may also suggest why a model does not fit well and how the fit might be improved.

Finally, the GIS analyst must be trained in the use of statistical methods, particularly the statistical theory of estimation and testing, multivariate analysis, linear models, and sampling, in addition to the training required for the GIS, if spatial data analysis is to be performed. The standard kriging and modeling features often found in GIS systems may not provide for the proper generation of the diagnostic information and alternative models. The analyst must be able to generate these reports externally. A GIS for spatial data analysis should include not only the GIS software, but also statistical computing software sophisticated enough for spatial analysis.

5.8. Conclusions

There is much more to spatial data analysis of GIS data than can be covered in a single chapter. The works of Cressie (1991), Haining (1990), and Isaaks and Srivastava (1989) provide wide coverage of many of the spatial data techniques. We have described a set of spatial data techniques that are of particular importance in GIS work, which is spatial prediction. Predictions are often needed to convert data that were sampled in a discrete manner into continuous data that may later be represented in the GIS as isoclines. Very often, predictions of new events are needed, as in the prediction of contaminant spread through a field and into the water table. These predictions could be based upon data that are already contained within the GIS data base.

We have shown the basic building blocks for constructing a spatial prediction model, and have demonstrated the difficulty of selecting between models where one emphasizes a trend structure and the other emphasizes a spatial dependency structure. It is very obvious that reliable spatial modeling will require interaction with the GIS, not only for data storage and retrieval, but also for the display of predicted values, residuals, cross-validation results, and other modeling diagnostics. It is also important that the analyst not depend upon a "built-in" estimation program that does not provide sufficient modeling diagnostics along with many alternative models and techniques. Rather, the GIS should provide ample flexibility to move between the GIS software and the statistical analysis software.

References

Clarke, G. P. Y. and J. H. Dane. 1991. A simplified theory of point kriging and its extension to co-kriging and sampling optimization. Bulletin 609, Alabama Agricultural Experiment Station, 44pp.

Cressie, N. 1985. Fitting variogram models by weighted least squares. Mathematical Geology 17, 563-586.

Cressie, N. 1989. Geostatistics. The American Statistician 43, 197-202.

Cressie, N. A. C. 1991. Statistics for Spatial Data. John Wiley & Sons, Inc., New York, 900pp.

Cressie, N. and D. M. Hawkins. 1980. Robust estimation of the variogram. Mathematical Geology 12, 115-125.

Emerson, J. D. and D. C. Hoaglin. 1983. Analysis of two-way tables by medians. Pages 166-210 in Hoaglin, D. C., F. Mosteller, and J. W. Tukey, eds. Understanding Robust and Exploratory Data Analysis, John Wiley & Sons, Inc., New York.

Haining, R. 1990. Spatial Data Analysis in the Social and Environmental Sciences. Cambridge University Press, Cambridge, 409pp.

Isaaks, E. H. and Srivastava, R. M. 1989. An Introduction to Applied Geostatistics. Oxford University Press, New York, 561pp.

Kitanidis, P. K. 1983. Statistical estimation of polynomial generalized covariance functions and hydrologic applications. Water Resources Research 9, 909-921.

Macchiavelli, R. E. 1992. Likelihood-based Procedures and Order Selection in Higher Order Antedependence Models. Ph.D. Thesis, The Pennsylvania State University.

Matheron, G. 1963. Principles of geostatistics. Economic Geology 58, 1246-1266.

Moser, E. B., R. E. Macchiavelli, and D. J. Boquet. 1994. Modelling within-plant spatial dependencies of cotton yield. Applied Statistics in Agriculture 6, 246-260.

Neter, J. W. Wasserman, and M. H. Kutner. 1990. Applied Linear Statistical Models: Regression, Analysis of Variance, and Experimental Designs, 3rd edition. Irwin, Homewood, IL, 1181pp.

Ripley, B. D. 1988. Statistical Inference for Spatial Processes. Cambridge University Press, Cambridge.

SAS Institute, Inc. 1992. The MIXED procedure. Pages 287-366 in SAS Technical Report P-229, SAS/STAT Software: Changes and Enhancements, Release 6.07, SAS Institute, Inc., Cary, NC.

Stroup, W. W. 1989. Why mixed models? Pages 1-8 in Applications of Mixed Models in Agriculture and Related Disciplines, Southern Cooperative Series Bulletin No. 343, Louisiana Agricultural Experiment Station, Baton Rouge.

Zimmerman, D. L. and M. B. Zimmerman. 1991. A comparison of spatial semivariogram estimators and corresponding ordinary kriging predictors. Technometrics 33, 77-91.

CHAPTER 6

GIS Needs and GIS Software

C. Collet, D. Consuegra and F. Joerin

6.1. Aim of the Chapter

The reader might be surprised at first to find in this chapter neither a GIS software directory, nor a comparative list of software performances. It is our conviction that such an attempt does not belong in this book for two major reasons. First, the interest for a software compilation decreases rapidly over time due to permanent progress in computer technology as well as in this particular domain of software development. Secondly and principally, a GIS environment is much more than just a cartographic tool used to produce maps or to display colourful maps and animated sequences of images. It is an information processing technology that requires substantial human and financial resources, with particular constraints on the manner by which tasks can be processed. Therefore the key issue is neither the selection nor the purchase of an adequate software package, but how such a technology can be integrated in an institution or a research group in order to satisfy both existing and planned tasks. The software selection should result from an evaluation of needs, the organisational context and resources. GIS technology embraces the whole spectrum of information processing tasks and thus influences the mode of handling information from its acquisition to the production of results. The critical issue is then to assess its ability and relevance in the field of hydrology and water resources management.

The evaluation and comparison of GIS software require a clear understanding of the nature of GIS and its related concepts. Furthermore, to consider GIS capabilities within any applied domain such as hydrology, it is assumed that needs are clearly stated. Conversely, familiarity with GIS offers a broader view of its potential, and leads to differentiate between specificity of hydrology and other scientific fields, as well as their common characteristics.

The goal of this chapter is to present and to briefly illustrate GIS capabilities in the field of hydrology. It describes the uses of GIS software within hydrological

V. P. Singh and M. Fiorentino (eds.), Geographical Information Systems in Hydrology, 115–174.
© 1996 *Kluwer Academic Publishers. Printed in the Netherlands.*

application requirements, such as information management, queries and mapping, spatial data preparation, modelling and simulation. Finally it provides keys for GIS software selection.

6.2. GIS Concepts and GIS Software

6.2.1. GIS OVERVIEW AND CONCEPTS

Nature and components of a GIS   Among the numerous and various definitions of a GIS, the following is retained as a global description: *a GIS is a computerised system designed to process information with a spatial dimension*. The three major characteristics shared by any GIS are its computerised environment in which the software component is integrated, its ability to process information and the nature of the information with a spatial content. This latter characteristic is the key feature that makes geographical information systems different from other information systems.

Information processing is composed of several stages including data acquisition, data preparation and data exploitation. Data acquisition produces information in computerised form (i.e. numerical and alphanumerical form). Data preparation organises information in a structure and format suitable for its storage within the database. Data exploitation consists of information retrieval in an appropriate form and information analysis (figure 6.1).

Since a GIS is an information system which processes spatial information (georeferenced information), it offers three similar processing stages. The three major components of a GIS are a geographical data base (GDB), a set of dedicated tools and a user-machine interface (figure 6.2).

A GDB is a database storing information in digital form and linking together the thematic and spatial contents of information.

The set of tools is made up of procedures for processing information at various stages:

- data acquisition tools;
- data preparation tools;
- GDB management tools, known as geographical database management system (GDBMS):
- data exploitation tools. They offer procedures for the retrieval of information from the GDB and for its representation either in graphical or text form. As information contains a spatial dimension, its graphical representation implies the use of cartographic tools;
- data analysis tools.

The user-machine interface is the component that permits the user to interact with the system. It offers a language to interrogate the GDB either interactively or not, to access analysis procedures and system resources with various degrees of ergonomy.

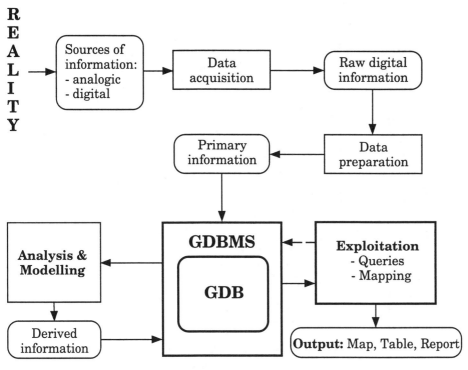

Figure 6.1. GIS as an information processing system

GIS types   GIS applications range from spatial information management to spatial analysis (Aronoff 1989, Burrough 1986, Laurini and Thompson 1992). Their respective developments followed a quite different path. On one hand the GIS environment was designed to survey and manage spatial features within a defined territory. Its development was aimed at cadastral survey purposes within a rigid legal context. Major contributors and users of such a so-called Land Information System (LIS) are surveying and planning offices. On the other hand, research institutions have developed a GIS environment to perform spatial analysis, modelling and simulation of spatial processes. It has been principally developed by earth and environmental scientists, using the same and therefore confusing name of Geographical Information System (GIS). Spatial information management and spatial analysis are two complementary goals corresponding to the two ends of a continuum illustrated in table 6.1. In practice, however, application needs are located somewhere along this continuum (Bédard 1987). Depending on GIS uses, the software component belongs to a very different context. Some of its major characteristics are described in this table, showing the interdependency between the software, the needs and the institutional infrastructure. They are examined in section 6.7. about GIS software selection.

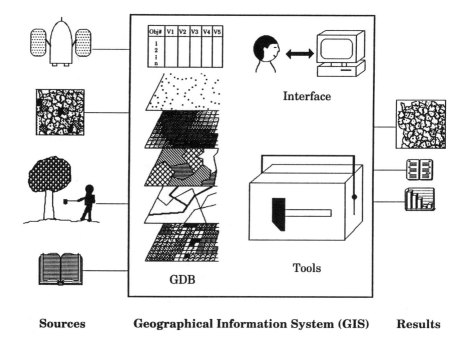

Obj#	V1	V2	V3	V4	V5
1					
2					
i					
n					

Interface

Tools

GDB

Sources   Geographical Information System (GIS)   Results

Figure 6.2. Three major components of a GIS

Today's needs are more ambitious and complex; one expects to work within an environment that integrates both management and analysis aspects. The management of resources and the planning of sustainable human activity development require such an integrated system.

Nature and content of the GDB   Information stored in the geographical data base is a model of the real world. It is a biased and simplified image of the reality, with respect to a thematic content (hydrology), a spatial content (scale of description, spatial objects) and a temporal content (period, time scale). This image of reality is biased because it is a specific view of the hydrologist, which is certainly different from the view of other Earth scientists. It is also a simplified image because of the limited number of information elements (variables, objects) and of the scale of description. Such statements lead to a very important remark: the content of a GDB is tailored for a clearly defined purpose (Goodchild 1992).

The construction of a GDB is aimed to serve different goals that can be grouped into two categories: the management of land resources and spatial analysis of phenomena and processes within a study area. The former has a general purpose at institutional level (public services, governmental agencies) as the second has a more specific purpose at research unit level (universities, engineering and planning offices, governmental research agencies). The GIS environment and more specifically the GDB structure and its content are distinct between the two categories.

The GDB is a set of descriptive units (entities, spatial objects or spatial features) with their attribute and spatial characteristics. Its organisation can be seen as layers describing spatial features of different kinds. A GDB can handle both spatially continuous distributions (i.e. elevation, precipitation) and spatially discontinuous distributions such as river network, watersheds, gauging stations, ... In order to ensure coherent spatial relationships between spatial features, they must be georeferenced within a defined projection system.

Spatial objects can be defined from two different points of view, such as *origin* (i.e. cadastral or administrative units) or *resulting* from the spatial distribution of a phenomenon (i.e. land cover, soil units). In the first case, spatial features are defined through their spatial dimension (geometry, limits, location) and then characterised by a thematic content that can be heterogeneous. In the second case, a spatial feature is an entity with a homogeneous thematic content with respect to the considered phenomenon.

Information stored in the GDB comes from various sources providing different spatial and thematic contents. Each content has a defined *quality* and *precision* related to the origin of the source document. Information quality expresses the intrinsic content of the source document and its conversion into digital form. It describes how well information described the real world in its thematic, spatial and temporal dimensions. From this point of view, it is useful but often difficult to distinguish between *primary* information documents containing direct field measurements or observations and *interpreted* information documents resulting from a complex interpretation such as a soil map or a landscape map. In this type of map, spatial objects are originated from an interpretation based on a limited set of sampled observations. On the other hand, precision is related to the amount of error contained in the information, both in its thematic and spatial dimensions. Errors present in the original document are combined with those generated during the acquisition stage.

Regardless of the manner in which a GIS processes information, the GDB is structured in several different forms. Basically there are two distinct structures: the vector and the raster structure. In a vector structure, information is organised into spatial objects such as points, lines, zones and facets; units of description are spatial objects. In a raster structure, units of description are regular and generally rectangular zones, called cells or pixels; spatial objects are made up of contiguous cells sharing the same attribute characteristics. In the raster structure, the description of the spatial dimension is mainly implicit, through the relative location of each cell within the image. The choice of a relevant structure is dictated by the nature of the source document, the nature of the phenomenon to be described, as well as the purpose of the GDB. However it is always possible to convert from one structure to another, assuming that there is a sufficient level of detail. As a simple rule, it is admitted that a raster structure is more efficient for the description of a continuous spatial distribution, while a discontinuous spatial distribution is more efficiently described with a vector structure.

6.2.2. GIS NEEDS IN HYDROLOGY

The mapping ability is from far not the strongest feature of Geographical Information Systems (GIS). The key issues are related to database organisation since the information displayed in the maps results from a retrieval and a spatial analysis of available data. It is also important to realise that even the most simple queries may involve very complex mathematical operations. A simple illustrative example may be to ask a given system if the urban areas over a given watershed are clustered, dispersed or uniformly distributed (quadrate and nearest-neighbour analysis). This information may be very important when subdividing the watershed into various sub-catchments each of them with uniform land-use. Therefore the analysis tools and the database organisation are with no doubt the most important components of a GIS system.

In many cases GIS applications in hydrology involve only input and output interfaces with given models. These systems do not guarantee full portability and flexibility of the various components. A change in the model or in the type of data required generally leads to a complete, or at least partial, regeneration of the whole GIS system. As indicated by Ball (1994) this situation is similar to that of early computer hardware where an upgrade or a change of manufacturer involved large modifications of current programs due to the narrow linkage between software and hardware.

The increased use of complex models requires the availability of huge amounts of data. This may be considered as a technical problem but it is important to realise that it is an administrative problem too. As a matter of fact, different countries or different regions in the same country do not have standards for data organisation and the selected databases are in many cases different and even worst, incompatible. This situation leads to data redundancy as indicated by Hoogendoorn et al. (1993). In the past, data collection was only achieved on the basis of the requirements of the particular model to be applied. If a few years later, another study has to be conducted in the same area, the previously gathered data will not be compatible with new computer codes and will lead to another data collecting process.

Another important problem is that the life span of hardware is very short (around 5 years). The innovation rate of GIS software is not that much longer (around 15 years). Any permanent GIS application is rapidly threatened by hardware improvements but also by software improvements resulting from market competition. These rapid changes result in financial investments but also in excellent skills from GIS staff and users. If the necessary investments are to be kept as cost effective as possible it is necessary to ensure easy implementations not only of the software but specially of the databases (Kaden, 1993).

Databases must also handle time components since most hydrological parameters present a temporal evolution. The storage of different object versions is necessary in many water management projects. For instance, the analysis of flood control alternatives may modify objects in the database on a temporary basis until the best solution is found. River conservation authorities often require the temporal evolution of cross section profiles in rivers to locate erosion or sedimentation problems. Land use is an important information that also requires periodic updates.

The tendency is now to move to relational common databases, with hierarchical structures of the information and pre-specified access keys according to the type of users and applications. For the above mentioned reasons, this chapter will be largely devoted to database management systems.

Schultz (1993) describes an interesting experience with the use of different GIS systems in hydrology. He distinguishes between four categories according to the level of complexity and ranging from self developed GIS to more sophisticated systems like ARC/INFO and GRASS. The simple self developed GIS systems offer the possibility of easy modifications and prototyping for inexperienced users. They obviously fail to analyse large scale problems and require the development of all the routines needed by all applications. The most sophisticated systems can overcome these problems but have lost flexibility with respect to changes or further software development. These packages often come with a lot of applications that may not suit the user objectives. The development of extra computer code can only be done by an expert GIS user. A solution to that problem can be to select a product with a large users-group involved in wide variety of applications. This creates a forum for code improvement and exchanges. However, this chapter will illustrate by means of simple examples that simple GIS systems like MapInfo and IDRISI linked with a database management system like DbaseIV can be used to tackle relatively complex hydrological problems.

6.2.3. GIS SOFTWARE CAPABILITIES

Among the large variety of existing GIS software in the market, they all share common features and particularly common components:

- data acquisition and preparation tools for the creation of the GDB;
- a geographical data base;
- a geographical data base management system (GDBMS);
- data exploitation tools for information retrieval from the GDB and its cartographic representation;
- data analysis tools.

Major differences between GIS software capabilities are in the weighting of those components as well as in the selected mode for processing spatial information. Depending on their purpose, GIS software offer more or less complex capabilities for the different processing stages, but they all offer some basic tools in these domains. More important is certainly the way that they process spatial information. The *object processing mode* considers spatial objects as units of description, while *image processing mode* is based on arbitrary and regular units of description, referred to as cells. The two approaches process spatial information in a very different manner and require a specific data structure, vector and raster respectively.

According to these distinctive characteristics, GIS software provide either management oriented capabilities with an object processing mode, a complex GDBMS and a palette of exploitation tools, or spatial analysis capabilities requiring rich and complex preparation and analysis tools but straightforward GDBMS and exploitation tools, generally with an image processing mode. Several software developers have attempted to offer a package combining the two types of GIS, but we are still far from having a truly integrated system that satisfies the two complementary purposes of management and spatial analysis.

6.3. Geographical Data Base Construction

Before building up the geographical data base, it is assumed that the model of reality is clearly defined. There are two distinct stages to the process of GDB construction: data acquisition and data preparation. They require specific pieces of hardware and software that may vary according to the type of source document.

6.3.1. SOURCES AND SPATIAL DATA ACQUISITION

Among the large diversity of source documents, only a limited number of them expresses a spatial content about units of observation. They can be organised into different types:

- field data (generally point measurements);
- cartographic documents (cadastral, topographic and thematic maps);
- analogic images (aerial photographs and satellite images);
- digital images (aerial and satellite data, scanned documents);
- existing GDB (layers of information from other institutions already in digital form).

The process of acquisition varies according to these types, making use of specific instruments and procedures. Furthermore acquisition instruments operate in different ways to produce digital information in either vector or raster structure. Figure 6.3 illustrates the process of acquisition from different source document types. GIS software should provide tools for acquiring spatial data from a digitising table, scanner and camera, as well as import procedures to read existing digital files in different vector and raster structures.

A digital file of spatial information without its related documentation file is like a crypted message (cryptogram) without its decipher code; it is useless. A documentation file must contain information about thematic, spatial and temporal dimensions of the referred digital information, including a history about data acquis-

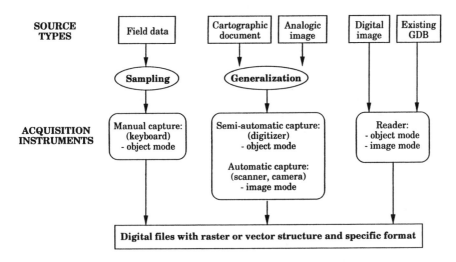

Figure 6.3. Spatial data acquisition process

ition and preparation process. A documentation file may contain the following relevant information:

- in the thematic dimension: content description, value unit, legend, quality, error;
- in the spatial dimension: coordinate reference system, reference system unit, georeference, position error, (cell size for a raster structure);
- in the temporal dimension: document date;
- for history: origin and scale of source document, person responsible for acquisition process, acquisition date and instrument, processing history (acquisition and preparation), recommendations and restrictions of use, ...

GIS software should offer the capability of creating and managing such documentation files, known as *meta-information* or *meta-data*.

6.3.2. SPATIAL DATA PREPROCESSING

Spatial data preparation consists of transformation processes for the production of georeferenced layers of information with suitable structure and form, as well as relevant thematic content. To fulfil these requirements, preprocessing tasks are the following:

— modifications of the thematic content;
— modifications of the spatial content (geometry, object limits, units of observation);
— modifications of the structure and format of data file.

Spatial data preparation can be a very tedious and complex process, depending
on characteristics of original data and those to be produced in the resulting files.
It includes transformations like interpolation, generalisation. Basic preprocessing
techniques can be divided into two categories, according to the resulting structure
they produce: raster or vector.

6.3.2.1. *Production of a Raster Structure*

Original digital information can be either in a vector structure or already in a raster structure but with inadequate thematic content, geometry or format. Figure
6.4 summarises standard transformations involved in the process of suitable raster

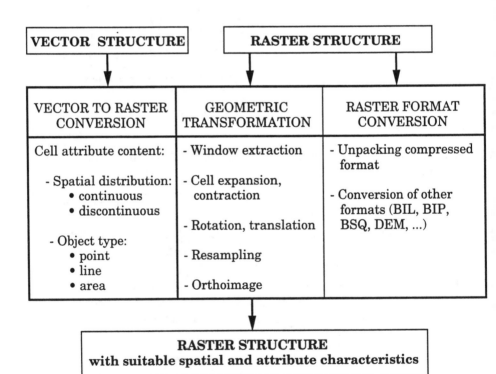

Figure 6.4. Production of a raster structure

structure production.

Vector to raster conversion   Vector to raster conversion is certainly the most
obvious and frequently applied transformation, particularly for spatial analysis.

However attention should be drawn to the nature of such a process, in which both thematic and spatial contents of information are modified. Transformation techniques applied to discontinuous spatial distributions are radically different from those applied to continuous spatial distributions.

Conversion techniques applied to discontinuous spatial distribution assign its thematic content to each cell inside a spatial feature. In most cases the conversion process is straightforward when applied to zones, as long as they are larger than the cell resolution. But what happens to linear and point features? In some instances the thematic content must be modified to express the presence or frequency within each cell, while in other circumstances it is suitable to split the object layer into several image layers in order to avoid such conflicts. In any case, it is essential to control the process in order to obtain relevant information.

For a continuous spatial distribution, conversion techniques estimate the attribute value of each image cell based on information available from the original vector information. Interpolation is the process involved in this type of conversion. A typical use in hydrology is the construction of a digital elevation model (DEM). Because a DEM is at the origin of the morphological description of the relief, great attention should be paid to the construction of this piece of information. The quality of morphological descriptors such as slope, aspect, watersheds and drainage networks relies on the consistency of the constructed DEM. One should keep in mind that a DEM is a model of reality, in other words it is an oriented construction build up from a limited set of measurements. Properties of an elevation model may vary according to its purpose; in hydrology for example, elevation distribution should maintain the continuity of water flow along the river system and be free of any local depression artificially created during the interpolation process. The most appropriate interpolation technique must be selected among numerous ones offered by software packages, such as weighted distance methods, kriging, rasterized contour lines and tessellation (Lam 1983, Gold 1989, Collet 1992, Delhomme 1978). As continuous surface values are estimated from a sample of point values or a set of contour lines, specific assumptions and rules control the process of estimating unknown values. In this chapter structure, the generation of a DEM or of any continuous spatial distribution is part of the data preparation stage, while the production of derived morphological indicators belongs to the spatial analysis stage. In chapter 7 different methods for generating a DEM and for producing morphological descriptors will be reviewed in much greater detail.

Vector to raster conversion makes use of very different transformation procedures which are suitable for distinct purposes and data types. Unfortunately in many software packages they are presented as black boxes. The user has little or no control over the process, and very often only limited information is provided about the algorithm and its related conditions of use. This is not only true for GIS software integrating such tools, but for software packages specialised in this process as well.

Image geometric transformation   Original images acquired in digital form may not fulfil the required geometric characteristics of the planned GDB. Geometric transformations allow one to modify the projection system of an image, to reduce or increase the resolution of units of observation (cell size), or to extract the relevant portion of the study area. Multiple geometric transformations can be simultaneously performed on the original image, using existing procedures offered by most image processing and digital photogrammetric software systems.

For photographic and satellite imagery, complex geometric transformations are involved such as rotation, warping and resampling in order to produce a properly georeferenced layer. Furthermore, subsequent exploitation of their thematic content implies thematic changes like classification and index transformations; they will be discussed in the analysis section of this chapter as they require supplementary information present in the GDB.

For other original images in digital form like imported layers of an existing GDB, necessary geometric transformations are more straightforward. They permit one to extract a study area and to modify the cell resolution or the projection system.

In a raster structure, geometric transformations operate by the construction of a new image in which cell values are re-estimated on the basis of original values. In other words the geometry of an image is transformed through an attribute value interpolation process. The selected interpolation procedure should of course be adapted to the nature of the thematic content.

Raster format conversion   Original information acquired in raster structure may not be in an appropriate format. Format conversion procedures provide tools to convert raster data files from and to standard formats. Raster format conversion is used for two different objectives:

— to compress or decompress raster data files. A compressed format is utilised for transmitting and archiving image files (run-length, block, chain and quadtree formats);
— to exchange raster data files between information processing systems. Although the number of standard formats is limited (BIL, BIP, BSQ, DEM, PCX, TIFF, GIF), many of them offer multiple variants.

In most GIS software numerous format conversion procedures are offered as part of the import-export module. The conversion process is normally straightforward, assuming the input format is known and supported.

6.3.2.2. *Production of a Vector Structure*
Depending on the type of equipment used to acquire information in digital form, its resulting structure may be in raster or vector form. Production of a vector structure with adequate spatial and attribute characteristics implies several different preprocessing stages. Figure 6.5 summarises two standard transformations involved in the process of producing suitable vector structure.

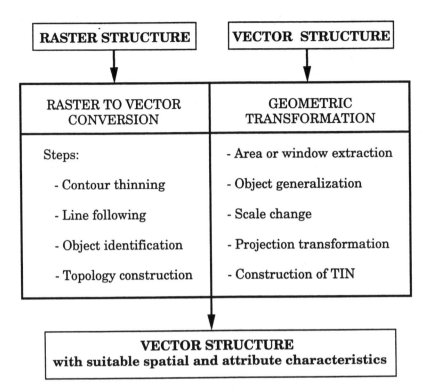

Figure 6.5. Production of a vector structure

Raster to vector conversion   This type of transformation generates a description of spatial objects from a layer in image form. Typically it is applied on digital documents acquired by scanning instruments such as a scanner or a camera. In order to delimit object boundaries, the considered distribution is assumed to be spatially discontinuous. The conversion process varies according to the type of spatial object considered (point, line or area) and the nature of the image content. Scanned images may contain lines expressing object limits or filled solid areas describing simultaneously object limits and their attribute content.

The conversion process requires several steps to produce a vector structure, assuming the input image is free of non relevant features such as text labels or other annotations present in many scanned map documents:

— contour thinning for images containing line drawings;
— line following to outline object limits;
— object identification derived from their attribute content or manually assigned;
— topology construction to build objects from the set of constructed lines.

The raster to vector conversion can be a very tedious task depending on the quality and the complexity of the original information in image form; this is particularly true with scanned map documents.

Geometric transformation   This type of transformation concerns vector files with an inadequate geometric content. Geometric transformations include change of projection, scale, and format as well as the extraction of features within the region of description and the construction of a surface.

Extraction of features contained in the region to be described with the GDB is a common task while the original information in vector form has been imported from an existing digitised file. Areal or linear features intersecting the edges of the selected region have to be truncated or clipped and therefore new objects need to be reconstructed. Projection transformations require the presence of projection functions with known parameters. Such cartographic transformations should be applied to produce a GDB with a common projection system.

Scale transformations are used to reduce the amount of information in the spatial dimension. This process is commonly called *generalisation* because it reduces the level of detail in the description of spatial features. A change in scale of description can transform areal features into points or even remove small objects. In order to preserve the topological integrity, common sampling procedures cannot be applied to line network or contiguous polygons, but specific algorithms must be used such as the one developed by Douglas and Peucker.

As the vector format of digital files is dictated by the acquisition environment (i.e. the equipment and its companion software), it may not correspond to the format supported by the current GIS in use. Format transformations allow one to convert vector files into various format standards and more specifically to construct the topology of spatial features from a non topological file structure. This is a requirement for object oriented GIS.

A vector structure can only handle bounded spatial objects such as points, lines and areas. Therefore a surface describing a continuous spatial distribution such as elevation or precipitation must be broken into pieces, generally triangular areas called *facets*. The original digital information of a continuous spatial distribution is a set of sampled data points. The surface is modelled through the process of *tessellation* producing a set of triangular plane facets known as a triangulated irregular network (TIN). Several tessellation algorithms are proposed offering simple Delaunay tessellation or more complex procedures handling break lines and other specific constraints (Gold 1989). From the constructed TIN, any point value from the whole surface can then be interpolated. Thus this process is the vector counterpart of the interpolation process in raster mode.

6.4. Geographical Data Base Management System

6.4.1. HISTORICAL OVERVIEW IN COMPUTER SCIENCES

GIS can be understood as the reunion of two units: a database management system (DBMS) unit plus a spatial data treatment unit. The latter is made of map-making functions that allow the management of topology and metrics, such as map overlays, area buffering and so forth ... The DBMS is where the data is stored, organised and treated. It provides the access to information through its questioning functions (SQL), but it is a powerful tool for a constant updating of the data, too. It also allows the creation of selective access paths (writing and reading).

The DBMS naturally manages information stored in the GIS. Even if in many cases, GIS are used as simple map-making software, their main purpose is to solve the problem set by the management of the huge volume of data necessary to describe the territory. Obviously, from this point of view, a DBMS is essential to GIS.

A short historical reminder about the evolution of computer sciences allows a better understanding of the importance of DBMS (fig.6.6). In the beginning of

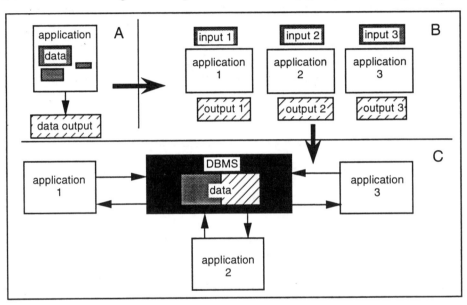

Figure 6.6.   The three steps in the computer methodology evolution

the computer era, when storage was accomplished on perforated cards, data were a part of the program itself. That is, both information and its treatment were stored in a same logical unit, the source code. Soon, technical improvements made possible the separation of data and its treatment. This way, programs got closer to so-called applications. As a matter of fact, computer programs were previously considered as computational routines created to solve one problem set by one particular situation. By getting separated from the data, instruction sets became tools that were useful to solve not only one particular problem but a whole type of problems. As a program then had many possible applications, the time spent to develop it was being paid back, and the use of computers improved a lot.

This information thought process is still widely used in hydrology. Most hydrologic models are designed according to these principles. Every model has its own data format. This situation is far from being the best. A case study, or a comparative research often require the use of several models. In this case, and even if the input data is mostly the same for all models, it can't be directly transferred from one model to another, because they don't share a same format. Consequently, either one has to duplicate the data, which means redundancy, or write small format transformation programs.

This situation generates many problems. First, when the data is numerous, it is nearly impossible to duplicate it, due to the lack of disk space. In addition, in this situation, the control of information is difficult to perform. For example, if a set of data is modified, the update has to be done wherever it can be found. It is then more than likely that one copy, or memory space won't be updated. On the other hand, the development of format translating programs (interfaces) is a heavy and painful work, because of the constant modifications made to the different applications.

DBMS were developed to solve all these problems. The basic assessment was that centralising the storage of information and developing systems to manage, organise and distribute them would be extremely profitable. The apparition of such systems made information management and transfer much easier than before.

6.4.2. SCOPE OF DATA BASE MANAGEMENT SYSTEM

Globally, DBMS deals with the management of information, that is :
- information distribution,
- information quality control,
- information consistency maintenance.

Access to the information   DBMS play the role of an interface between data and its users (either man or software). Besides the fact that the whole set of data is centralised, DBMS offer query tools that allow a faster and more efficient access to necessary information. The burst of communication networks even increased the accessibility to information. As DBMS were quickly adapted to this new way, the user now has access not only to one database, but to a network of databases. The confidentiality of some information is taken into account by DBMS, that restrict the access to confidential data only to authorised persons.

Consistency   The centralisation of information allows an easier and more efficient control of data. The consistency of the database is created by integrity rules. Then controlling the integrity of the database sums up into controlling that all integrity rules are respected. An integrity rule may specify that any river gauging station has to have an owner. Then, if one wants to insert a new station in the database, the DBMS will see that, according to the rule, the new station does have an owner.

Diminishing the degree of redundancy of the database is an indirect way to preserve its consistency. As seen previously, without a DBMS, many information are stored in different places at the same time. This redundancy threatens the consistency of information, as updates are not performed with all the care they require, and therefore induce contradictions.

Maintenance   Query functions, redundancy reduction and integrity rules contribute to the maintenance of the database. Query functions make a selection of the data to be updated, integrity rules verify that updates are consistent and redundancy reduction makes those operations faster and easier.

In addition, DBMS were created to handle such operations as updates. The user can be sure that an update will be either fully, or not at all, completed. For example, if there is a system failure, the system automatically erases uncompleted operations. On the other hand, two people cannot modify one set of data at the same time. DBMS also takes care of update authorisations. Only users formally identified by a password are allowed to modify the content of the database.

6.4.3. DESCRIPTION OF A DBMS

Different kinds of DBMS   There are many kinds of DBMS. The first ones were hierarchical. They are based on "father-children" relations between information, like in an organisation tree. For example, the department of physical engineering (father) is made of three labs (children): the nuclear physics lab, the astronomy lab, and the plasma lab. Every lab is under the direction of one professor, who hires several teaching assistants.

This sort of DBMS does have a few disadvantages. First, to get information, the user has to follow the path imposed by the way the data is organised. Then, this hierarchical model is not always suited to the reality to be represented. In particular, the fact that one child cannot have more than one father is a strong limiting factor. To bypass this limitation, one has to introduce artificial redundancy, with all the regrettable effects it brings.

Finally, the use of such a database demands the knowing of its organisation tree. The integrity rules imposed by the hierarchical structure make the representation of the reality not very sensible and not intuitive at all.

After that came the so-called network system. This structure can be considered as a development of the hierarchical structure. The basic assessment of the concept is still the organisation tree, but here, children are allowed to have more than one father. This improvement makes queries easier. By densifying the relations between the elements of the database, the DBMS creates different access paths for one piece of information. In addition, network DBMS give a better representation of real systems. However, next to these two points, the disadvantages attached to the use of this kind of DBMS stay the same as in the previous case.

The apparition of relational DBMS was the great step ahead in this field. The main innovation is that all attributes of an entity make one access to the information. The structure of this DBMS does not specify the relation between two entities. The user of the DBMS has to create them according to his needs and wills. This structure is evidently extremely adjustable and provides an unlimited potential of application. However, as the user himself has to set the relations used in his queries, the use of the query language for this kind of DBMS is a little bit touchy.

Anyway, relational DBMS are the most commonly used in GIS, even if few attempts are made towards more revolutionary concepts, such as object oriented DBMS. This is why the focus of this section is on relational DBMS.

Description of the relational DBMS Every information that can be considered as an object, or entity (such as monitoring stations) is stored as a TABLE in a relational DBMS. Every table has several occurrences, or tupples. A table is made of several fields, each of them containing an elementary set of data about the entity. Figure 6.7 presents this concept: tupples correspond to real-life objects, and

Monitoring station

Table name

Fields

Number	Owner	Watershed	Starting period
128	SHGN	Broye	12/10/1968
129	Mauvoisin	Val de Bagne	4/7/1967
...			

tupples

Figure 6.7. Relational table

the fields are the descriptors of these objects. In the example, the gauging station # 128 is the real-life object. It is described by four fields that successively indicate its number (128), its owner's name (SHGN), its watershed (Broye) and its starting period (12/10/1968).

6.4.4. ORGANISING A DATA BASE

The objectives assigned to the DBMS (access to information, coherence and maintenance) can only be achieved if the database is organised correctly. Organising a database consists in, on the one hand, choosing the information that will be contained in each table, and on the other hand, which information will be common to one or more tables. Simplifying matters somewhat, one may assume that the tables are the computerised expression of objects belonging to the real world, whereas information common to several tables conveys the relationship between these different objects. The organisation of the database should fulfil two aims. First of all, it should avoid the duplication of information in order to minimise memory space utilised and facilitate maintenance. Furthermore, it should reflect the real world to be as close as possible to the user's intuition.

This last objective is important, as it allows any user to have access to information, without prior knowledge of the database schema. If we retake the example of figure 6.7, at the position where the address of an institution may be found in the Monitoring Station table, it is unlikely that someone looking for this address would consult this table, one would, however, intuitively look for a table describing the institutions.

The conception or organisation of a database is probably the most difficult and most important step in setting up an information system. Numerous research has been carried out in this field. Consequently, many different methods exist in planning the organisation of a database. The method shown here is the one most currently used to conceive a database in relational mode. It comprises three fundamental stages which are materialised by the realisation of a schema. Each stage contributes to realising the gradual transition between the real world and its representation through a database. However, this approach was developed for the conception of information systems and not geographical information systems and, this difference sometimes creates problems. Conception methods such as ModulR (Gagnon 1993) or Me.Co.S.I.G. (Pantazis 1994) put forward solutions for the SIG conception. Nevertheless, the approach is globally the same, and we will not go into these specificities here.

The stages of the conception of a database are: first the conceptual level, then the logical level and finally the physical level. They respectively lead on to the realisation of the conceptual data model, logical data model and physical data model. Figure 6.8 illustrates this method of conception.

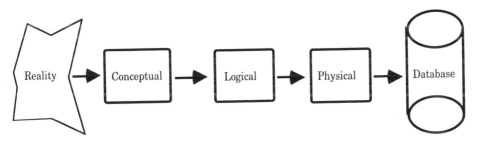

Figure 6.8. The three stages of the conception of a database

Conceptual level   The realisation of the conceptual data model is generally based on a systemic approach. This approach is based on the theory of systems for modelling reality (Le Moigne 1990). The real world is perceived as a system and the database is conceived as a model of this system. The conceptual data mode is an exact image as possible of the real system. It is independent of technological aspects and is consequently not influenced by the hardware and software constraints by which the database will be affected. These constraints will be taken into consideration during the next stages of conception of the GIS (logical and physical).

The conceptual data model is by definition a model, that is to say a simplified representation of reality. Even if the conceptual model aims to represent the real system as closely as possible, it does not constitute an objective image. This representation corresponds to the user's point of view with regard to his perception of reality, or to application objectives. Simplifications carried out, must concord with the user's needs, who will be directly implicated in the realisation of the conceptual data model.

The conceptual schema is expressed though a formalism. The most common formalism is "entity-relationship". It is based on two concepts, entities (objects) and relationships. Figure 6.9 demonstrates what could be the conceptual schema of the Figure 6.7 example.

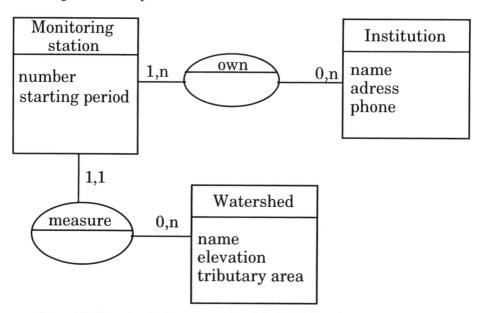

Figure 6.9. Example of a data conceptual model with the formalism entity-relationship

It should be noted that Monitoring Station and Institution entities are only described by attributes which are particular to them. The attribute "owner" which can be found in the Monitoring Station table (Figure 6.7) is expressed by the relationship "own" that links the limnigraphic stations to the Institutions. The cardinal point (1,n) of the relationship between Monitoring Station - Institution expresses the fact that a limnigraphical Station has at least one owner, but may have several (co- owners). Whereas the cardinal point (0,n) of the relationship Institution - Monitoring Station, expresses that an Institution may possess none or several limnigraphical Stations.

We have seen that the aim in organising a database is not only to project a true and intuitive image of reality, but also to ensure that there is no redundancy. At the conceptual level, this aspect is taken into account by the application of rules tending to normalise the model. The first stage in the realisation of a conceptual model conceives an initial schema without taking into consideration the question of redundancy. This first schema is then standardised in order to eliminate all redundancies. The optimisation of the conceptual schema results in the properly so called conceptual data model.

Logical Level   Contrary to the conceptual level, the logical level is adapted to the software on which the database will be implanted. As the objectives of representativity and normality have already been achieved by the conceptual model, they no longer form a part of the objectives of the logical level. The aim of the logical level is thus, on the one hand to adapt the conceptual model to the type of DBMS used and on the other hand to offer an efficient organisation with regard to the future utilisation of the database. In other words, it is a question of finding the right compromise between the total elimination of redundancies and the speed at which operations may be carried out by the database user.

The same conceptual models take on different logical forms for different DBMS's. In compliance with the choice previously made we will examine the case of relational DB's.

Briefly, the transition from the conceptual stage to the logical stage is carried out in the following manner: the entities of the conceptual model become tables, whereas relationships are expressed either by tables or by duplication of the attribute allowing for the joint to be made. These two cases may be distinguished by cardinal points of the relation.

The conceptual model in Figure 6.9 is represented in Figure 6.10. Both cases (Case A and Case B) illustrate the two ways in which a relationship in the logical model may be expressed. Case A represents the situation, in conformity with the conceptual model, where a limnigraphic stations may have one or more owners. In this situation, where one person may own several limnigraphical Stations, it is necessary to create a table to establish this relationship. Case B only allows for a limnigraphical Station to have one owner and the relationship may therefore be more easily established by adding the owner's name to the attributes of the limnigraphical station. This second type of figure corresponds to the situation portrayed by figure 6.7.

Physical Level   The physical level is the last step before the database itself. This stage not only takes into account the type of DBMS, but also the software chosen. The language of the physical model corresponds to the programming language of the DBMS software.

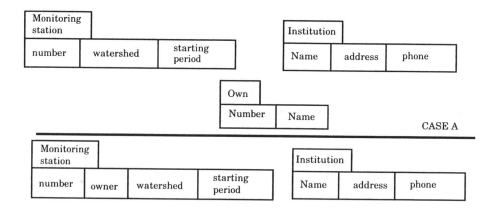

Figure 6.10. Example of a logical data model

The physical model is an image, distant from reality. It is quite difficult to read and in the absence of a logical or conceptual model, it is almost impossible to comprehend the organisation of a database from this single schema.

6.4.5. SPECIFICITY OF A GDBMS

The differences between a GDBMS and a DBMS are for the most part, the same as those existing between a GIS and an information system (IS). Nevertheless, we will not consider the functionalities of these two systems separately. We will rather emphasise the essential characteristics of geographical information and the management of these characteristics by the GDBMS.

The organisation of databases is based on the description of real systems through two concepts: entities and relationships. Entities and relationships are described by attributes. When the system to be modelled is georeferenced, if for example the territory is to be modelled, the entities, as well as the relationships or the attributes may be spatial. A spatial entity is an entity possessing a form and a position, a spatial relationship is said to be a relationship of proximity (is next to, under, near to ...), whereas a spatial attribute is a characteristic that varies in space (e.g. altitude, depth of underground water-level).

Spatial Entities   Four different kinds of spatial objects may be distinguished: points, lines, polygons, and continuous surfaces. Thus a limnigraphical station may be represented by a point or a watershed may be represented by a polygon type of object. The polygon associated with the watershed describes its form and position within a known cartographical projection system. This enrichment of the description of entities allows, for the example of watersheds, to analyse the occupied ground or their geomorphology. Lines are often used to represent networks such as electricity networks, road or hydrological networks.

Spatial Attributes An entity may be described by spatial and/or non-spatial attributes. Taking the example of a limnigraphical station (point type of spatial object), attributes such as the name of the owner, the name of the watershed or the starting date are not spatial for they are determined solely by the station they describe. On the other hand, a watershed (represented by a polygon type of spatial object) has a relief that can obviously be expressed by measures of altitude. One may, therefore, consider that the object watershed may be described, among other things, by its altitude, even though this characteristic is not constant on the polygon surface. The altitude, may therefore be considered to be a spatial attribute of watersheds.

Whereas a spatial object may be recognised by its form and its position, a spatial attribute is defined by the fact that its value is not only determined by the object it describes, but also by its spatial position (eg. temperature of a building).

Spatial relationships Spatial relationship are, at least partially, managed by GIS's. These systems, in particular, offer spatial operators allowing the user to carryout queries integrating spatial relationships. Two types of spatial operators may be distinguished.

Metric operators are used for taking measurements such as the distance between two points, or the length, the perimeter or the surface of an object. Topological operators deal with proximity relationships named topological relationships. This kind of relationship may be defined as a spatial relationship, independent of the projection system. Adjacency is, for example, a topological relationship, for if object A is adjacent to object B, this relationship remains true whatever the projection system.

Topological operators are used in the same way as relational operators. The main topological operators are "intersect", "contains" and "within". Nevertheless, a large number of different topological relationships exist and these operators, alone, cannot totally manage them. In most GIS's, the topological operators are integrated into the queries of the type "select-from-where". This SQL extension is generally identified by the term GQL: Geographical Query Language (see section 6.5.1).

6.4.6. ILLUSTRATIVE EXAMPLE

The GESREAU project is presented here to illustrate the role of an GDBMS in a GIS application. This project developed by the County of Vaud/Switzerland and the "Institut d'Aménagement des Terres et des Eaux" (Land and Water Management Institute) of EPFL, aims to conceive and realise a computerised tool for water management (Crausaz and Musy, 1996).

Under one form or another, water management falls under the responsibility of public institutions. This covers, in particular, the upkeep of watercourses, authorisations for taking samples or discharging, approbation for constructing all works modifying a section of the river, acquisition and handling of hydrometeorological measures. This management should also establish prospective studies of the foreseeable behaviour of watersheds with regard to Parish and District development plans.

The usual procedures for managing both technical and administrative files, no longer enable these services to comply with the expected speed and quality required by the complexity of the phenomena dealt with. Data processing and GIS softwares, combined with structured and high-capacity databases are apparently becoming an indispensable and highly efficient tool.

GESREAU is a software and database intended to carry out the functions of several public services, whose activities mainly consist in the management of surface (and occasionally underground) watercourses. Thus, GESREAU is a data processing tool for civil servants specialised in the field of water management (hydrologists, engineers specialised in the correction of watercourses, biologists and chemists responsible for inspecting water quality, etc.). It is therefore an integrated system. The final version should comply with functionalities ranging from the administrative stage up to assisting in decision making. The application was constructed within the ARGIS software, distributed by UNISYS. The GIS is coupled with the database ORACLE.

Within the framework of the GESREAU application, the description of the hydrographical network is of capital importance. Most of the information indexed within this application is linked with watercourses. The modelling of this network, the real backbone of the data information, is of utmost importance for the success of the project.

Apart from the hydrographical network, GESREAU manages information concerning the environmental quality of watercourses, watersheds and their ground occupation, limnigraphical and pluviometrical measuring stations and authorisations to intervene with regard to the hydrographical network under district control. The diversity of the information, together with the specific expectations of the administrative services (the future users of this database) have resulted in an extremely complex database organisation. The schema shown thereunder (Figure 6.11) is an extremely simplified extract of the GESREAU conceptual data model.

Modelling of the hydrographical network pursues several aims. First of all, it satisfies the administrative management of watercourses by, for example, indicating for each section, the owner (therefore responsible) or the state of the river bed: natural or corrected... Secondly, modelling the hydrographical network allows for the simulation of the hydrological behaviour of watercourses. These watercourse simulations are carried out, depending on the needs, either by the help of exterior hydrological models or directly with the assistance of the network analysis module

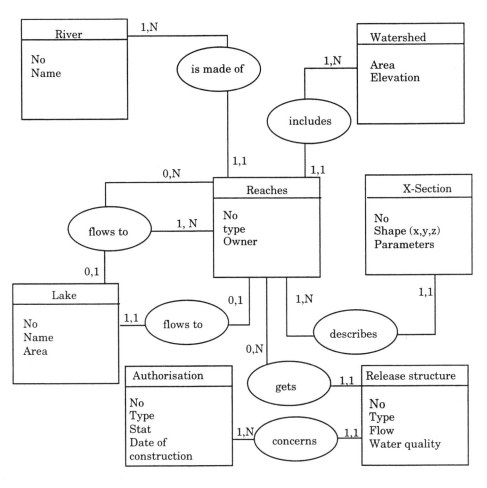

Figure 6.11. Simplified extract of the GESREAU data conceptual model

furnished by the GIS software.

In order to illustrate an utilization of this database, we shall examine a case history for a licence to construct discharge works. If a company wishes to discharge industrial water into a river, it must apply for a licence. This application will be introduced into the database under the object "Licence". That is to say, a tupple will be created in the table describing licences. The tupple will necessarily have an identifier, and the value of its "state" will at first be "non evaluated". Parallel to the creation of this tupple, an occurrence should also be introduced for the object "discharge works" in order to technically describe the construction project. Finally, in order to determine which section is affected by this application, tupples are also created for the relationships "concerning" and "takes place on".

From this point onwards, the database user may have access to all the applications for a licence, either in progress, or already granted, for any given section. Therefore, if a licence has already been requested for the same section, he knows he will have to take into consideration the coupled impact of both discharges.

The licences whose attribute "state" has the value "non-evaluated", are being dealt with. In other words, the service granting licences evalues what impact the discharge will have upon the quality of the water and/or the river flow. This evaluation generally calls for hydraulical or hydrological models. Once the impact has been calculated, it is introduced into the attribute "impact" of the object "discharge works" and the value of the state of the licence becomes "evaluated".

The responsible service now has all the necessary information at its disposal to take a decision. If it judges that the licence may be granted, the state of the licence becomes "granted" and the discharge works associated by the relationship "concerns" may be updated to describe the works constructed (and no longer the construction project). If the application is refused, the tupples describing the licence and the discharge works, as well as the relationship tupples "concerns" and "takes place on" may be destroyed. They may however be preserved for administrative or legal reasons (possibility of appeal).

The last stage in the case history of a discharge licence comes about if the validity is out-of-date. This last event results in the modification of the state of the licence which takes on the value "expired". The owner of the discharge works then has two alternatives. He can re-apply for a licence and the above- mentioned steps are renewed or he can take the decision to destroy the discharge works and in this case, once it is in actual fact destroyed, all the concerned information (licence discharge works, the relationships "concerns" and "takes place on") are eliminated from the database.

Synthesis   To sum up, it appears that the GDBMS is a fundamental component of GIS's. It ensures the management and maintenance of data information and should be capable of handling the spatial characteristic of information contained in the geographical database. The differences between commercial GIS's at the GDBMS level, can essentially be located in the strength of the exploitation languages and in the richness of spatial operators with regard to the exploitation language. Most of the commercial GIS's are associated with existing DBMS's and integrate all, or at least a part of their functionalities.

6.5. Exploitation

The exploitation of geographical information includes two processes: information retrieval and mapping. The selection of spatial features makes use of a query language while result output often requires mapping capabilities. Since units of observation are spatial features, their selection can also be achieved from a displayed map. Mapping capabilities should therefore be integrated into the query process as an interactive cartographic interface and offer a complete mapping environment for the production of suitable map form documents.

6.5.1. INFORMATION RETRIEVAL

Information retrieval is the process of extracting existing information from the GDB based on selection criteria. The latter can combine specifications from the three considered dimensions: thematic, spatial and temporal.

6.5.1.1. *Query Language in Object Mode*
As stated in section 6.4, the system-operating language used in spatial databases usually is SQL (Structured Query Language) or GQL (Geographical Query Language) if spatial operators are added. The database is where the data is stored, and the query language allows the operator to get the information he needs. The DBMS produces the information that "answer" the queries made by the user.

The basic structure of SQL-GQL is the statement: "Select...From...Where". The "select" specifies what kind of an output the query will generate. The "from" indicates the location of the desired (selected) information. The "where" allows the setting of conditions for selecting the data. The query is formally stated in the "where" statement.

There are two types of operators: relational and topological. The six basic relational operators are: intersection, union, difference, joint, relational projection, restriction and division. The most widely used topological (or geographical) operators are: intersect, contains, and within (cf. sections 6.4.3. and 6.4.5.).

This section will present the use of these languages with a GIS in object mode through query examples. We shall first talk only about relational operators, then we will integrate the topological operators to our query examples. The examples presented further are based on the logical model discussed in section 6.4. (Figure 6.10, case B). To that model is appended a new entity "Parish" described by a name, the townhall address and phone number (Figure 6.12).

Relational operators   If the user wishes to know the name of all the owners of a (river) gauging station, he has to write the following query:

Select Number, Owner
From Monitoring_station

Notice that this query does not contain a "where" statement. As a matter of fact the user wants to know the name of all gauging station owner. Therefore, he does not set any condition to the selection of the "name" field. This operation is a projection, and its result is shown in Figure 6.13.

Now, the user wants to know the watershed and the start-up date of station #128, the query will be:

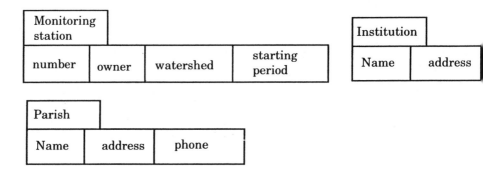

Figure 6.12. Logical data model for this example

Number	Owner
128	SHGN
129	Mauvoisin
...	

Figure 6.13. Result of the projection type of query (attribute)

Select Watershed, Starting_period
From Monitoring_station
Where Number = 128

This query is a restriction, illustrated in Figure 6.14.

Another crucial operator is "joint". It is through its use that the database can be structured. In the logical model, there is a relationship between a gauging station and the institutions. This relationship is set by the owner's name and the institution's name.

Then, the user who wishes to know the address and the phone of the owner of gauging station #128 has to join the tables representing (or containing...) the gauging station and the institution. The query will be:

Select Monitoring_station.number,Institution.address, Institution.phone
From Monitoring_station, Institution
Where Number = 128 and Monitoring_station.owner = Institution.name

Watershed	Starting period
Broye	12/10/1968

Figure 6.14. Result of the selection type of query with condition (attribute)

Figure 6.15 shows the joint made between the attribute Owner of table Monitor-

Monitoring station

Number	Owner	Watershed	Starting period
128	SHGN	Broye	12/10/1968
129	Mauvoisin	Val de Bagne	4/7/1967
...			

Institution

Name	Address	Phone
SHGN	2 rue du Rhône	36 21 12
Mauvoisin	5 rue Valère	53 18 11
...		

Number	Address	Phone
128	2 rue du Rhône	36 21 12

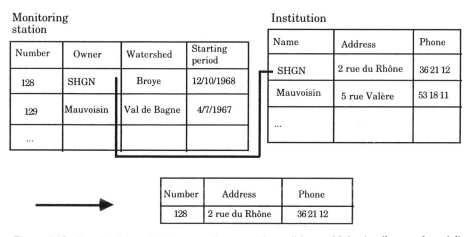

Figure 6.15. Result of the selection type of query with condition and joint (attribute and spatial)

ing_station and the attribute Name of table Institution. The last table is the output of that query. Information concerning the name of the institution, such as the address and the job could very well be found in the table Monitoring_station. However, if one owner owns more than one station, his address and job will be redundant data, because repeated in every Monitoring_station occurrence.

This solution allows the optimisation of data access speed, even if the data volume increases *de facto*, and complicates the maintenance of data. If the one owner of many stations moves out, the address change has to be made in all Monitoring_station tables, whereas if this information is stored in the institution table, it has to be updates only once. That's how jointing operators can help structuring a database.

Topological operators   If one wants to know what gauging stations can be found
in the parish of Geneva, one has to write the following query:

    Select Monitoring_station.number
    From Monitoring_station, Parish
    Where Parish.name = "Geneva"
    and Monitoring_station.object <u>within</u> Parish.object

Notice that a query can be made of both relational and topological operators
(within). To perform this request, the GIS will select the points that describe
gauging stations and the polygons (areas) representing parishes. Then, for every
gauging station, it will see if the point positioning the station is within the polygon
describing the parish. This is a very simple topological query, because there is no
ambiguity in the overlay of a polygon and a point. However, there are many different spatial relationships, and the choice of the right spatial operator and a reasonable interpretation of the answer given by the GIS is a critical stage in object mode
spatial analysis.

Figure 6.16 shows which conditions are respected when using spatial operators

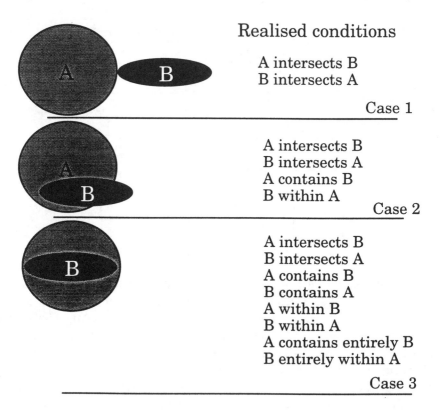

Figure 6.16. Example of topological relationships

with a specific commercial GIS. Notice that in case #3, "A contains B" is true, as well as "B contains A". This to show the fact that relationships like "within", "contains", "intersect" are not univocal.

Moreover, this example takes into account only a few relationships between polygons (2D). By considering also points (0D), lines and all the possible combinations of these elements, Champoux & Bédard defined that in a three-dimensional space, there are sixteen possible combinations of the following spatial relationships: "touch", "include", "be adjacent", and of the following states : "not at all", "partially", "totally". At the moment, there seems to be no GIS managing all these spatial relationships. Therefore, quality and diversity of spatial operators are primary criteria when buying a GIS in object mode. The user who intends to perform spatial analyses must be extremely demanding on that particular point.

6.5.1.2. *Query Language in Image Mode*

Image mode GIS do not provide such an elaborate query language. The retrieval of information results from a sequence of processes exploiting available thematic and spatial tools. As units of description are cells but not spatial objects, many queries imply the reconstruction of considered features. Like analysis tasks, information retrieval in image mode is based on a query model describing the processing stages. Simple queries can be achieved within a single stage, whereas more complex queries require several consecutive processing stages. In many instances a query process in image mode is made up of two distinct steps: first the construction of a layer containing the selected units with the use of thematic and spatial tools, and then the extraction of information from this derived layer, expressed in tabular or map form.

6.5.2. MAPPING

Map production is certainly one of the expected capabilities of any GIS software. It is one of the most efficient analogical forms to describe the spatial distribution of phenomena. However, many GIS packages, particularly those in image mode, fall short in two ways: the inability to produce complete maps rather than simple image printouts and the lack of consistent sets of patterns, tones and symbols. Cartographic language as means of communication is aimed to present a clear and understandable message. Like any other language, cartography has developed rules designed to produce efficient map documents. A set of basic rules accompanied by relevant computer mapping capabilities can help a non professional cartographer to make quality maps (Buttenfield et al. 1991).

Mapping capabilities of a GIS software should include the following:

- production of a complete map including a title, a legend caption, a map orientation, coordinate references and a map scale;
- an interactive map display and map printing;
- sets of patterns, tones and symbols able to handle all types of spatial feature (point, line, area and continuous surface) graphically for colour or black and

white printing;
— scalable map display and printing.

6.6. Spatial Analysis and Simulation

The two basic aims of a GIS are to manage georeferenced information for specific queries and retrievals and to analyse spatial relationships and interactions between phenomena in a defined study area. This first task will be discussed in the next section on the exploitation of the GDB. The second task includes the processes of modelling and simulation within a GIS environment. The reason for discussing analysis processes first is because it produces new layers of information that can be stored permanently or temporarily in the GDB rather than simply as output results. There is no rigid limit between analysis and exploitation tasks; they should instead be seen as successive stages of information processing.

Analysis process produces derived information from that which is already stored in the GDB. GIS provide basic tools to process information in the two spatial and thematic dimensions. The analysis is performed through the construction of analysis models making use of available tools. The major difference between object mode and image mode GIS is the ability to perform spatial analysis. The former approach is more effective for processing information in the thematic dimension, while the second offers equal ability in spatial and thematic dimensions.

The evaluation of GIS packages for analysis purposes raises the following essential question: Can existing models be run within the GIS environment? Two positive answers would be either to be able to reconstruct those models with the available tools, or to undertake data exchange between the external model and the GDB, assuming that efficient import/export procedures are provided. One cannot expect any GIS software to provide numerous specialised models for the wide range of scientific applications, unless specific modules were developed for a particular domain of application. However, the number and the quality of proposed processing tools remain a key factor when selecting a GIS package. This section first covers basic analysis tools required to process spatial information; more specialised tools in the field of hydrology will then be discussed. Finally, issues about use and integration of hydrological models within a GIS environment will be raised.

6.6.1. GIS OPERATORS

Basic processing tools available in GIS software for analysis purposes can be grouped into two distinct categories according to the dimension they process: thematic and spatial tools. In many instances they are combined in a sequential manner to produce quite complex spatial analysis. They make use of transformation operators.

6.6.1.1. *Thematic Tools*

Thematic tools transform the attribute content of descriptive units or spatial objects without any change to their geometrical characteristics. Depending on the mode of analysis involved, the transformation process is different.

Object mode Since in object mode thematic and spatial dimensions are clearly separated, thematic transformations operate only on attribute table(s), in a manner similar to a database or a spreadsheet transformation process. Individual values are derived for each object which are considered as simple observation items. Frequent transformation operators are:

- logical operators to group attribute values into categories or classes;
- mathematical operators and functions to produce univariate and multivariate transformations:
- unit change, classification, indices, principal components ...
- statistical descriptors to summarise distribution of values among observations.

Image mode In image mode thematic and spatial dimensions are embedded to constitute a set of georeferenced layers of information. Thematic transformations process image values on a cell-by-cell basis, without respect to their relative location within the image. Transformation results are stored as derived images or spaceless statistical descriptors. Typical transformation operators are identical to those applied in object mode. In many GIS software, this process is called *overlay*, but this is only a generic term; more important is to be aware of what operators are available to transform the superimposed images (Collet 1992).

6.6.1.2. *Spatial tools*

Spatial tools allow one to simultaneously process information in its thematic and spatial dimensions. This is the strength of a GIS, and it makes spatial analysis possible. Spatio-thematic transformations fall into three categories according to the spatial characteristics they consider:

- *geometric measurement operators*: description of size and shape of selected spatial features,
- *proximity operators*: measure of distance between units of description (Euclidian or non Euclidian), measure of access, path analysis or allocation.
- *context operators*: take into account the neighbourhood of units of description to derive thematic characteristics such as morphological descriptors, and to filter or to relate units with their neighbourhood (watershed, ...).

Geometric measurement operators Spatial features can be described according to their geometric characteristics. As point features have no dimension, geometric measurements are applied on areas and line features. For lines, length, sinuosity and network connectivity are the typical descriptors, as for areas, perimeter, area and compactness are used (Unwin 1981). In object mode, such descriptors are stored as attributes (also known as spatial attributes) in the GDB. The precision and accuracy of measurements depend on the scale and the level of generalisation of digital information.

In image mode the description of spatial features implies first to group cells into corresponding spatial features (also called *regions*); geometric measurements are then derived and assigned to each feature cell, producing a new image. Due to the stepwise description of feature limits, length and perimeter measurements tend to be overestimated, as the calculated area matches the one estimated in object mode. The cell size influences of course the precision and accuracy of measurement.

Proximity operators   Proximity analysis is based on distance measurements to and from units of description. Distance can either be expressed with a geometrically horizontal measurement (Euclidian distance), or in a more flexible manner as an accessibility measurement, with the use of weighted distance. Based on distance measurements, path and allocation analyses can then be performed.

In object mode, proximity analysis is usually limited to the horizontal distance measurement to and from spatial features, as well as to the construction of buffer zones.

The image mode offers a more appropriate environment for proximity analysis. Spatial units under consideration can be either the cells or regions corresponding to spatial features. Many GIS software offer operators to derive horizontal and weighted distance measurements, to estimate the shortest path between two features or to outline areas of proximity. The structure of the image mode allows one to consider the *anisotropic* nature of numerous spatial distributions.

Context operators   The spatial dimension gives information about the arrangement of thematic content within the study area. Context operators relate the content of spatial units with that of their surroundings. This process generates either new attributes or new spatial features. Object mode capabilities are restricted to contiguity analysis for aggregating contiguous objects, to topological overlay and network analysis. This latter capability is certainly the strength of the object mode.

In image mode, context transformations operate at two different levels: the cell and the region (Tomlin 1990). At cell level, the neighbourhood is defined within a window moving across the image. According to the transformation applied to window cells, various indices are produced: morphological indices (slope gradient, aspect, illumination, terrain shape, ...), textural indices (diversity, dispersion, ...), as well as other processes such as low or high pass filtering. A more particular use of the moving window is applied for producing more complex spatial transformations such as watershed and viewshed delineation. At a region level, similar transformations are applied within region cells or between contiguous regions, producing regional indices.

6.6.2. REMOTELY SENSED DATA PROCESSING

Remotely sensed information needs to be processed in order to produce relevant information. This process includes both thematic and geometric transformations. The latter was already discussed in the section on spatial data preparation. As remotely sensed information is by nature in image form, only image mode based systems are able to process such information. Image processing tools can be either provided as an independent software package or integrated into an image mode GIS software. In addition to radiometric preprocessing procedures and geometric transformation tools, the image processing software should include several multispectral classification and transformation procedures as well as digital photogrammetric capabilities for DEM and orthoimage construction.

In the field of hydrology, derived information from remotely sensed data is mainly of two kinds: land surface cover content and terrain topography content.

Land surface cover Digital airphotos and satellite images are major sources of information for producing the landcover and landuse of a study area and to monitor seasonal and permanent changes on the Earth's surface. However, to derive such information, remotely sensed data must be integrated within a set of data containing supplementary information such as ground truth or other relevant land surface characteristics. Extraction of landcover and landuse results from a sensitive and frequently complex classification process (Jensen 1986). Unfortunately, proposed classification procedures do not make fully integrated use of contextual operators and therefore the classification process is principally based on spectral characteristics of cells, leading to erroneous results. The accuracy assessment of derived information is one of the most critical stages in this process; it requires dedicated tools. In addition to landcover characteristics, remotely sensed data provides valuable information in hydrology about water quality, flood extension and soil surface roughness. This latter characteristic can be obtained from various derived vegetation indices.

DEM construction Digital photogrammetry offers additional processes to derive a digital elevation model from remotely sensed information. In many instances available topographic maps do not provide an elevation model with a sufficient precision for hydrological uses. From stereo pairs of digital airphotos or satellite images it is then possible to generate a DEM with a precision dependent on the scale of digital stereo pairs. Such processing tools are now available on micro stations or personal computers. Some image processing packages or even GIS software offer such photogrammetric tools.

6.6.3. MORPHOLOGIC MODELLING

Many terrain characteristics can be derived from a digital elevation model (DEM) –either in vector (TIN) or in raster structure– with the use of context operators.

While several of them are common to any Earth science discipline, some are more specific to the field of hydrology. The latter for the most part require dedicated procedures involving the use of context operators. Typical derived morphologic indices can be grouped as follows:

— general terrain descriptors: slope gradient, slope aspect, terrain shape;
— hydrology related descriptors: catchment areas, drainage network, waterflow and steepest flow path, isochrones, Strahler ordering.

A detailed description of procedures and algorithms generating morphologic indices can be found in numerous published papers (Band 1986, Moore and Grayson 1991), as well as in chapter 7 of this book.

General terrain descriptors are expected to be found in most image mode GIS packages, but more complex hydrological descriptors are proposed as optional specialised modules integrated or not into the GIS package.

6.6.4. DYNAMIC MODELLING

This section describes an application of dynamic modelling applied to the assessment of agricultural damages produced by flooding. This example is taken from another study (Consuegra and Joerin, 1994) describing an overall methodology for flood delineation and impact assessment. Hydraulic computations rely on polygonal meshes and the solution of a two dimensional diffusive approximation of the Saint-Venant equations. For each polygon the hydraulic model computes an average depth of flooding and the overall flow exchanges with adjacent nodes through the interfaces which are the facets of each polygon. From the hydraulic calculations, it is also possible to derive the duration of submersion for each polygon and produce the corresponding map.

Flood damages to agriculture result from more or less prolonged submersion affecting the normal growth rate of plants. According to the period of flooding and the duration of submersion, the farmer may decide to accept either a partial, even complete damage or to seed a replacement crop if the season and/or the soil conditions allow for it. The duration of submersion is defined as the time span between the beginning of flooding and the drying out of the soil. Until then, the soil and the plants may be damaged by labour or heavy machinery.

6.6.4.1. *Evaluation of Damages*

The proposed damage assessment procedure can be applied to a wide variety of agricultural schemes in humid temperate climates. However, yield losses and economic figures result from on site inquiries involving local farmer associations in the Basse Broye region (Consuegra 1992). It was then possible to identify crop types, rotation characteristics and to draw a map indicating the areas where similar crop rotations are followed. Profit margins and yield losses could also be estimated from local sources of information.

6.6.4.2. *Crop Rotations and Crop Sensitivity*

Figure 6.17 (A) shows a typical crop rotation in the study area. The cycle covers a period of 6 years. At a given plot, winter wheat will be followed next year by sugar beet and by winter wheat again two years later. Winter wheat is then followed by potatoes, winter wheat again and finally corn for grain. The profit margins for each individual crop are also shown. This crop rotation corresponds to average weather conditions and guarantees adequate overall profit margins. The crop sensitivity changes during the growing season from seeding to harvest periods. For each single type of crop, the total damages were estimated on the basis of the duration of submersion and the period of the year. For instance, figure 6.17 (B) shows the time evolution of total losses for winter wheat according to the season and the duration of submersion without considering replacement options.

Damages have been assigned to the period where the flooding occurs even if the consequences may appear later during the year (indirect costs). For duration between 5 and 15 days the crop sensitivity is higher at the beginning of the growing season and decreases until harvest. From April to harvest, duration of submersion higher than 15 days lead to a total loss equivalent to the gross product. The latter includes profit margin, seeding and soil treatment expenses, marginal fees as well as indirect costs, which vary from one season to another.

6.6.4.3. *Seeding Delays and Crop Replacements*

Flood damage computations must also account for eventual replacement crops which are intended to minimise yield losses. Flooding in the study area does not occur, on average, more than once a year. More frequent flooding would prevent economically feasible agricultural practice. For this reason, the crop replacement policy shown in figure 6.17 (C) can only be derived on the basis of one single flooding event per year. However, there are no restrictions regarding the period of occurrence of submersions. Obviously, each crop replacement is associated with a yield loss. The growth cycle of winter wheat in the study area starts at the beginning of October (seeding) and finishes at the end of July (harvest). If winter wheat seeding was done in normal conditions the crop will be sensitive to submersions from the beginning of April until harvest (critical period in figure 6.17 (C)).

Submersion duration higher than 5 days between October and March will postpone sowing to the end of March or the beginning of April at the latest. Related yield losses are equivalent to 30% of the profit margin plus the re-seeding expenses. During the critical period, seeding of winter wheat can not be postponed and farmers in the study area have to implement a replacement crop. During the first two weeks of April, submersion for a duration between 5 and 15 days will ruin the winter wheat seeded in October and force replacement with corn for grain sowed during the last two weeks of the same month. During the second fortnight of April, submersion duration higher than 5 days will force the replacement of the crop and the seeding of corn for silage in early May. During the month of April, submersions no longer than 5 days will not generate substantial yield losses. Figure 6.17 (C) illustrates the remaining crop replacement options until the harvest period. It is always necessary to compare the damages to the actual plant with those involved with the crop replacement. Figure 6.17 (D) shows that for winter wheat and submersions between 5 and 15 days, the farmer will prefer to sustain losses instead of sowing a forage crop. A different decision will be taken for submersions higher than 15 days.

6.6.4.4. *GIS Implementation*

Figure 6.18 illustrates the overall set-up within the GIS framework of the flood impact methodology. The procedure requires the flood map describing the duration of flooding for each polygon and that showing the spatial distribution of individual crops. Both maps are on a vector format. However a raster support can also be used. In the DBMS, the following tables are also needed: a) the "Plot" table describing for each plot the current crop in the field and b) the "Potential Losses" table indicating the yield losses per hectare for each crop type and for several submersion duration at the moment of flooding. The "Potential Losses" table includes yield losses for both the replacement and the no replacement options. The "Plot" table is updated at the beginning of each agricultural season. For each plot, the crop type is determined according to rotation rules similar to that shown in Figure 6.17 (A). The "Potential Losses" table is updated every 15 days to follow the evolution of crop sensitivity to flooding.

The procedure starts with a "polygon overlay" operation between the map showing the duration of submersions for each control volume and that describing the spatial distribution of plots and crop types. Each plot will be broken down into smaller polygons covered by one single duration of submersion. For instance, Figure 6.18 shows that plot # 1035 will be broken down into two polygons # 5425 and # 5426. Polygon # 5425 corresponds to the non flooded surface of plot # 1035 while polygon # 5426 relates to the surface of that same plot which is submerged for 3 days. This splitting procedure leads to a new table "Flood-Plot" indicating the number of the newly created polygons, the plot number to which it belongs, the corresponding submersion duration and the area flooded. The "polygon overlay" operation can be done on both the raster and the vector GIS.

The computation of flood damages uses the "Flood-Plot" table as the main source of information and is entirely achieved within the DBMS. For each newly created polygon (# 5425 for instance) and on the basis of the plot identifier (ex: #1035) it is possible to find the crop type on the field at the beginning of flooding in table "Plot". On the basis of the crop identifier, the duration of submersion and the surface flooded, damages can be computed with the information included in table "Potential losses".

Computations are done for both the replacement and the no crop replacement options. The resulting economic damages are stored in the "Losses-New Polygons" table (Figure 6.18). To obtain the flood damages for each plot (#1035) it is necessary to cumulate economic losses for all the corresponding sub-polygons (#

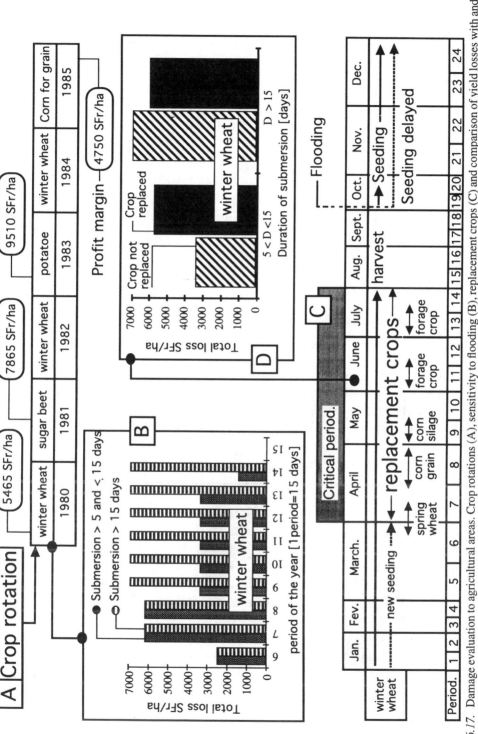

Figure 6.17. Damage evaluation to agricultural areas. Crop rotations (A), sensitivity to flooding (B), replacement crops (C) and comparison of yield losses with and without replacement (D)

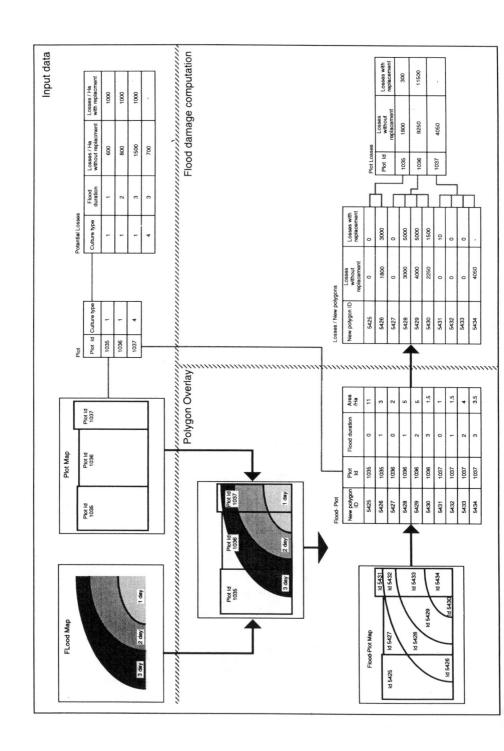

5425, 5426, etc.). The results are stored in the Plot-Losses table for both the replacement and the no replacement options. The cheapest alternative is selected. In case of replacement, it is assumed that the entire plot surface will be seeded again with the new crop.

Flood damages to all plots can be easily mapped with the vector GIS. Figure 6.19 illustrates a typical output from MapInfo. The first map illustrates the spatial distribution of plots and crop types. Polygonal lines indicate the limits of the plots while the colours relate to a particular crop type. The flood map with the duration of submersions has been overlaid. Dark indicates long duration of flooding while the lighter shades correspond to shorter submersions.

The second map illustrates the spatial distribution of flood damages. Interactive query information can be gathered directly from the second map to visualise the calculated flood damages for each single plot.

6.6.4.5. *Dynamic Data Base Modelling*

To model the dynamic character of the database, we propose to use the static and dynamic conceptual approach REMORA (Rolland et al. 1988). This method is based on three concepts linked together with three relationships. Those concepts are 'objects', 'events', and 'operations'. The relationships which link them are 'an event activates an operation', 'an operation modifies an object', and 'an object modification is noticed by an event' (figure 6.20). Using those three concepts, the database evolution can be modelled and managed. The dynamic conceptual model illustrates for all events which updating operation(s) have to be activated, and which object(s) will be modified. A simple example is shown in Figure 6.21.

In the context of agricultural practice, time may be represented by the following hierarchy of events as illustrated in Figure 6.22 (Joerin, 1994). First, time is constituted by a sequence of crop seasons (Sc). Each crop season begins with a winter season, during which farming is limited to field work. Then, as spring approaches, the agricultural activity starts (A) and crops grow. When a crop reaches the optimal state of growth, it is harvested (H). This period is critical, because at that time crops have the greatest worth. Finally, after the harvest, new crops are sowed and the spatial distribution of crops is again modified (S). Crop season duration is not fixed. Indeed, it depends on the regions climate and the specific meteorological conditions of the period. In Switzerland, a crop season lasts for one year, in warmer countries its duration can be just a few months. Moreover, the harvest period might be delayed for a few weeks if meteorological conditions are not favourable. Nevertheless, the proposed time structure is not modified by these fluctuations because the sequence of events is always the same.

This illustrates the interest of a time reference system that is not based on exact time positioning. It is worth noting that this time perception is well suited to a non gradual evolution (with abrupt changes), but not to a phenomena presenting gradual changes like crop sensitivity to flooding. Crop sensitivity is of interest during the agricultural activity period (A). To follow this gradual behaviour, it is

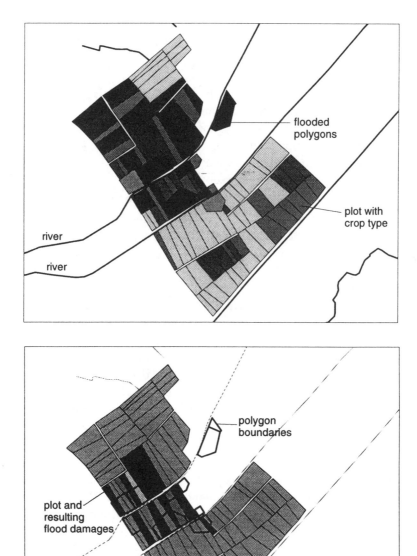

Figure 6.19. MapInfo outputs illustrating the duration of flooding for concerned polygons, the plot and crop type spatial distribution and the resulting yield losses

necessary to discretise the time evolution of crop sensitivity and define each interval as a pseudo-event. This concept is illustrated in figure 6.23. This may not be the optimal solution since it is not always easy to define events on the basis of a

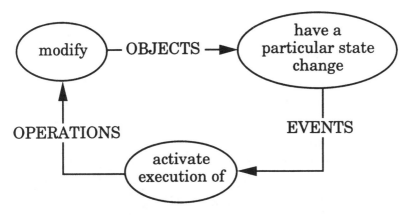

Figure 6.20. Concepts used for dynamic modelling

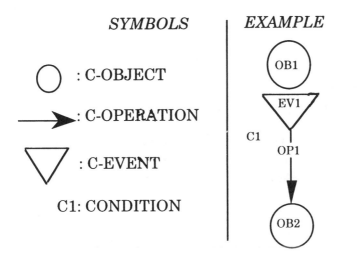

Figure 6.21. Symbols for dynamic modelling

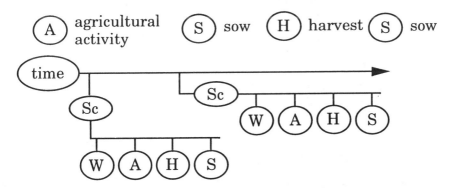

Figure 6.22. Time perception in agricultural practice

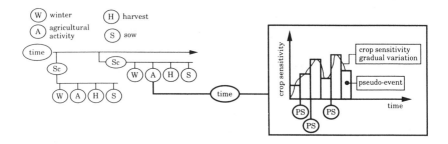

Figure 6.23. Time representation of crop sensitivity

gradual evolution of a given phenomena. Using this hierarchical time structure it becomes possible to clearly identify the most important events, which will determine the evolution of the geographical database. Each of those events correspond to a modification in the real world. Obviously the database needs to be modified accordingly, this is achieved using updating operations.

Modifications resulting from operations are governed by rules. In this application, three rules determine the database evolution. The first expresses the rotation process, the second models the crop sensitivity evolution, while the last manages the replacements of crops after a flood. Figure 6.24 represents the dynamic conceptual model which translates these three rules of evolution. The first one describes the rotation process, the second models the crop sensitivity evolution while the third one manages the replacement of crops after a flood. The latter indicates that when flooding occurs the modification of OBJ4 is noticed by EV3 and activates the execution of OP3 that updates OBJ5 introducing the resulting flood damages in his attributes. EV3 also activates OP4 that updates OBJ1 introducing an eventual replacement crop. It is also noted that operation 2 (OP2) is activated by a temporal event. The latter occurs regularly (in our case every two weeks) independently of outside conditions as opposed to an external event such as a flood. In this application the temporal event is used to update the crop sensitivity data with a regular time step.

6.6.5. OBJECT ORIENTED MODELLING, APPLICATION IN HYDROLOGY

Hydrological studies for water management studies very often imply the use of hydrological models. The choice of a particular hydrological model is always a com-

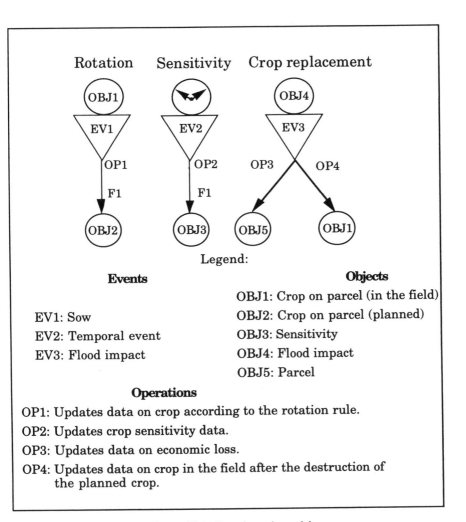

Figure 6.24. Data dynamic model

promise between the scope of the study, the availability of data, the experience of the user and the costs involved with the application. Available methods to compute design floods or droughts seldom indicate application ranges and many engineers are wondering if these specifications will be ever available. For these reasons, hydrologists and practitioners need to interpret and analyse themselves the available hydrometeorological data. These studies go beyond traditional approaches such as regional estimates of observed maximum specific floods or mapping the spatial distribution of annual rainfalls. The main interest is to test different computation methods, to calibrate and validate different models and derive the main rules describing catchment behaviour.

For this purpose, the concept of object oriented modelling provides an interesting alternative. The latter is characterised by a great deal of flexibility in:

— the choice of the basin elements to be modelled (type and level of detail);
— the selection of an adequate model describing the behaviour of a particular element;
— the definition of relationships between watershed elements.

However object oriented modelling requires some hydrological expertise because the user will have to construct a specific model (or set of models) suited to his particular catchment. The following paragraphs describe an application of object oriented modelling for watershed schematisation and runoff computation. These developments are taken from a more comprehensive study on GIS applications to hydrology (Abednego 1988).

6.6.5.1. *Hydrological Objects*
The concept of hydrological object can be easily illustrated on the basis of the grid used in some physically based sophisticated models. A single grid is characterised by:

— a series of pertinent parameters describing its hydrological characteristics (size, slope, land use, soil type, etc.),
— a set of equations or models using these parameters to quantify the time evolution of hydrological response;
— a topological situation describing the surrounding and inter-acting grids.

The concept of hydrological object can be easily extended to a sub-watershed level provided the user is capable of describing the governing equations and the related parameters. A sub- catchment can be seen as a surface object. The latter can also be considered as a single entity. Another advantage of the hydrological object concept is that is can be applied to non surface objects like channels and hydraulic structures such as detention basins or derivations. In all cases, an object will be defined by:

— a set of parameters describing its hydrological behaviour;
— a set of governing equations quantifying its hydrological response;
— a topological network defining links and interactions with other objects.

6.6.5.2. *Application to Watershed Hydrological Modelling*
A given watershed can be described by a set of objects. The list and sub-list structures defined by a programming language like PROLOG (Giannesi et al. 1985) is an efficient way to schematize a river basin. By definition a list is characterised by a head which is the first object and a queue which includes the remaining ones. Figure 6.25 illustrates some list examples. The first case (1) is a simple list with two objects: [a] the head and the queue [b]. The second case (2) shows an example with 4 objects: [a] is the head of the list and the sub-list [b.c.d] is the queue. The third case (3) includes 5 objects: the sub-list [a.b] is the head and the sub-list [c.d.e] is the queue. A PROLOG list can have a queue with an empty element called 'Nil'.

A watershed like that shown in figure 6.26 can be represented in objected ori-

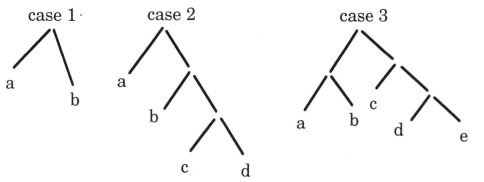

Figure 6.25. Examples of PROLOG lists

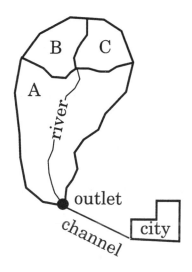

Figure 6.26. Typical watershed schematization

ented programming as follows if the hydrograph of sub-basin [A] is assumed to contribute directly to the outlet.

In this PROLOG list, all hydrological objects must be defined as heads of individual lists. The corresponding queues can be the 'Nil' element or another sub-list representing an input hydrograph to the object. The types of hydrological objects represented by simple elements in the PROLOG list include:

a) Surface objects (watershed, sub-watersheds, urban and/or rural areas, etc.);
b) Vector objects (river reaches, channels, conduits, etc.);
c) Point objects (outlets, detention reservoirs, flow diversions, etc.).

The PROLOG list described in Figure 6.26 includes every hydrological element of the watershed and the necessary topological information for runoff generation and hydrograph routing to the outlet. In the PROLOG language, it is necessary to define the type of mathematical operations to be performed between the head and the queue of a given list. Figure 6.27 illustrates the resulting PROLOG

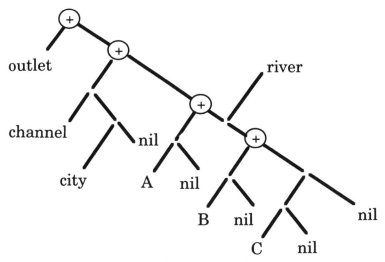

Figure 6.27. Watershed schematization using PROLOG

scheme for the watershed shown in Figure 6.26. Two types of operations can be defined:

- A series of equations describing the hydrological behaviour of the object if the head is a single element. This operation is directly related to the type of object the head of the list represents. In Figure 6.26, the surface object B is associated with a model [B] including rainfall excess computation and convolution with a unit hydrograph. Object [river] is a river reach associated with model [r] to route the sum of the computed hydrographs from the surface objects [B] and [C].
- A summation of hydrographs if the head is a sub-list (more than one element) or an outlet.

6.6.5.3. *PROLOG Implementation*

The PROLOG implementation requires the set up of a list of instructions including a predicate and some arguments. The general form of such instructions is as follows:

hydrological-object (n-ame, t-ype, m-odel, t-o-name)

The predicate is *"hydrological object"* and the terms between the brackets are the arguments: *'n-ame'* is the name of the hydrological object, *"t-ype"* is the type (surface, vector, point), *"m-odel"* is the type of operation to be performed on that particular object and *"t-o-name"* is the name of the connected hydrological object. The watershed scheme shown in Figure 6.26 leads to the following PROLOG instructions:

1. hydrological object(city,surface,model-city,channel),
2. hydrological object(channel,vector,model-channel,outlet),
3. hydrological object(area[A],surface,model-[A],outlet),
4. hydrological object(area[B],surface,model-[B],river),
5. hydrological object(area[C],surface,model-[C],river),
6. hydrological object(river,vector,model-[r],outlet),
7. hydrological object(outlet,point,sum,outside).

In the first instruction, the hydrological object is called city, the type is surface, the model to be used is model-city and the next object is the channel. In other words, the hydrograph from the object city will be routed through the object channel. The flood routing method is that included in the model-channel.

6.6.5.4. *Advantages of Object Oriented Modelling in Hydrology*

The object oriented approach provides interesting advantages in the field of hydrological modelling. The instructions described in the previous paragraph, can be easily changed. The user can easily modify the attributes of the object and the associated model. Several model combinations can be tested and analysed in terms of sensitivity. The structure of the schematization can also be changed easily. A retention reservoir at the outlet can be integrated in the watershed scheme by modifying instructions No: 2, 3 and 6:

2. hydrological object(channel,vector,model-channel,reservoir),
3. hydrological object(area[A],surface,model-[A],reservoir),
6. hydrological object(river,vector,model-[r],reservoir),

and by adding a new one:

8. hydrological object(reservoir,point,model-[S],outlet).

Instruction No: 8 defines the new point object reservoir associated with model-[S] which can include a simple Storage Indication Method. Obviously the object reservoir needs a certain number of attributes.

Abednego (1988) developed an expert system linked to PROLOG for easy and flexible manipulation of objects and topology. One of the main advantages of this system is that it is able to detect where object attributes have been modified and can efficiently recalculate the contributions of the modified objects only and recall those of the non modified ones. This leads to a significant gain in computation time, specially for long continuous simulations.

6.6.6. SIMULATION APPLICATIONS

This paragraph describes two applications of GIS technology to assess flood damages to built- up areas and traffic network conditions. These examples are taken from a comprehensive study for flood mitigation and control (Consuegra and Jo-

erin 1994). The prototypes described below, have been implemented in the Data
Base Management System DbaseIV and the raster GIS IDRISI.

6.6.6.1. *Evaluation of Damages to Built-up Areas*
Economic damages to buildings are based upon standard stage-damage relations
similar to that shown in figure 6.28. These relations only consider submersion ef-

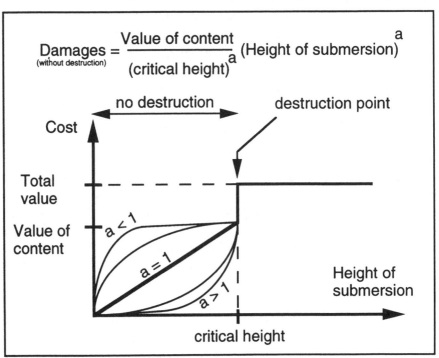

Figure 6.28. Typical stage-damage relation for built-up areas

fects and do not account for nearby velocities or debris transport which may be an
important cause of destruction. The relationship between these hydraulic variables
and the economical damages are not easy to establish (Torterotot, 1994). However,
if future studies derive such kind of relations, it will be relatively easy to include
them in the prototype described hereafter. The flood damage computation is based
on a evaluation of the depth of water around each building ("Height of submersion"). If this value is higher than a critical height, the building is completely destroyed. Losses are equal to the value of the structure and that of the building content. If the water depth is less than the critical value, losses are proportional to the
value of the building content. The exponent "a" allows to parameterise the stage-
damage relation according to the building type. If a<1 small flood depths will be
strongly penalised while a>1 indicates that the expensive contents of the building
are located in the upper stories.

  For a given flood episode, the pre-processing routine in the DBMS identifies
for each node the maximum water depth during the entire flood event. This inform-

ation is transferred to the GIS to produce a map showing the spatial distribution of maximum water surface elevations. Figure 6.29 illustrates the subsequent operations to assess flood impacts on built-up areas.

The flood map mentioned above is first overlaid with that delineating the boundaries of individual buildings (Building map) to identify all the constructions affected by flooding (Buildings and Flood map in figure 6.29). The GIS is then able to compute the average depth over the surface covered by each single building. This information is stored in the "Building and flood depths" table.

In this example, the building #105 is partially flooded. According to the "Building" and the "Type of building" tables, building #105 belongs to a commercial area and the corresponding stage-damage relation can be approximated by a polynomial equation in which the exponent "a" is equal to 0.5. Using the "Type of building" and the "Building and flood depths" tables damages can be computed for each single building. These damages are then transferred to the GIS to produce a map showing the spatial distribution of economic losses. Cumulating individual damages leads to the global economic flood impact on built-up areas.

Typical output maps produced by IDRISI are shown in Figure 6.30. Figure 6.30 (left) shows the spatial variability of maximum depths for each polygon and indicates all the buildings affected by flooding. Figure 6.30 (right) illustrates the resulting damages expressed as a function of the total value of each single concerned building.

6.6.6.2. *Flood Impact on Traffic Conditions*

Traffic damages relate to flow interruptions and to reduced circulation speeds. In case of road cut-off, the search for the fastest alternative path between two points is definitely an important question for rescue services. It is also required to derive an overall index representing the spatial distribution of traffic density resulting from road cut-off. These evaluations can be done on the basis of minimum travel times between departure and arrival points in the road network (D_{opt}).

The road network is subdivided into a series of stems (Figure 6.31). A stem represents a piece of road between to cross-roads. Each stem can be characterised by a travel time according to traffic conditions and speed limits. The network analysis operators in the GIS system allow to determine the fastest link between two connected points which are typically considered to be major urban agglomerations (Figure 6.31).

The set of stems describing the fastest links between two cities is called an axis (table "Stems and axes" in Figure 6.31). Theses axes are also characterised by an average daily traffic flow derived from regional development plans commonly available in Switzerland (table "Axes" in Figure 6.31). The proposed methodology assumes that travel times are equal in both directions of circulation. The average flow per axis is equal to the sum of those in each traffic direction.

In case of flooding, the system identifies the road sections where circulation is no longer possible. Traffic cut-off occurs if the depth of submersion on the road is

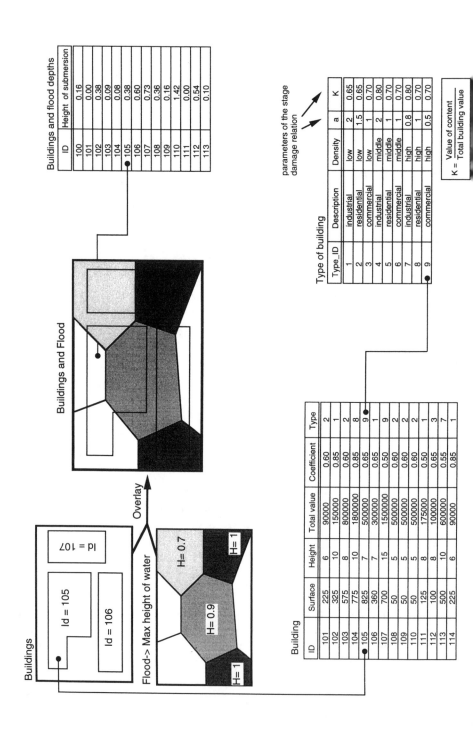

Buildings and flood depths

ID	Height of submersion
100	0.16
101	0.00
102	0.38
103	0.09
104	0.08
105	0.38
106	0.60
107	0.73
108	0.36
109	0.16
110	1.42
111	0.00
112	0.54
113	0.10

parameters of the stage damage relation

Type of building

Type_ID	Description	Density	a	K
1	industrial	low	2	0.65
2	residential	low	1.5	0.65
3	commercial	low	1	0.70
4	industrial	middle	2	0.80
5	residential	middle	1	0.70
6	commercial	middle	1	0.70
7	industrial	high	0.8	0.80
8	residential	high	1	0.70
9	commercial	high	0.5	0.70

$$K = \frac{\text{Value of content}}{\text{Total building value}}$$

Buildings and Flood

Buildings

Id = 105

Id = 106

Id = 107

Overlay

Flood-> Max height of water

H= 0.7

H= 0.9

H= 1

H= 1

Building

ID	Surface	Height	Total value	Coefficient	Type
101	225	6	90000	0.60	2
102	325	10	150000	0.85	1
103	575	8	800000	0.60	2
104	775	10	1800000	0.85	8
105	825	7	500000	0.65	9
106	360	7	300000	0.65	1
107	700	15	1500000	0.50	9
108	50	5	500000	0.60	2
109	50	5	500000	0.60	2
110	50	5	500000	0.60	2
111	125	8	175000	0.50	1
112	100	8	100000	0.65	3
113	500	10	600000	0.55	7
114	225	6	90000	0.85	1

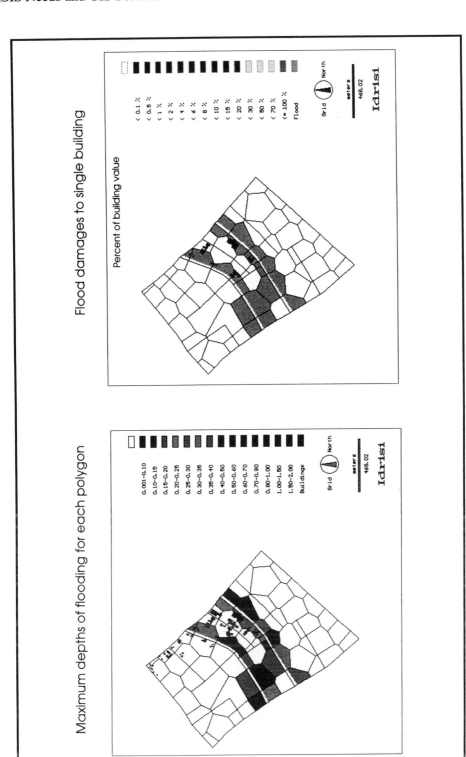

Figure 6.30. Typical IDRISI outputs showing the extent of flooding and the resulting damages to each single building

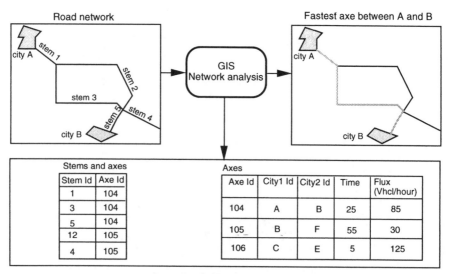

Figure 6.31. Road network description

higher than a given value and lasts longer than a pre-specified interval.

For selected departure and destination points, the system re-computes new travel times and identifies the fastest links between connected cities. For each axis the impact of flooding on traffic flow can be computed with the following formula:

$$N = (1 - (D_{opt}/D_{ino}))F * D \tag{6.1}$$

where D_{ino} is the travel time in case of flooding, F is the average vehicle flow in normal conditions and D the duration of flooding. N is computed for each axis. It represents the number of vehicles that could not reach destination during the flooding period. If $N = 0$, the axis is not concerned with flooding. The maximum value of N is equal to FD indicating that the axis between the two concerned cities can not be used during the entire flooding duration.

The overall impact of flooding is derived by cumulating the individual values of N for each axis. The system will also compute the new spatial distribution of traffic density resulting from flooding. This allows to identify overloaded stems and to suggest alternative paths.

Figure 6.32 (A) illustrates the database architecture to identify the major axis concerned with traffic cut off due to flooding.

Two cities are connected by one single axis (the fastest one). The latter is described by the attributes shown in table "axes". The fastest travel time (D_{opt}) is determined within IDRISI using the COST and PATHWAY functions. COST computes travel times from a given origin to all destination points. PATHWAY identifies the fastest paths between selected departure and arrival points. The stems corresponding to a given axis are stored in table "Stems-axes". If flooding prevents normal circulation through a given stem (table "Flooded-stems" coming from ID-RISI), dBaseIV looks for all the concerned axes in table "Stems-axes". For each selected axis, the concerned cities can be found in table "Links-Cities". City and axes identifiers are stored in table "Concerned axes".

For each disturbed axis, IDRISI calculates the new fastest travel times (D_{ino}) in a similar manner as that used to compute D_{opt} and stores the results in the table "New travel time" shown in figure 6.32 (B). With query type operations, dBaseIV finds in table "Axes" traffic flow and D_{opt} for all concerned axes and calculates the effects on traffic flow (N) according to equation (1). For each single axis, the corresponding value of N is stored in the table "Impact on axis". The new spatial distribution of traffic is computed by cumulating for each single stem the vehicle flow of all axes passing through it. Figure 6.33 shows a typical output produced by IDRISI comparing traffic densities for both normal and flooding conditions.

6.7. GIS Software Selection

The selection of an appropriate GIS software is a key issue for many institutions or research groups interested in the integration of spatial information processing within a computerised environment. However, from various points of view this selection is not the main issue, neither from its financial involvement, nor from a structural issue. This selection is a component of a multicriteria and multiobjective decision making process that notably involves the institutional context and computer requirements as well.

The introduction of a computerised GIS environment within an organisation raises several critical issues to be considered during the decision process: the integration of current and new tasks in the organisation, the compatibility between the existing information system and the one required by the GIS technology, and the rapid developments in technology that make both hardware and software quickly obsolete. Regardless of the complexity of tasks to be supported by the planned GIS environment, the decision making process should include the following sequential stages:

- identification of tasks currently carried out manually or outside an integrated computerised environment;
- identification of tasks and procedures to be integrated within the future GIS environment;
- evaluation of changes in the institution (work-force and infrastructure) resulting from the GIS implementation;
- evaluation of existing computer capabilities;
- selection of a suitable GIS software integrated into a suitable computer hardware structure.

This five stage list outlines major components involved in the decision making process. It shows the final position of the software selection. Both the complexity and the time span of each stage are obviously related to the type of task

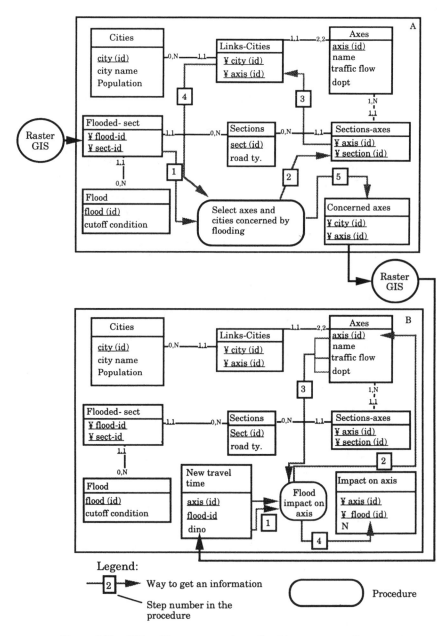

Figure 6.32. GIS architecture to evaluate flood impacts on traffic network

and the size of the institution. For multidepartmental institutions with management oriented tasks, the first three stages are certainly the most significant; whereas for research groups the software selection remains an important issue. Table 6.1 illustrates the interdependence between organisation tasks, software capabilities and

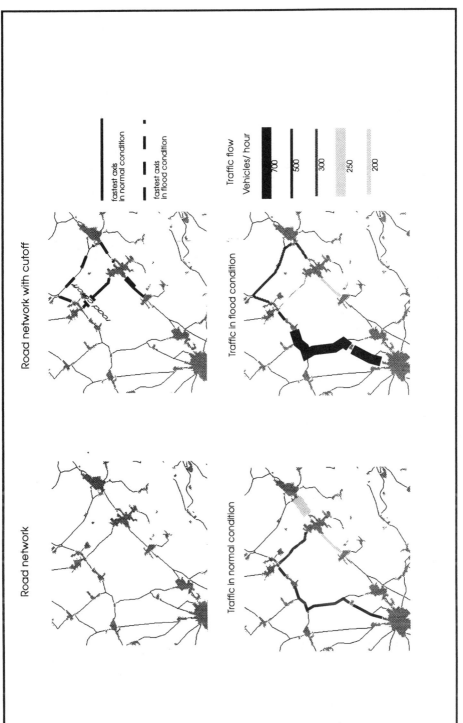

Figure 6.33. Typical IDRISI output showing increase in traffic density due to road cutoff

the nature of information to be processed. It shows some fundamental differences between information management and spatial analysis tasks and consequently the type of appropriate GIS environment.

TABLE 6.1. Major characteristics of GIS applications and their environment (adapted from Thériault 1992 and Bédard 1987).

Spatial information management	CONTINUUM	Spatial analysis, modelling, simulation
Land information system LIS	CHARACTERISTICS	Geographical information system GIS
Institutional	Organization	Research group
Management	Application tasks	Research, planning
Territory	Model of reality	Natural components and human activities
Legal	Context	Decision making support
Query, mapping	Uses	Spatial analysis, modelling, simulation
Regular	Frequency of use	Project dependent
Fully defined	Information processing	Partly undefined
Rigid	GDB structure	Flexible
Continuous	GDB update	Irregular
Cadastral	Description scale	Local, regional
High	Precision of location	Variable
Object (vector)	Spatial description mode	Image (raster) or object
Important	Topology	Essential
Required	Information access control	Optional
Permanent	GDB lifetime	Project dependant
Complex, organized	Computer environment	Simple
Required	Computer staff	Optional

6.7.1. SELECTION KEYS

Management and retrieval of spatial information require software capabilities that are different from those requested for spatial analysis. However, in both domains of application the task load will determine on which computer system the software should be installed: a mainframe, a workstation or a personal computer. Thus the GIS type, the computer system and available operating systems form a first set of selection keys. Other capabilities discussed throughout this chapter can be used as selective criteria. At the final stage of the decision process, the selection of a suitable GIS software can be made, based on periodicals such as The 1994 European

GIS Yearbook or Mapping Awareness 1994 that offer a short description of software capabilities and a list of local distributors.

For management and retrieval of spatial information, an *object oriented GIS software* should offer the following characteristics:

— spatial data acquisition modules including projection transformations;
— an access capability to a standard relational database;
— a powerful standard query language (SQL) with supplementary spatial operators to offer a real geographic query language (GQL);
— a true mapping system including an interactive mapping display for spatial data selection and retrieval;
— local distribution support offering training courses;
— a long lifetime.

For spatial analysis, an *image oriented GIS software* should offer:

— spatial data acquisition modules including geometric transformations;
— a flexible module to import and export spatial data in both vector and raster structures;
— several complementary interpolation procedures;
— a geographical database that can manage documentation files (meta-information) and access relational databases;
— the ability to handle very large size georeferenced images for spatial analysis purposes;
— a rich set of thematic and spatial operators;
— an open environment to integrate user-defined transformation models;
— an interactive mapping display and user-friendly interface.

References

Abednego B., 1988: Apports de la télédétection à la conception de modèles de simulation en hydrologie. PhD thesis No:806 Institut d'Aménagement des Terres et des Eaux du Département de Génie Rural de L'Ecole Polytechnique Fédérale de Lausanne, Switzerland.

Aronoff S., 1989: Geographic information systems, a management perspective. WDL Publications, Ottawa, Canada.

Ball J.E., 1994: Hydroinformatics-Are we repeating past errors?. Hydroinformatics94, Proceedings of the first international conference on Hydroinformatics; Balkema, Rotterdam, ISBN 905410 -515 1 and -516 X, the Netherlands.

Band L., 1986: Topographic partition of watersheds with digital elevation models. Water Ressources Research, Vol. 22, No. 1, pp. 15-24.

Bédard Y, 1987: Sur les différents types de SIRS. Actes du congrès conjoint de Carto-Québec et de l'ACC. Québec, Canada, pp. 73-87.

Blenheim, Mapping Awareness (Ed), 1994: The 1994 European GIS Yearbook. London, UK.

Burrough P.A., 1986: Principles of geographic information systems for land ressources assessment. Clarendon Press, Oxford, UK.

Buttenfield B., McMaster R. (ed), 1991: Map generalization: making rules for knowledge representation. Longman, London.

Champoux P., Bédard Y., 1992: Notions fondamentales d'analyse spatiale et d'opérateurs spatiaux. Revue des sciences de l'information géographique et de l'analyse spatiale. France, Vol.2, pp. 187- 208.

Collet C., 1992: Systèmes d'information géographique en mode image. PPUR, Lausanne, Switzerland, pp. 46-57.

Consuegra D., 1992: Concept de Gestion des Eaux de Surface: Aspects méthodologiques et application au bassin versant de la Broye en Suisse. Thèse No 1064. Département de Génie Rural, Ecole Polytechnique Fédérale de Lausanne. Switzerland, 200 pp.

Consuegra D. and Joerin F., 1994; Flood plain hydraulic modelling and GIS applications for impact assessment. Report submitted to the Fonds National Recherche Scientifique Suisse, in the framework of the AFORISM project. Contract EPOCH-CT90-0023 (TSTS), Bern, Switzerland.

Crausaz P.A.C. an Musy A., 1996: GESREAU: un outil d'aide à la gestion des eaux, application au canton de Vaud. Vermessung Photogrammetrie Kulturtechnik/ Mensuration Photogrammétrie Génie Rural, mai 96, Sigwerb A.G., 5612 Villmergen, Schweiz.

Delhomme J.P., 1978: Application de la théorie des variables régionalisées dans les sciences de l'eau. Bulletin du BRGM, Section III, No. 4, Paris.

Giannesi F., Kanoui H., Pasero R. and Van Caneghem M., 1985: Prolog. Inter Editions. Paris.

Gagnon P.D., 1993: Modul-R, version 2.0. Centre de Recherche en Géomatique, Faculté de Foresterie et de Géomatique, Université Laval, Sainte-Foy, Canada.

Gold C., 1989: Surface interpolation, spatial adjacency and GIS. Three dimensional applications in GIS, Raper J. (Ed), Taylor & Francis, London, UK, pp. 21-35.

Goodchild M., 1992: Geographical data modeling. Computers & Geosciences, UK, Vol. 18, No. 4, pp.401-408.

Hoogendoorn J.H., Van der Linden W. and Te Stroet C.B.M. 1993: The importance of GIS in regional geohydrological studies. Applications of GIS in Hydrology and Water Resources Management; Edited by Kovar and Natchtnebel; IAHS publication No:211. ISBN 0-947571-48-5.

Jensen J., 1986: Introductory digital image processing, a remote sensing perspective. Prentice Hall, New York, USA.

Joerin F. and Claramunt C., 1994: Intergrating the time component in a GIS: Application to assess flooding impact on agriculture. Proceedings of EGIS/MARI'94, Paris, Volume I, page 524.

Kaden S.O., 1993: GIS in water-related environmental planning and management: problems and solutions. Applications of GIS in Hydrology and Water Resources Management; Edited by Kovar and Natchtnebel; IAHS publication No:211. ISBN 0-947571-48-5.

Lam N., 1983: Spatial interpolation methods: a review. The American Cartographer, Vol. 10, No. 2, pp. 129-149.

Laurini R., Thomson D., 1992: Fundamentals of spatial information systems. Apic Series, Academic Press, London, UK.

Le Moigne J.L., 1990: La théorie du système général, théorie de la modélisation. Presses Universitaires de France, Paris.

Moore I., Grayson R., 1991: Terrain-based catchment partitionning and runoff prediction using vector elevation data. Water Ressources Research, Vol. 27, No. 6, pp. 1177-1191.

Pantazis D., 1994: La méthode de conception de SIG Me.Co.S.I.G. et le formalisme CON.G.O.O. EGIS/MARI'94, 5th European Conference and Exhibition on GIS, Paris, France, pp.1305-1314.

Rolland C., Foucaut G. and Benci G., 1988: Conception des systèmes d'information: la méthode REMORA. Edition Eyrolles, Paris, France.

Schultz G.A., 1993: Application of GIS and remote sensing in hydrology. Applications of GIS in Hydrology and Water Resources Management; Edited by Kovar and Natchtnebel; IAHS publication No:211. ISBN 0-947571-48-5.

Thériault M., 1992: Systèmes d'information géographique, concepts fondamentaux. Course notes, LATIG, Dept. of geography, Laval University, Québec, Canada.

Tomlin C., 1990: Geographic information systems and cartographic modeling. Prentice Hall, Englewood Cliffs, New Jersey, USA.

Torterotot J.P., 1994: Les coûts des dommages dûs aux inondations: Estimation et analyse des incertitudes. Thèse de Doctorat de l'Ecole Nationale des Ponts et Chaussées, spécialité Sciences et Techniques de l'Environnement, Paris, France.

Unwin D., 1981: Introductory spatial analysis. Methuen, London, UK.

CHAPTER 7

Digital Terrain Modelling

A. Sole and A. Valanzano

This chapter presents a review of available techniques for generating Digital Terrain Model (DTM). It starts by analyzing data capture methods and shows differences between them in terms of accuracy and precision of the derived DTM and of optimal strategies for capturing morphologic features of study area. Two methods for creating DTM are analyzed, regular grid and triangulated irregular network, and the peculiarities of each one are illustrated. Algorithms for calculating information relevant to hydrological modelling, such as slope, aspect, watershed and drainage networks are presented. Finally, a brief review of available software packages for generating DTM and hydrological parameters is presented.

7.1. Introduction

The spatial distribution of topographical and geomorphological features of drainage basins and of drainage networks is relevant for determining the spatial variability of some hydrological phenomena.

Such phenomena are usually analysed with hydrological distributed models which require some parameters. Among these the most important are: catchment area, topographic form, slope and aspect for the basin; and topologic structure and slope of links for drainage network.

The quantification of these parameters is tedious and time-consuming when accomplished manually, therefore there is a need for an automatic determination of these parameters. This is possible with Geographical Information Systems (GIS) and digital elevation models (DEM) or digital terrain models (DTM), which together provide an ideal support for hydrological distributed models.

DTMs allow the determination of topographic parameters relevant for many hydrological models. TOPMODEL, developed by Beven and Kirkby (1979), is a physically-based, topographically driven flood forecasting model. It allows the prediction of distributed soil moisture status on the basis of spatial indices which

V. P. Singh and M. Fiorentino (eds.), Geographical Information Systems in Hydrology, 175–194.
© 1996 *Kluwer Academic Publishers. Printed in the Netherlands.*

depend on slope and cumulative upslope area derived by flow pathways, (Quinn et al., 1991).

The Geomorphic Instantaneous Unit Hydrograph-GIUH (Rodriguez-Iturbe and Valdez, 1979, and others), is a commonly used method for determining the rainfall-runoff response of catchment, it uses Horton's ratios in order to calculate its characteristic parameters. Horton's ratios can be determined from digital channel networks derived from DTM.

DTMs and GIS have been used by Djokic and Maidment (1991) for urban storm water modelling. In particular they have used DTM for calculating parameters necessary for time of concentration. Lanza et al.(1993) developed a set of automatic procedures for flash flood forecasting using a DTM and an hydrologically oriented GIS. Copertino et al. (1991) used DTM and GIS for calculating geomorphological parameters useful in modelling flood forecasting.

7.2. Data Source for Generating DTM

Data for DTM should be observations about elevation and shape of terrain surface with particular attention to surface discontinuities (breaklines) and special location (passes, pits, peaks, points of change in slope, ridges, stream channels etc.). These data can be collected using different techniques: ground surveys, photogrammetry using aerial photographs or satellite images, digitizing or scanning existing maps (Weibel and Heller 1991).

The choice of data source is dependent on a combination of several factors such as the extent of the study area, desired precision and accuracy of DTM in relation to specific applications. Accuracy is a measure of how much an estimated value differs from the true value, while precision, in statistical terminology, is a measure of the dispersion of observations about a mean value and is not related to the meaning of the data but to equipment used to collect data. None of the previous collection methods is however error-free and the extent of error is dependent on the specific technique.

Ground surveys
This technique allows the creation of very accurate DTM because surveyors usually tend to capture the elevation of discontinuities and/or special location that are characteristics for the area under observation. However, it is relatively time consuming and therefore is usually applied to particular projects which involve small study areas. Recently, a new technique called Global Positioning System (GPS) has been developed. GPS consists of a constellation of satellites which transmit continuous time and position information, enabling users of GPS receivers to plot their positions on Earth to within metres of their true location.

Photogrammetry
This technique is suitable for study areas greater than those covered by the previous technique as well as nation-wide areas. In this case the accuracy of DTM depends

on the sampling method adopted. Using stereoscopic aerial photographs or stereo-scopic SPOT images and suitable equipment, it is possible to collect elevation data using different sampling methods. There are three main sampling techniques: regular pattern (fig.7.1), progressive sampling(fig.7.2) and selective sampling (fig.7.3). The first consists of sampling based on a fixed interval in two directions (grid

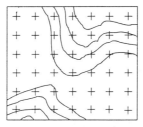

Figure 7.1. Regular pattern

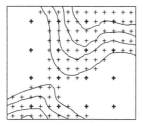

Figure 7.2. Progressive sampling

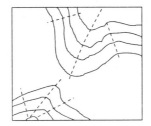

Figure 7.3. Selective sampling

sampling). This method is the simplest and can be fully automated, however it presents some disadvantages. for example the amount of data collected, depending on the chosen interval, may be redundant for flat or homogeneous areas and not optimal for areas with sharp changes.

Progressive sampling is an iterative process. The process starts from a low density sampling grid and, if necessary for better accuracy, recursively adapts the density of sample points to the complexity of terrain surface. This method, which can be automated, collects fewer points than regular sampling and produces more accurate DTMs.

Selective sampling is a manual process which is used for rough terrain with many elevation discontinuities and special locations. The correct identification of such discontinuities is relevant to the accuracy of DTM. The combination of the last two methods is called composite sampling. This method is used for areas which present different terrain structures. Topographic breaks are captured with selective sampling and the rest of the terrain data are captured using progressive sampling. Thus the composite sampling techinique cannot be fully automated.

Digitizing or scanning existing maps
In manual digitizing, topographic maps are positioned on a digitizing table and a pointing device is used to trace contour lines or to pick elevation inter contour points. In scanning, a topographic map is positioned in a special equipment and an electronic detector moves across the map to produce a digital image. Subsequently a manual editing operation is necessary in order to assign the elevation value to each line and point. The accuracy of DTM generated from data captured using such techniques depends on the quality and scale of original source maps.

Digital data sets are becoming available in most of the developed countries to satisfy a wide range of users. These data are produced by private or public organizations using one of the above mentioned methods.

7.3. Methods for Creating DTM

Once data have been collected using one of the previously mentioned methods, which as shown before can produce regularly or irregularly spaced data , it is necessary to build a model capable of approximating the surface behaviour. Basically two models are used more commonly for representing terrain surface (fig.7.4): regular grids (elevation matrices) and triangulated irregular networks (TIN).

The first model represents the surface as a regular grid, superimposed on the sample data, for which the mesh point elevation values need to be calculated. The second model represents the surface as a network of adjacent triangles whose vertices are the sample points.

7.3.1. REGULAR GRIDS

In regular grids the data structure consists of a matrix of elevation values. These values refer to equally spaced points (mesh points). Three different cases need to be considered in order to calculate mesh point values from sample points. (1) If sample points are regularly spaced and the distance between them is coincident with the desired grid size then no interpolation is needed. (2) If sample points are regularly spaced but with a distance between them not coincident with the desired grid size then usually either a bilinear interpolation method (fig.7.5) or a cubic convolution method can be used to resample the grid.

With bilinear interpolation a new value for a mesh point is calculated from the four input mesh points surrounding it. The first two linear interpolations along x direction are performed for calculating elevation values for points A and B, and then a linear interpolation along the y direction is performed for calculating the P elevation value. The same procedure can be performed interpolating first in y

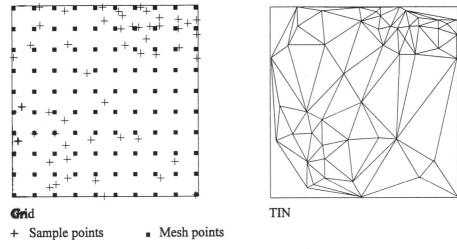

Grid   TIN

+ Sample points   ■ Mesh points

Figure 7.4. Grid and TIN models

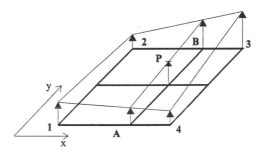

Figure 7.5. Bilinear interpolation from four input points (1-4) to compute a new value for the mesh point P

direction and then in x direction. Cubic convolution uses sixteen neighbour points and a non linear weight for the distance. (3) If sample points are irregularly spaced then several methods of interpolation can be used for estimating elevation values at positions different from sample point locations. The most commonly used are Inverse Weighted Distance and Kriging.

Inverse Weighted Distance
The basic assumption of this method is that the influence of one data point on another declines with distance from the point being estimated. The estimated value

can be obtained from the following expression:

$$Z(x) = \frac{\sum_{i=1}^{n} Z(x_i)d_i^{-m}}{\sum_{i=1}^{n} d_i^{-m}} \tag{7.1}$$

where x is the point where the surface is to be interpolated, x_i are sample data points, and d_i are distances from each of the sample points to the point being estimated, x; m is the weighting power.

The above expression can be rewritten as

$$Z(x) = \sum_{i=1}^{n} Z(x_i)\lambda_i \tag{7.2}$$

where $$\lambda_i = \frac{d_i^{-m}}{\sum_{i=1}^{n} d_i^{-m}} \tag{7.3}$$

are called weights.

There is a relation between the chosen value of m and the influence of neighbour points. The higher the value of m, the lower the influence of further points will be. For example, suppose there is a need to estimate the elevation at point P of fig.7.6, surrounded by six sample points. Table 7.1 shows the effect of the inverse distance weighting power on the sample weights and the relative P estimated elevation values. As m decreases, the weights become more similar and therefore the estimated value tends to approximate the mean value. As m increases, the difference between weights tends to increase.

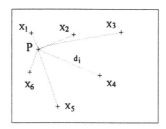

Figure 7.6. Distribution of sample points

In particular, for m=10, the furthest sample point receives a very small weight while the nearest one a very high weight, therefore the estimated value tends to approximate the nearest neighbour value. The most common value for m is 2, although estimates obtained using such value are not necessarily better than those obtained with different value of m (Isaaks and Srivastava 1989).

TABLE 7.1. Estimated Z values for point P using different weight powers m.

| | Z_i | d_i | λ_i | | | |
			$m=0.5$	$m=1$	$m=2$	$m=10$
1	700	20.59	0.2132	0.2630	0.3628	0.8029
2	650	28.18	0.1823	0.1922	0.1937	0.0349
3	600	51.66	0.1346	0.1048	0.0576	0.0001
4	625	52.48	0.1336	0.1032	0.0559	0.0001
5	675	48.05	0.1396	0.1127	0.0666	0.0002
6	700	24.17	0.1968	0.2241	0.2634	0.1620
Z(P)			663.92	669.35	678.70	698.24

Kriging

This method is based on the regionalized variable theory. For DTM applications the variable is the altitude value. Kriging assumes that the spatial variation of the variable is best represented by a stochastic surface rather than by a mathematical function. Kriging, in order to obtain an estimated z value for an unknown point, calculates weights to be applied to each sample point using a random fuction model rather than using a simple function of the distance, as with the previous method. The distinctive characteristic of this method compared to others is that it calculates weigths which are derived from the distance expressed in statistical terms rather than in purely geometrical terms (Isaaks and Srivastava 1989). In more detail, the spatial continuity is measured using the semivariance that is half of the average squared difference in z values between pairs of input sample points which are separated by a fixed distance h called lag.

$$\gamma(h) = \frac{1}{2n} \sum_{i=1}^{n} [Z(x_i) - Z(x_i + h)]^2 \qquad (7.4)$$

where n is the number of pairs of sample points separated by distance h. Using different values of lags a diagram called semi-variogram can be produced for a defined data set (fig. 7.7).

A sample semi-variogram is usually characterized by three elements: range, sill and nugget effect. At relatively short lag distances the semi-variance increases with the distance, but beyond a distance called range, a, it assumes a relatively constant value called sill, s. For distances greather than the range, the variation in Z values is no longer spatially correlated. Although the value of the semi-variogram for $h = 0$, according to the equation (7.4), should be zero, several factors, such as sampling errors and spatial variations, that occur over distances much shorter than the sample spacing, produce a value of $\gamma(h)$ greater than zero. This value is called

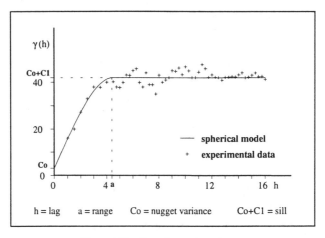

Figure 7.7. Sample semi-variogram

nugget effect. There exist however some cases in which the sill does not exist and the semi-variance increases with the disctance without reaching any plateau. In this case pairs of points located at whatever distance between them are spatially related. A sample semi-variogram can be fitted with one of several mathematical models (spherical, exponential, linear and others, fig.7.8) in order to be able to describe the way in which semivariance of sample points changes with lag.

The user can choose the model which better fits the semi-variogram of sample points and then calculate the desidered weigths by applying the previously mentioned theory. For both methods, Inverse Weighted Distance and Kriging , there is the problem of choosing the number of points or search radius to be considered in interpolating an unknown value at a mesh point. While using the first one the user has to choose an arbitrary value of the nearest points or search radius without any analytical criterion, using Kriging, and looking at the range, if it exists, of the semi-variogram, the user has a clear indication of this value and therefore can better control the interpolation process. Regular grids have been used for a long time because of simplicity of handling their data structure. The choice of the grid size is critical. A small grid size allows a more accurate representation of uneven terrain but generates a large amount of redundant data for representing uniform terrain inside the same study area. On the other hand, a large grid size, which would be able to effectively represent uniform areas without redundancy, is not capable of accurately representing complex topographic features.

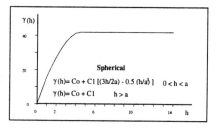

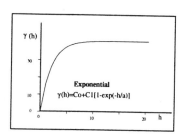

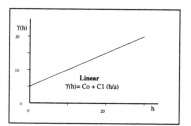

Figure 7.8. Examples of some mathematical models

7.3.2. TRIANGULATED IRREGULAR NETWORKS

In TIN the data structure is based on two basic elements: sample points with their x, y, z values and a series of edges joining these points to form triangles. Such an operation of connecting points is called triangulation. The network of all these triangles represents the terrain surface. The method allows the representation of areas with a complex topography using fewer points than altitude matrix. It efficiently uses the information carried by sample points in terms of location and elevation (i.e. the elevation and location of triangle vertices are those of sample points), thus minimizing errors due to interpolation. With altitude matrix the number of processed points is a function of grid size while with this method the number of points used to build the model is either coincident with sample points or lower when redundant points are eliminated. Several algorithms, Fowler and Little (Fowler and Little 1979), Drop Heuristic (DH) (Lee 1989), Adaptive Triangular Mesh (ATM) (Heller 1990), Hierarchical Triangulation (HT) (De Floriani et al. 1984), exist which select significant points. These methods and Very Important Point (VIP) (Chen and Guevara 1987) are also used when there is the need to reduce the number of points of a regularly spaced data set (original sample points or regularly interpolated grid).

Lee (1991) presents a comparison of some methods. He points out that there are two key problems in differentiating methods between them: "a) the establishment of the importance of elevation points, b) the establishment of the rules used to stop the selection procedures". The conclusion of the comparison is that each of the analysed methods has its advantages and disadvantages. However all methods require the user to specify a tolerance and a stopping rule. Therefore in order to obtain better results trial and errors experiments are needed in relation to the specific application.

The most common method of triangulation is the Delaunay triangulation. This algorithm produces a network of triangles that are as equiangular as possible and therefore any point inside the triangle, for which the elevation is unknown, is as close as possible to a vertex (sample points) and its elevation can be more accurately determined. In formal terms a triangulation conforms to the Delaunay criterion if, and only if, the circle which passes through 3 nodes of a triangle contains no other point.

There exist, however, some critical cases, such as breaklines and faults, which cannot be properly handled by pure Delaunay triangulation. Breaklines are linear features which represent a particular surface behaviour, e.g., shorelines, dams, roads, embankments, etc.. Shorelines mark a sharp transition between the planar surface behaviour of the water and the terrain surface (fig.7.9). All the other mentioned linear features do not mark any sharp transition but establish elevation values along them, which can be constant (dams) or variable with a known law (roads and embankments). Faults represent a discontinuity in surface behaviour e.g. geological faults in which for a given x, y location there are two z values (fig.7.10). When breaklines exist, the Delaunay triangulation must be adjusted in order to represent these linear constraints, i.e. some triangle edges must be coincident with part of these features (fig.7.11). These surface behaviour peculiarities cannot be easily handled by an elevation matrix.

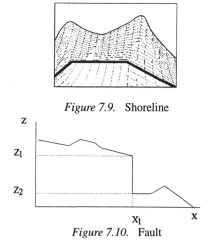

Figure 7.9. Shoreline

Figure 7.10. Fault

Several interpolation methods exist for obtaining the elevation for points different from sample points. Most commonly used methods refer to a planar surface (linear interpolation) or a curved surface (polynomial interpolation with higher order functions) within each triangle.

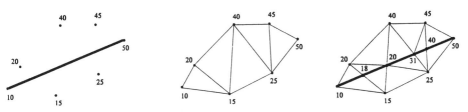

Figure 7.11. Breakline and relative modification of triangulation

7.4. Examples of Products that can be Derived from DTM

A great deal of information can be derived from DTM, based on elevation matrix or TIN, and some of this is important in hydrological modelling. In particular, in this paragraph methods and algorithms for determining slope, aspect, watershed and drainage networks starting from DTM are illustrated. Watershed and drainage network calculation algorithms are illustrated only for DTM based on regular grids because these algorithms are more commonly used. However Palacios-Velez and Cuevas-Renaud (1986), Gandoy-Bernasconi and Palacios-Velez (1989) and Jones et al. (1990) proposed algorithms for determining the above mentioned information using DTM based on TIN.

7.4.1. SLOPE AND ASPECT

Slope is defined by a plane tangent to the surface as modeled by the DTM at any given point and comprises two components: gradient, the maximum rate of change of altitude, and aspect, the compass direction of this maximum rate of change (Burrough 1986). Slope is expressed in decimal degrees or percents and aspect in decimal degrees. Generally speaking, however, the term slope is used to mean gradient and this terminology will be used in this chapter.

Regular grid
The slope is defined as:

$$\tan S = \sqrt{[(dz/dx)^2 + (dz/dy)^2]} \qquad (7.5)$$

where z is altitude and x and y are the coordinate axes.
  The aspect is defined (Burrough 1986) as:

$$\tan A = -(dz/dy)/(dz/dx) \qquad -\pi < A < \pi \qquad (7.6)$$

Most of the methods for calculating slope and aspect use a three by three moving window that is successively moved over the grid (fig. 7.12), and calculate the partial derivates, defined in eq.7.5 and 7.6, as finite differences.

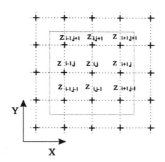

Figure 7.12. Three by three window

The simplest way to calculate finite differences is to account for only the neighbours above, below and on either sides of the central point of the moving window. Therefore the dz/dx and dz/dy finite differences can be calculated as

$$dz/dx = (z_{i+1,j} - z_{i-1,j})/2Dx \qquad (7.7)$$
$$dz/dy = (z_{i,j+1} - z_{i,j-1})/2Dy \qquad (7.8)$$

where Dx and Dy are the distances between cell centres in x and y directions. However this method is very sensitive to local errors in terrain elevation.

Skidmore (1989), comparing several methods, including the above mentioned, finds that the one proposed by Horn (1981) gives a good accuracy and does not require extensive CPU time. It is based on numerical analysis and the expressions used for calculating finite differences are

$$dz/dx = [(z_{i+1,j+1} + 2z_{i+1,j} + z_{i+1,j-1})$$
$$-(z_{i-1,j+1} + 2z_{i-1,j} + z_{i-1,j-1})]/8Dx \qquad (7.9)$$
$$dz/dy = [(z_{i+1,j+1} + 2z_{i,j+1} + z_{i-1,j+1})$$
$$-(z_{i+1,j-1} + 2z_{i,j-1} + z_{i-1,j-1})]/8Dy \qquad (7.10)$$

Fig.7.13 shows an example of slope and aspect classes calculated using the Grid model.

Triangulated irregular network
Almost all methods based on TINs for calculating slope and aspect represent the surface within each triangle as a plane. The equation of this plane, determined by the three vertices of each triangle, is

$$z = Ax + By + C \qquad (7.11)$$

where $A, B,$ and C are constant and can be calculated by simultaneous solution of the equation (7.11) at three vertices. The slope is given by:

$$\text{Slope} = \arctan\sqrt{(A^2 + B^2)} \qquad (7.12)$$

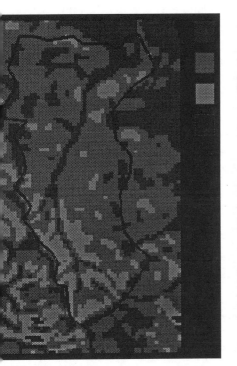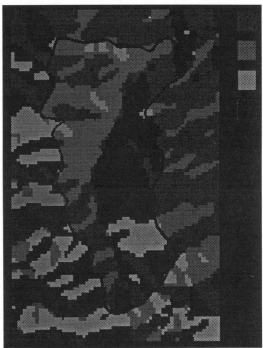

Figure 7.13. Example of four slope and aspect classes calculated using the Grid model for Lapilloso basin, Southern Italy

and the aspect is given by

$$\text{Aspect} = 180 - \arctan(B/A) + 90(A/|A|) \qquad (7.13)$$

These values can be seen as average values because they are constant for all points within each triangle.

7.4.2. WATERSHED

Several algorithms have been developed in the past for demarcating watershed. Most of them are however combined algorithms which delineate both watershed and drainage network and are based on elevation matrices. One algorithm that delineates only the watershed has been proposed by Marks et al. (1984). A pre-processing of the elevation matrix is needed in order to remove pits as much as possible which could influence the basin delineation.

The algorithm delineates the drainage basin using slope and aspect matrices, calculated as shown above, and an outlet cell interactively selected by the user. In addition, the user has to assign a threshold value of slope for the definition of flat regions. Aspect values are coded using 8 intervals each one of 45 degrees, (fig.7.14). The algorithm uses a 3 × 3 moving window initially centred on the outlet selected cell. It marks this cell as belonging to the basin and sequentially checks the adjacent eight neighbours. A neighbour cell is considered upstream (i.e. within the same basin) if its aspect faces toward the centre cell (fig.7.14) or if it has a slope lower than the threshold value chosen by the user.

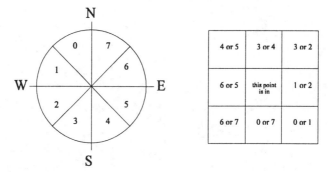

Figure 7.14. Aspect classes and values of these classes for which a neighbour cell belongs to the same basin of the centre cell

The algorithm is recursive in that, once an upstream neighbour is found, it is marked as belonging to the basin and the 3 × 3 window is moved and centred on it. At this stage the algorithm analyzes neighbour cells and again starts checking for upstream cells. The algorithm is repeated until it encounters a basin boundary or an edge of the grid. It then backtracks repeating the procedure until all cells belonging to the basin have been marked.

This algorithm is simple and has the advantage of analyzing only grid cells that are in or just adjacent to the drainage basin and ignoring the rest of the elevation grid. In addition, during the recursive process cells already marked as upstream are ignored.

7.4.3. DRAINAGE NETWORKS

One of the first methods for determining drainage networks from gridded DTM was proposed by Peuker and Douglas (1975). The authors developed an algorithm for identifying concave-upward portions of a DTM, where surface runoff tends to be concentrated. The algorithm flags the highest pixel in a window of 2 × 2 grid cells. At the end of this process all the unflagged cells represent the estimated channel network. Networks derived by using this algorithm are not necessarily connected and present many discontinuities.

Band (1986) proposed a set of thinning and connection procedures, based on image processing techniques, which, starting from networks derived using the previous algorithm, chosen because of its simplicity, determine fully connected networks and watershed boundaries. The approach used by Peucker and Douglas (1975) can be considered a geometric approach which considers only local surface concavity. A different approach, classified as hydrologic, has been used by several authors, Mark (1984), Jenson and Domingue (1988), Carrara et al. (1989). This approach is based on the assumption that "the drainage represents those points at which runoff is sufficiently concentrated that fluvial processes dominate over slope processes. If the spatial concentration of surface runoff is simulated, then those points at which this runoff exceeds some threshold can be considered to be the drainage network" (Mark, 1984).

Mark (1988) reviews previous works and formalizes a strategy to be followed in order to determine drainage network using the hydrologic approach. The author delineates several phases: (1) calculation of drainage direction matrix, given the elevation matrix; (2) pit identification and removal; (3) definition of weight matrix; (4) calculation of drainage accumulation matrix; (5) application of a threshold criterion in order to produce a channel network. A drainage direction matrix is a matrix which stores the flow direction for each cell.

All methods assume that each cell can drain only to exactly one neighbouring cell. The flow direction can be calculated using either aspect information or steepest descent direction. The latter method is recognized as being hydrologically more appropriate (Mark 1988). The steepest descent or maximum drop is calculated as

maximum drop = change in z value / distance
where distance is determined between cell centres.

If a 3×3 window is considered, the central cell of this window has eight neighbours. If the cell size is 1, the distance between two orthogonal cells is 1 and the distance between two diagonal cells is 1.4142, the square root of 2. Very often in grid based DTM cells exist, called pits, for which all elevation values of neighbours are higher. These are most of the time due to errors in DTM generation and only in special geomorphic environments (glaciated terrain or karst topography) they represent real features. For these points the flow direction cannot be calculated and therefore a modification of the elevation matrix is needed in order to compute a flow direction. Several authors proposed methods for eliminating pits, most of them are based on local filtering or smoothing operation. Mark (1984), Carrara (1988), Jenson and Domingue (1988) proposed a method based on filling operation.

In gridded DTM cells exist, different from pits, for which the steepest descent is equal for more than one neighbour cell and therefore the flow direction in undefined. This happens in flat areas. The flow direction for these cells can be calculated using a combination of neighbourhood techniques and iterative region growing procedures, which use not only information on elevation but also information on flow direction of surrounding cells, Jenson and Domingue (1988).

In order to model different drainage basin phenomena, it is necessary to define a weight matrix. This matrix influences the calculation of the drainage accumulation matrix, in which each element represents "the sum of the weights of all elements in the matrix which drain to that element" (Mark 1988).

As stated before different weight matrices produce different values in the drainage accumulation matrix. A short list of examples of different weight matrices will follow: (1) if each element of the weight matrix is set to one, then each cell of the drainage accumulation matrix will represent the number of cells that flow into it; (2) if each element of the weight matrix is set to the cell area, then each cell of the drainage accumulation matrix will represent the area of the cells that flow into it; (3) if only one cell of the weight matrix is set to one, with all the rest zero, then each cell of the drainage accumulation matrix with a value equal to one will represents the path from that cell to its outlet; (4) if the weight matrix represents average rainfall during a given storm, then the drainage accumulation matrix will represent the amount of rain that would flow through each cell, assuming complete runoff.

Once the drainage accumulation matrix is obtained, then by applying a threshold value to the minimum number of contributing cells or minimum drainage area, depending on the chosen weight matrix, it is possible to extract the drainage network.

Drainage density of extracted channel networks varies with the chosen threshold value. Small threshold values will generate more detailed channel network with higher drainage density. High threshold values will generate coarser channel networks (fig.7.15).

Tarboton et al. (1991) suggest that the network extracted from DTM should have properties traditionally ascribed to channel networks and have as high a resolution as possible. They present two analytical procedures for determining an appropriate threshold value, based respectively on a constant drop property and a power law scaling of slope with area. Mark (1988) reports a method proposed by Band which uses, as threshold, a critical value of the following parameter:

$$P = \ln(a/\tan b) \qquad (7.14)$$

where a is the drainage area contributing to the cell, and b is the surface slope.

For both methods the threshold value is determined as follows. For several arbitrary threshold values a number of different digital channel network are generated; network properties are calculated for each one of these digital networks; optimal threshold value is that which has generated a digital network for which, are still valid laws or network properties derived for river networks .

7.5. Software

Three main classes of software products are available for generating DTM, each one of these classes has its own peculiarities in terms of analytical capabilities and costs.

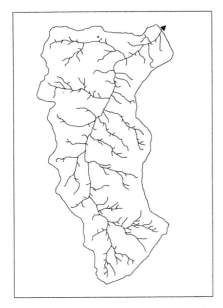

(a) Drainage network automatically extracted by
DTM with flow accumulation value equal to
100 cells (cell dimension = 25 m x 25 m).

(b) Drainage network automatically extracted by
DTM with flow accumulation value equal to 150
cells (cell dimension = 25 m x 25 m).

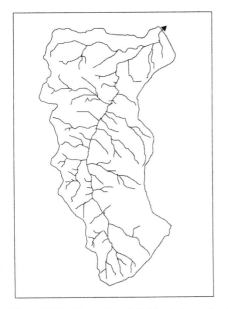

River: Lapilloso (Southern Italy)

Catchment area: 28.548 km2

(a) Drainage density: 2.68 km/km2

(b) Drainage density: 2.06 km/km2

(c) Drainage density: 2.33 km/km2

(c) Blue Lines digitized from map at
scale 1: 25.000.

Figure 7.15. Examples of digitized channel network and digital channel networks generated using
different threshold values

The first class includes software packages which usually can generate elevation matrices starting from a set of regularly or irregularly sample points. They use different interpolation alghoritms such as kriging, inverse weighted distance and minimum curvature. Once the DTM is obtained, they can produce profiles, contours and elevations for points different from the original data set. One example of this class is SURFER, developed by Golden Softvare , Colorado, USA, which runs on PC based systems.

The second class includes software packages which usually can generate DTM, both grids and TIN and related parameters such as slope and aspect. They meet the need of numerous disciplines and mainly of civil engineering and surveying but do not allow the calculation of watershed or drainage network. They usually operate inside CAD software: one example is QUICKSURF, developed by Schreiber Instruments inc., Denver Colorado, USA, which runs within AutoCAD on PC based systems.This package accepts input from point data sets and contour lines without any limitation. It uses the Delauney criterion for generating TIN and a proprietary gridding algorithm for generating elevation matrix. It allows the incorporation of surface discontinuities such as breaklines and faults into DTM and can perform volume and profile calculation .

The third class includes GIS software packages which usually have modules or commands which generate DTM (grids and/or TIN) and some or all the previous described parameters. One advantage of using this class of software is that results of DTM analysis can be combined with other geographic data in order to obtain new relationships between them.

One example is IDRISI, a grid based GIS developed by Clark University, Worchester, Massachusetts, USA, which runs on PC based systems. It allows the creation of a grid based DTM starting from points or lines. It, in addition to the calculation of slope and aspect, delineates watershed once assigned one or more target cells.

Another example is Arc/Info, a GIS developed by ESRI, Redlands, California, USA, which runs on several platforms ranging from PC to workstation and mainframe. It supports both vector and grid data, thus a user can choose the more appropriate data model (vector or raster) depending on the specific problem. Arc/Info can generate DTM based on TIN or grid models and calculate slope and aspect. The TIN module of Arc/Info allows the modelling of highly variable terrain surfaces which include discontinuities and breaklines. The GRID module of Arc/Info enables the generation of watershed boundaries and drainage networks using the algorithm proposed by Jenson and Domingue,1988. However in order to generate an appropriate drainage network, the user still has to specify a threshold value for the contributing area. Once a digital drainage network is generated, GRID allows the ordering of such a network using the methods proposed by Strahler (1957) and Shreve (1966). TIN and GRID modules are present on all platforms except for PC based systems. To this class belongs also GRASS, a public domain GIS software developed by U.S. Army Corps of Engineers. It runs on workstation under UNIX operating system and on PC under Linux operating system. It supports both vector and grid data and can generate DTM based on grid models, calculate slope and aspect, and enable the generation of watershed boundaries and drainage network - for a more detailed description of GIS software packages please refer to Chapter 16 in this book.

7.6. Conclusions

This chapter has introduced some of the basic concepts involved in terrain analysis using DTM and has shown the usefulness of Digital Terrain Models in hydrological application. Methods and algorithms have been illustrated and compared in different sections of the chapter. An attempt to show limits and potentialities of each one has been made. More common and simple algorithms have been illustrated with more detail, in order to make clear the principles on which terrain analysis is based. For this reason, the chapter is not intended to be exhaustive of surface modelling methods and the reader should refer to references for a more detailed description of such methods and for research topics.

References

Band L., 'Topographic partition of watershed with Digital Elevation Models', Water Resource Research, vol.22, 1986.

Beven K. J., Kirkby M. J., 'A phisically-based variable contributing area model of basin hydrology', Hydrol. Sci. Bull., 24, 1979

Burrough P. A. Principles of GIS for land resource assessment , Oxford Clarendon Press, 1986.

Carrara A., 'Drainage and divide networks derived from high-fidelity digital terrain models'. In Chung C.F. et al., Quantitative analisys of mineral and energy resources, D.Reidel Pub. Co, Dordrecht, 1988.

Carrara A., Detti R., Federici G., Pasqui V., 'Reticoli idrografici e parametri morfologici di bacino da modelli digitali del terreno', GNDCI-CNR, Roma 1989.

Chen Z. T., Guevara J. A., 'Systematic selection of very important points (VIP) from digital terrain model for costructing triangular irregular network', Proceedings of AUTO-CARTO 8 Baltimore, MO, U.S.A., edited by Chrisman N. (Falls Church:American Congress of Surveying and Mapping), 1987.

Copertino V. A., Fiorentino M., Sole A., Valanzano A., 'Il Sistema Informativo dei Bacini Idrografici Pugliesi' in Copertino V. A., Fiorentino M., Valutazione delle Piene in Puglia, GNDCI-CNR, Final Report, Roma 1991.

De Floriani L., Falcidieno B., Nagy G., Pienovi C., 'A hierarchical structure for surface approximation', Computer Graphics, vol.8, 1984.

Djokic D., Maidment D. R., 'Terrain analysis for urban stormwater modelling', Hydrological Processes, vol.5, 1991

Fowler R. J, Little J.J., 'Automated extraction of irregular network digital terrain models', Computer Graphics, vol.13, 1979

Gandoy-Bernasconi W., Palacios-Velez O., 'Automatic cascade numbering of unit elements in distributed hydrological models, Journal of Hydrology, vol.112, 1990.

Heller M., 'Triangulation algorithms for adaptative terrain modeling ', Proceedings of 4th Int. Symposium on Spatial Data Handling, Columbus Ohio,1990

Horn B. K. P., 'Hill shading and the reflectance map', Proc. IEEE , vol 69, 1981

Isaaks E. H., Srivastava R. M., An introduction to applied geostatistics, Oxford University Press, 1989.

Jenson S. K., Domingue J. O., 'Extracting Topographic Structure from Digital Elevation Data for Geographic Information System Analysis', Photogrammetric Engineering and Remote Sensing, vol.54, 1988.

Jones N. L., Wright S. G. and Maidment D. R., 'Watershed Delineation with Triangle-Based Terrain Models', Journal of Hydraulic Engineering, vol.116, 1990.

Lanza L., Conti M., La Barbera P., 'Automated modelling of flashfloods in the mediterranean area. The september 1992 event over the Liguria region.', Proceedings of the 12th IASTED International Conference, Innsbruck, Austria, 1993.

Lee J., 'A drop heuristic conversion method for extracting irregular networks for digital elevation models' Proceedings of GIS/LIS '89, Orlando, FL, U.S.A. 1989.

Lee J., 'Comparison of existing methods for building triangular irregular network models of terrain from grid digital elevation models', International Journal of GIS, vol.5, 1991.

Mark D. M., ' Network models in geomorphology'. In Anderson M.G., Modelling Geomorphological Systems, John Wiley & Sons Ltd.,1988.

Mark D. M., 'Automated detection of drainage network from digital elevation models', Cartographica, vol.21, 1984.

Marks D., Dozier J., Frew J., 'Automated basin delineation from digital elevation data', Geoprocessing, vol.12, 1984.

Palacios-Velez O. L., Quevas-Renaud B., 'Automated river-course, ridge and basin delineation from DEM', Journal of Hydrology, vol.86, 1986.

Peuker T. K., Douglas D. H., 'Detection of surface specific points by local parallel processing of discrete terrain elevation data', Comput.Graph.Image Proc.,vol.4,1975.

Shreve R. L. 'Statistical law of stream number', Journal of Geology, vol.74, 1966.

Strahler A. N., 'Quantitative analysis of watershed geomorphology', Transactions of the American Geophisical Union, vol.8, 1957.

Quinn P., Beven K., Chevallier P, Planchon O., 'The prediction of hillslope flow paths for distributed hydrological modelling using digital terrain models', Hydrological Processes, vol.5, 1991

Rodriguez-Iturbe I., Valdez J. B., 'The geomorphological structure of hydrologic response', Water Resources Research vol.15,1979

Skidmore A. K., 'A comparison of techniques for calculating gradient and aspect from a gridded DEM', International Journal of GIS, vol.3, 1989.

Tarboton D.G., Bras R. L., Rodriguez-Iturbe I., 'On the extraction of channel network from digital elevation data', Hydrological Processes, vol.5, 1991.

Weibel R., Heller M. 'Digital terrain modelling', in Maguire D. J., Goodchild M.F., Rhind D.W., Geographical Information Systems, vol.1, Longman Scientific Technical, 1991.

CHAPTER 8

GIS for Distributed Rainfall - Runoff Modeling

C. Colosimo and G. Mendicino

Abstract. In this chapter the rainfall-runoff phenomenon is considered emphasizing the role assumed by the runoff production. Through the use of GIS, capable of managing and storing a great amount of data, different mechanisms of runoff production are analyzed. In particular a modified and spatially distributed version of TOPMODEL, which is capable of modeling both infiltration excess and saturation excess runoff and incorporating the spatial variability of soil, topography and rainfall, had been applied to simulate the hydrological behavior of an experimental basin (Turbolo Creek - Southern Italy). Finally, through geostatistical analysis, the spatial distribution of soil parameters is considered to obtain useful information on the characteristic dimension of the elementary drainage area.

8.1. Introduction

The principal aim of modern hydrological research is the correct definition of the interconnected relationships between the different hydrometeorological quantities within a basin and their representation by mathematical models of varying complexity.

A basin, defined by the whole of its measurable attributes (climate, temperature, precipitation, humidity, infiltration, transpiration, evaporation, runoff, lithological, pedological, geomorphological characteristics, etc.), can be assumed to be a natural system which at whatever action (rainfall) creates a corresponding response (runoff) depending on the actual state of the system. The state of the system can be defined by state variables (water level, discharge, soil humidity, etc.) a knowledge of which allows the precise identification of the nature of the outflow in relation to a determined inflow.

V. P. Singh and M. Fiorentino (eds.), Geographical Information Systems in Hydrology, 195–235.
© 1996 Kluwer Academic Publishers. Printed in the Netherlands.

A knowledge of the measurable attributes of the basin system is necessary for the correct definition of its state, with the aim of reducing all the derived sources of imprecision to a minimum: from time variations in the basin's geomorphological characteristics caused by natural and anthropic processes; from errors in hydrological measurements (velocity of infiltration, evapotranspiration, peak flow); to the imperfect knowledge of the real spatial distribution of inflow, outflow and state variables. Such uncertainties can never be entirely eliminated. Consequently, the identification of the basin system, taken to be the precise determination of the effective nature of the existing relationship between the inflow and outflow variables, would appear to be manifestly impossible. For this reason, different phases undertaken on a "prototype" hydrological system can only provide an approximate idealized representation in the model form (Spriet and Vansteenkiste, 1988).

Mathematical models which simulate the process of transformation from rainfall to runoff in a basin, are developed in three main directions variable according to their structure:

- black-box models for which the hydrological transformation process is approached by a closed black box analysis, which independently of the physical representation of the phenomenon, defines the transformation agent of rainfall into runoff;
- lumped conceptual models which describe the basin's response to precipitation by conceptual relationships, based on the representation of the physical process, by means of suitably integrated simple structures (linear reservoir, linear channel, geomorphological scheme related to the concentration of local contributions through the river network);
- physically-based distributed models, which through an analysis based on principles of conservation of mass, on quantities of motion and on energy, describe the evolution of the physical system, subdividing the domain into simple interconnected cells, upon which hydrological quantities can be hypothesized homogeneous.

The lumped-model types, as is well-known, require a long parameter calibration period which depends on the availability of a high number of events (rainfall-runoff). Parameter calibration will therefore be so much more precise in relation to the reduction in data acquisition error and to the type of integration of the recorded data at the basin scale. Since such models consider the basin system as a single homogeneous unit, they imply that the corresponding lumped variables (of inflow, outflow and state) are averaged out at the basin scale. These latter, being functions of time only, t, do not provide indications of their spatial variability within the interested domain. In any case, even if, for several factors which control the transformation process, it is possible to define their spatial variability in a more restricted range, further subdivisions of the main basin into homogeneous sub-basins is almost always precluded by the absence of hydrometric records at the outlet of the sub-basins.

Noticeable improvements can be obtained instead, through physically-based distributed models since they allow an evaluation, on the one hand, of the non-uniformity of precipitation and on the other, because they take the heterogeneity of the topography, vegetation and soil into account. These models such as SHE (Abbott et al., 1986a, b; Bathurst, 1986a, b), TOPMODEL (Wood et al., 1988), IHDM (Rogers et al., 1985) have no need, in theory, of precautionary calibration since the relative parameters offer a clear physical significance which makes an estimate of their values possible in relation to a knowledge of the basins characteristics.

However, the great attention given to this type of model has been partly revised according to Beven's observations (1989), which stressed the following limits :

- the differential equations concerning the various hydrological processes (overland flow, infiltration, percolation etc.) are solved for the single cells in which the basin is subdivided, introducing a conceptualization of the phenomenon itself for them (the heterogeneity of the hydrological quantities inside the cell are ignored);
- conceptualizations introduced in this way, can result in different performances in the models themselves with variations in the assumed scale;
- the applicability of such models is limited only to smaller basins since the differential equations which govern the various transformation phenomena require a high spatial and time discretization which, because of their non-linearity, can cause problems of numerical instability;
- do the methods of stochastic interpolation used to extend punctually acquired information to the whole basin, which suppose the assumed area to be a predominantly stationary stochastic field, allow an effective representation of the spatial variability of hydrological quantities?

Despite Beven's well-founded arguments (1989) the criterion on the basis of which the greater the level of detail in the representation of a basin and of the hydrological quantities relative to it, the higher are the performances provided by the different simulation models, remains valid.

An excessive level of detail introduced into the simulation could however lead to a computational load, which in some cases, would not be justified by the improvements in performance of the models. Useful indications with regard to the aggregation levels of single areas, for the identification of a characteristic dimension for a representative elementary area (REA), can be drawn from the study carried out by Sivapalan et al. (1987), Beven et al. (1988) e Wood et al. (1988).

The REA concept, while producing extremely interesting results regarding the identification of a representative scale for the hydrological process, results however in a simplification of reality, since it does not enable any explanation of non-stationary spatial phenomena associated with long or complex scale correlations.

For a clearer understanding of such problems, as well as for their computational evaluation, on the one hand, there emerges the need for a considerable amount of spatial data to elaborate and, on the other, for their availability in digital form.

These demands are currently satisfied by advances in computational systems capable of more rapid elaboration of the input and output parameters of models which simulate spatially distributed processes.

Geographic Information Systems (GIS) represent integrated systems aimed at the running, storage, elaboration and visualization of geographic information regarding topography, soil use, territorial features, climatic conditions etc. Their integration with simulation models allows a better definition of the scale of detail required by the correct parameterization of the model. The improvement brought about by GIS with respect to a knowledge of the local properties of hydrological quantities, implies a reduction of uncertainties caused by the spatial extrapolation of information by means of average operators.

GIS can be classified in two distinct categories, vector or raster, variable according to how the spatial data are represented.

A vector data model uses lines or points to identify positions. In this model the form of the spatial entities is represented by a set of lines, each of which is defined by points at its extremities and by some connective form (polygons, networks, irregular triangle networks). Among the various vector GIS the most-used is ARC-INFO (Environmental Systems Research Institute).

The raster model, instead, subdivides an area into a grid of regular cells ordered in a specific sequence. In this model the form of spatial entities is represented by a set of cells, located on a grid. Each cell is cross-referenced by a row and column number and contains a single type or value attribute. Among the raster GIS available today, GRASS (U.S. Army Corps of Engineers, Construction Engineering Research Laboratory, Champaign, Illinois) is the one which offers the greatest number of facilities. There have been many studies carried out in the field of GIS integration with rainfall-runoff transformation models. Essentially they are based on hydrological evaluation, identification of model parameters, design and construction of hydrological models within GIS and on the interface GIS-models. Hodge et al. (1988) integrated GRASS with the distributed physically-based ARMSED model. Vieux et al. (1988) and Vieux (1991) using the vector GIS ARC-INFO set up a finite element model capable of providing a distributed estimate of surface runoff in relation to determined precipitation. Wolfe and Neale (1988) also interfaced a distributed finite element model (FESHM) with the raster GIS GRASS. Cline et al. (1989) instead making use of a vector GIS-like (Computer Aided Design) software, automatically generated input files for the lumped rainfall-runoff model HEC-1. Stuebe and Johnston (1990) by means of the arithmetic expressions contained in GRASS, automatically estimated the amount of surface runoff produced by a precipitation. Also using modules contained in GRASS, Vieux and Kang (1990) developed a "Waterworks" program capable of automatically generating topographic parameters for use in concentrated-type hydrological models. Further methodological contributions, even if in the field of nonpoint source pollution models, have been provided by De Roo et al. (1989) Engel et al. (1991) Panuska et al. (1991) and Sasowsky and Gardner (1991), regarding the integration of AN-

SWERS, AGNPS, SPUR models with some of the GIS mentioned above. Nor should the contributions made by Fedra (1993), Maidment (1993) and Moore et al. (1993), be overlooked, which in analysing the problem of GIS integration with hydrological models, have provided a thorough framework regarding different experiences carried out in that sector. While emphasizing the many benefits derived from integration with GIS models, they however, define some conceptual differences which, on the one hand, consider GIS as an instrument of static and discrete reality representation and, on the other, hydrological models as instruments for the simulation of phenomena which evolve dynamically and develop in a continuous domain. In fact, in the case of GIS, basic concepts exist which concern geographic location, spatial distribution and relationships. Vice versa, in the case of hydrological models the fundamental concept is that of state, expressed in terms of transfer of mass and energy, interaction and variability of quantities. Such considerations, while limiting the potential offered by GIS because of their inability to dynamically manage the time component of hydrological quantities, should not however, introduce false convictions regarding the countless benefits brought about by GIS in the understanding of the complex rainfall-runoff transformation phenomena.

The aim of the present chapter, is to emphasize the potentiality offered by GIS (particularly the raster type), both in direct and indirect identification of the basin's measurable attributes, and in checks at a local scale of the surface and subsurface response as a result of a rainfall event.

In particular, only the more complex part of the transformation phenomenon is considered, or rather, that relative to the production of surface runoff due to Hortonian-type and surface saturation mechanisms, disregarding the routing of generated runoff over the hillslopes and along stream channels.

Finally, with the aid of tools contained in GIS, the spatial distribution of some soil and topographic parameters is examined with the aim of identifying a representative area which enables the carrying out of simulations no longer at a local scale, but at a sub-basin scale.

8.2. Computed and Observed Data

8.2.1. TOPOGRAPHIC PARAMETERS

The Turbolo Creek Basin, tributary of the River Crati in Southern Italy (Fig 8.1), was selected for hydrological simulations. Its catchment area is about 29 km^2. The basin, whose main river reach measures about 13 km, has an average altitude of 290 m above sea level and an average hillslope of 26%. After having completed the digitalization of the contour lines taken from an official 1:25,000 map scale using stochastic interpolation techniques (Matheron, 1971; Delfiner and Delhomme, 1975; Delhomme, 1978) a digital terrain model was obtained (DTM) on a regular 50 m grid (Fig. 8.2). The use of stochastic rather than deterministic interpolators was considered necessary, not only because it is possible to check the efficiency of results provided by estimators which minimize error variance, but also to re-

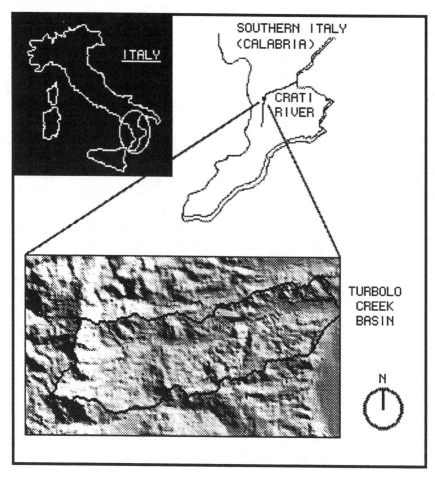

Figure 8.1. Shaded relief of the study area in Southern Italy

duce the number of flat zones and pits, in the DTM. This became evident through a comparison of this latter altimetric matrix with a previous one obtained on the same study area and on a regular 100 m grid, by means of deterministic method based on 8 points inverse distance interpolation (Colosimo and Mendicino, 1992). From the new altimetric matrix, made up of 135 rows and 247 columns, all the primary attributes were extracted using different functions, some internal to GIS and others integrated to it in the form of auxiliary software procedures. It should be emphasized that the structure of GIS is not a single software package, but an integrated expanding system, made up of a range of procedures which, from time to time, are used to solve different problems.

In this study, the primary attributes locally obtained from the DTM concerned the estimate of the following parameters: average slope; maximum slope; aspect; specific catchment area; flow path length; profile curvature and plan curvature.

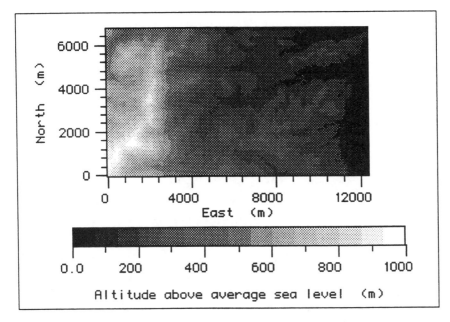

Figure 8.2. Spatial distribution of altitude over the study area

Some of these, although being easy to calculate, are not included in the GIS evaluation system (understood as a single software package).

Therefore, it was necessary to create some auxiliary procedures capable of interacting with the main program to increase available evaluation capacity. Among these, the format is that which enables the elimination of all flat zones and pits from the DTM; their presence leads to application failure in whatever simulation model concerning the theoretical concentration of surface runoff. Some authors suggest the removal of such holes by means of numerical filter algorithms which for DTM, with low levels of precision, could lead to noticeable variations in the altimetrics of the area under consideration (O'Callaghan and Mark, 1984; Band, 1986). Instead, in this study, a correction to DTM was preferred assuming a flooding of the holes and successively evaluating the best water course from them in the overflowing phase (Jenson and Domingue, 1988; Mendicino, 1989; Martz and Garbrecht, 1992).

The unpitted DTM was therefore elaborated by means of a procedure which allowed for the estimate of some of its morphological and hydrological parameters. This procedure analyzes the altimetric matrix by means of a moving window of 3x3 points centred on the point of analysis (i, j) (Fig. 8.3), which, moving along the directions of maximum slope, sequentially shifts from the highest DTM levels to the lowest. This algorithm, thoroughly described by O'Callaghan and Mark (1984), Jenson and Domingue (1988), Martz and De Jong (1988), Mendi-

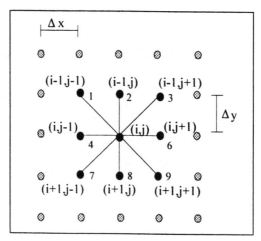

Figure 8.3. Moving window of 3 × 3 points centred on the point of analysis (i, j)

cino (1989), Martz and Garbrecht (1992), is indicated in short as D8 (Deterministic 8-neighbors method). At the end of this elaboration three new matrices were generated georeferenced to that of the altimetric levels. These matrices respectively define the distribution of surface runoff (Fig. 8.4), their theoretical course (Fig. 8.5) and the local surface slope, $\tan(\beta)$, along the direction of maximum slope (Fig. 8.6). To each element (i, j) of the altimetric matrix the contribution of the upslope drainage area of the element under consideration, the theoretical course taken by the surface runoff up to it and the value assumed locally by the topographic gradient, can therefore be associated.

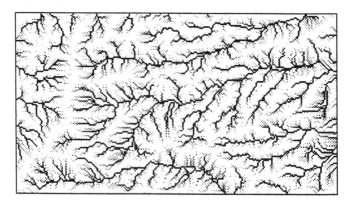

Figure 8.4. Spatial distribution of upslope drainage area over the study area

A higher level of detail can be obtained by introducing a stochastic component into the D8 algorithm along the N-E, S-E, S-W and N-W directions (Farfield and

Figure 8.5. Spatial distribution of flow direction over the study area

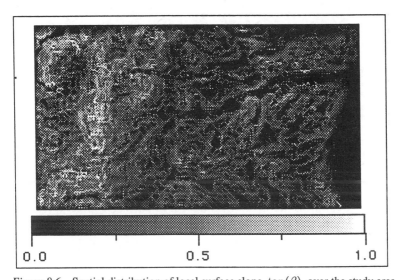

Figure 8.6. Spatial distribution of local surface slope, $\tan(\beta)$, over the study area

Leymarie, 1991). This improvement, known as Rho8 algorithm, is mainly found in those slightly sloping areas along which the automatically extracted water course would tend to runoff in parallel along the preferential directions, according to the D8 approach.

Both the D8 and Rho8 approaches, even if much used in the automatic generation of river networks and for the identification of the corresponding sub-basins, do not wholly represent the course of surface runoff, especially in those areas typified by divergent surfaces. If the accumulated upslope area for any one cell is distributed amongst all those of the downslope direction, according to weighted percentages relative to the value assumed by the slope in the same directions, further improvement is introduced into the evaluation of the theoretical concentration of surface runoff.

A multiple flow direction approach defined as FD8 algorithm, was initially proposed by Freeman (1991) and successively, analyzed by Quinn et al. (1991). The same authors have, however, stressed that the FD8 algorithm cannot simulate well in topographic conditions such as those found in alluvial plains. In these circumstances a pronounced expansion of surface runoff along the alluvial plains is noticeable instead of well-delineated stream channels (Fig. 8.7). The FD8 algorithm therefore has to be modified according to the main channels of the river network.

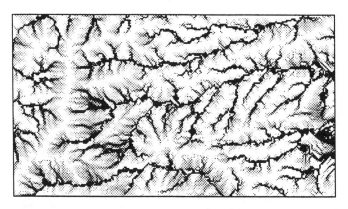

Figure 8.7. Spatial distribution of upslope drainage area over the study area using a FD8 approach

In particular, it is necessary to refer to a mixed scheme FRho8 algorithm, which allows the evaluation of the theoretical course of surface runoff in relation to the permanent drainage network. Runoff coming from the slopes (FD8 algorithm), after having reached one of the main channels of the river network, has to remain in it (Rho8 algorithm) until it reaches the basin outlet (Fig. 8.8).

To better understand the improvements introduced by the algorithm examined, it can be observed in Fig. 8.9 how surface runoff, produced by an instantaneous unit evenly distributed precipitation, is concentrated along an impermeable cone-shaped surface.

The computational load implied by the algorithms so far observed is more than compensated for by the benefits produced from the improved morphological, but also hydrological, representation of the surface runoff concentration phenomenon (as will be thoroughly discussed in the following paragraphs).

The use of these procedures, suitably integrated with others (Mendicino, 1990; Martz and Garbrecht, 1992) have allowed the automatic extraction of a connected

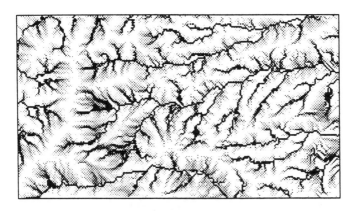

Figure 8.8. Spatial distribution of upslope drainage area over the study area using a FRho8 approach

and codified river network, and of its corresponding sub-basins, for the area under consideration.

8.2.2. SOIL PARAMETERS

Information relative to the composition and use of the soil was subsequently added to that of a morphological nature. This information was obtained by means of numerical elaboration techniques from satellite images and aerial photographs. Images acquired in digital form by means of an optical scanning device, were initially subjected to dynamic expansion aimed at heightening their contrast (Colosimo and Mendicino, 1993). Afterwards, the equalized images were processed with different numerical filters (linear, non-linear and morphological), variable depending on the attribute to be extracted. The homogeneous areas derived from it (vector information-polygons with associated attributes) were georeferenced and converted into the raster format with the same resolution of DTM. Regarding soil data, the sample area was stored inside GIS according to four different classes (A, B, C and D), following the classification proposed by the Soil Conservation Service (1968) (Fig. 8.10). In the case of soil use the same area was stored in GIS according to twelve different classes (Fig. 8.11).

Since no information regarding the hydraulic characteristics of the soil was available for the basin under consideration reference was made to studies of a hydraulic nature carried out previously by Morel-Seytoux and Verdin (1982) and by Colosimo and Mendicino (1993). Integration through GIS of the results obtained in these studies at a basin scale, allowed an estimate at a local scale (with necessary caution) of the values assumed by the hydraulic conductivity of the soil at natural saturation, K_s, and of the storage suction factor, S_f. The latter should be seen as a product of effective capillary drive (in relation to the wetting front), Ψ, and the

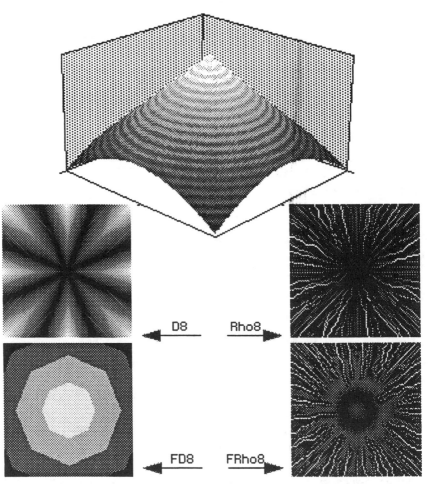

Figure 8.9. Shaded drainage images for cone-shaped surface obtained through D8, Rho8, FD8 and FRho8 algorithms

initial moisture deficit of the soil , $\Delta\theta$.

The procedural steps which enabled this estimate were developed beginning with the definition at a local scale of the CN parameter proposed by the Soil Conservation Service (1968). This estimate was carried out, rather than on tables which are difficult to manage automatically, by means of a procedure proposed by Giorgini (1990). This latter, beginning with the value taken from the CN parameter for a soil class A, and for a standard antecedent moisture condition AMC=II, defined as $CN_{II,A}$, allows the determination of the CN parameters for the other

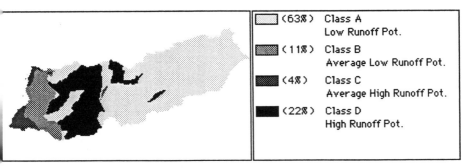

Figure 8.10. Soil classes according to the classification proposed by Soil Conservation Service

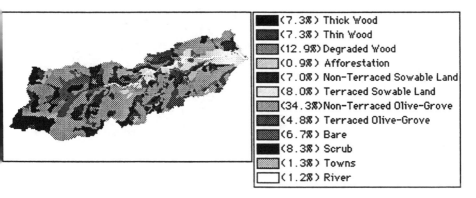

Figure 8.11. Soil use

soil classes, CN_{II}, through the equation:

$$CN_{II} = 100 - \frac{1.4}{x + 0.4}(100 - CN_{II,A}) \tag{8.1}$$

in which x is equal to 1, 2, 3 and 4 respectively, for the soil classes A, B, C and D. Simple reclassification procedures carried out using LUT (Look Up Table), associated to some algebraic operations carried out on matrices, were done inside GIS with the aim of defining the spatial distribution of the CN_{II} parameter. Since 29 mm of rain were measured in the five days preceding the simulated event, and considering the dormant season in which the precipitation occurred, it was necessary to calculate the new local CN values because of the high soil moisture. The new

CN_{III} parameter obtained by means of the following formula (Chow et al., 1988)

$$CN_{III} = \frac{23\ CN_{II}}{10 + 0.13\ CN_{II}} \tag{8.2}$$

gave an average value equal to 88 for the whole basin (Fig. 8.12).

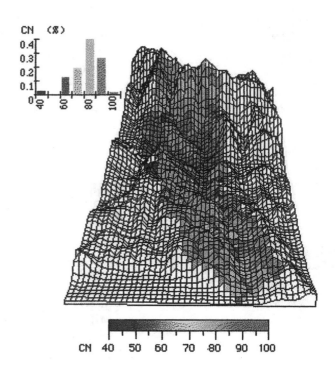

Figure 8.12. Spatial distribution of CNIII parameter superimposed on an isometric projection of the 50 m DTM of the Turbolo Creek basin

In previous research carried out on the same basin about the effective determination of net rainfall, Colosimo and Mendicino (1993) analyzed the performance of infiltration methods present in model HEC-1 (1990). For the rainfall event here considered, by means of the optimization parameters technique present in the HEC-1 model, the same authors fixed the best value assumed by CN_{III} parameter for the SCS-CN method at equal to 85. This result, even if obtained by means of research carried out at a basin scale, appears to confirm the estimate made by GIS at a local scale.

Referring moreover, to the conversion scheme proposed by Morel-Seytoux and Verdin (1982), about the relationship between the CN parameter, the hydraulic conductivity of the soil at natural saturation, K_S, and the storage suction factor, S_f,

(Fig. 8.13), the matrices of the CN values were reclassified by two LUT ($CN \rightarrow K_s$; $CN \rightarrow S_f$). The average values typical of the basin were extracted from the two matrices obtained, which represent the spatial behavior of the coefficient, K_s, and of the factor, S_f, and were respectively fixed at equal to 2.18 mm/hr and at 33.65 mm.

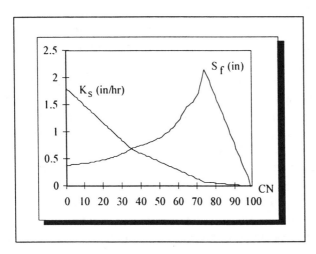

Figure 8.13. LUT conversion

These latter values, analogous to the CN parameter, were compared to those obtained by means of the optimization technique present in HEC-1, related to the Green-Ampt method (1911). In the case of the latter, for the rainfall event under consideration, the best values assumed by the parameters were $K_s = 2.50$ mm/hr and $S_f = 1.33$ mm. The analysis of the results, on the one hand, confirms the estimate made at a local scale about the values assumed by parameter, K_s, on the other, suggests a reflection with regard to the estimate of factor $S_f = \Psi \Delta \theta$, obtained with HEC-1. The discrepancies shown in the latter case could be due to the initial soil moisture condition, in this specific case, near to saturation. The estimate of $\Delta \theta = 0.04$ at a basin scale (HEC-1), owing to its noticeably low value, might not be explained by a local scale analysis.

8.3. Mechanisms of Runoff Production

The relative importance of the components in the rainfall-runoff transformation process depends both on its working scale and on the geographical, climatic and environmental conditions of the site under consideration. The necessity therefore arises to operate a classification of the models on the basis of their descriptive capability. In particular, referring to the single process models, whose descriptive capability is limited only to surface runoff, can be opposed those complete process models, which offer a detailed description of the rainfall-runoff transformation. The models which belong to the first class are suitable for those basins whose hydrological response is predominantly controlled by the routing phenomenon. Instead, in the other case, the models have to reproduce the interactions between one phenomenon and another which contribute to the surface runoff production.

Dunne (1978), through different field experiences, underlined that surface runoff production along basin slopes is the result of an already saturated surface layer of the soil. The saturation phenomenon can come about by means of two fairly definite mechanisms: infiltration excess runoff and saturation excess runoff. The former mechanism, evidenced for the first time by Horton (1933) (hortonian mechanism or infiltration excess mechanism), is found every time rainfall intensity exceeds the saturated hydraulic conductivity of the soil. As the rainfall event proceeds, the moisture content at the surface increases up to a saturation limit value reached in relation to the ponding time. After that moment a moisture profile is formed which, propagating in the lower layers of the soil, leads to a rapid diminution in its infiltration capacity. When the infiltration rate is less than rainfall intensity, a surface runoff production is found. This mechanism is particularly evident in those basins lacking vegetation cover and anyway, typified by low values of hydraulic conductivity. In fact, consolidation and compacting of the soil caused by rainfall and repeated agricultural activity, leads to a reduction in hydraulic conductivity with a consequent increase in hortonian-type surface runoff.

From the other hand surface runoff production according to Dunne (1978) (saturation excess mechanism) comes about from a saturation of the surface from below because of rainfall input and downslope subsurface flow. This mechanism is particularly evident on near-channel wetlands, which, tend to expand and contract during and between rainfall events. This mechanism therefore, is more suitable to surface runoff production in those wet areas rich in vegetation, typified by convergent topography and with shallow water tables.

In both cases it is evident that the spatial evolution of the runoff tends to be concentrated on precise basin locations (contributing areas) and therefore, cannot be considered to be distributed in areas with continuity. A confirmation comes from research carried out by Betson (1964), with regard to an analysis of the hydrological response of a great number of American basins. In particular, using a nonlinear mathematical model which incorporated Horton's (1940) infiltration capacity function, Betson showed that the percentage of runoff contributing area for 14 basins varied from 5%-36% with an average of 22%. The partial area concept introduced by Betson was successively studied more deeply by numerous authors such as Ragan (1968), Engman and Rogowski (1974), Dunne and Black (1970), Dunne et al. (1975), Beven and Kirkby (1979), O'Loughlin (1981), Beven (1986), etc.

These studies contributed to an explanation of the spatial-time evolution of the partial area runoff with regard to the two foregoing mechanisms. In one case (infiltration excess mechanism) the contraction and/or expansion of the partial area depends on the spatial variation of the rainfall, on the soil features and on the initial soil moisture content. In the other case (saturation excess mechanism) the surface topography (convergent or divergent hillslopes) and the position of the water table take on a dominant role. Many models at a basin scale, hillslope scale and local scale, were developed with the aim of simulating runoff by means of the two mechanisms, taking into account spatial variation of the hillslopes topography, of the soil and of the rainfall. Many of them are based on the spatial distribution of the topographic index (Beven and Kirkby, 1979) to predict the variation of the contributing areas. The topographic index, $\ln(a/\tan\beta)$, where a represents the upslope drainage area per unit contour length and $\tan\beta$ is local surface slope, results in agreement with the real distribution of the saturated surfaces in the basin. Beven (1986) moreover modified this index with the aim of considering, in addition to the spatial variations of the topography, those derived from the transmissivity of the subsurface zone, T. Therefore, when the scales of the spatial variabilities of topography and transmissivity are comparable, it is correct to combine the two effects into a single topographic-soil index, $\ln(a/T\tan\beta)$.

The use of GIS for the local definition of these parameters is undoubtedly useful. The advantages deriving from it relate, not only to typical map algebra operations carried out on matrices which describe the spatial distribution of hydrological quantities (for example the natural logarithm of the ratio between the matrix of the upslope drainage area values and the matrix of the $\tan\beta$ values), but above all to the improvement in the estimate of the runoff contributing area by means of FRho8 approach of the topographic index. The latter, as is well-known, can be expressed in the more general form by means of the following index:

$$\ln\left[\frac{A}{L\tan\beta}\right] \tag{8.3}$$

in which, in the case of D8 or Rho8 approach, A represents the whole upslope drainage area of the element under consideration, $\tan\beta$ indicates the local topographic gradient in the direction of maximum slope and finally, L represents the grid step (Fig. 8.14). In the case of a FD8 approach, referring again to Fig. 8.14, one has:

$$\tan(\beta) = \sum_{j=1}^{n}(\tan(\beta_j)L_j)/\sum_{j=1}^{n}L_j \tag{8.4}$$

$$L = \sum_{j=1}^{n}L_j \tag{8.5}$$

in which the subscript j represents the number of downslope directions. The results obtained can vary according to the approach adopted, and to each of them different hydrological forecast can be related. These differences, highlighted in Figs. 8.15 and 8.16, can be quantified by means of the cumulative distributions of the topographic index (Fig. 8.17).

The distributions of $\ln(a/\tan\beta)$ can be directly considered in terms of storage patterns, since they give the corresponding amount of area which directly contribute to runoff to an assigned value of the topographic index. Therefore, for a given

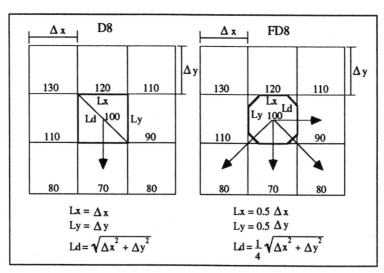

Figure 8.14. Flow partitioning using the single and the multiple flow direction approach

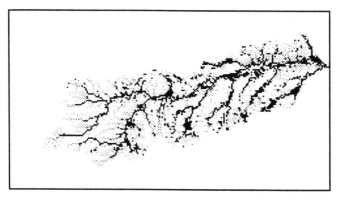

Figure 8.15. Spatial distribution of topographic index over the study area using a Rho8 approach

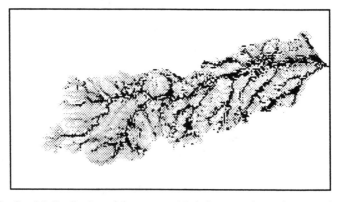

Figure 8.16. Spatial distribution of the topographic index over the study area using a FRho8 approach

opographic index, a greater quantity of contributing areas can be seen as an increase in available runoff at the basin outlet.

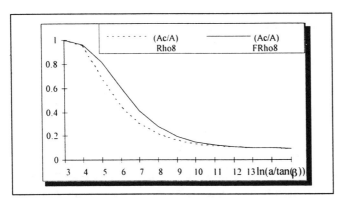

Figure 8.17. Cumulative distributions of topographic index for Rho8 and FRho8 approaches

8.4. Infiltration Excess Models

In this section some models are analyzed which enable an evaluation of surface runoff by means of a hortonian-type mechanism, or the excess runoff due to exceeding of the infiltration soil capacity. One of the best-known and most-used rainfall-runoff models is the Curve Number method, proposed by the Soil Conservation Service (1968). The development of this model is a result of its simplicity, which derives, on the one hand, from the use of a unique CN parameter and, on the other, from the possibility of obtaining that parameter from simple descriptive indications of the vegetation cover and soil characteristics of the basin. The SCS-CN method assumes the following equation :

$$S'/S = Q/P \tag{8.6}$$

in which Q (mm) represents the cumulative runoff volume up to the instant t, per unit area, P (mm) is the cumulative rainfall volume up to the same instant, per unit area, S' (mm) is the entirely lost volume, per unit area, and finally S (mm) is the soil potential maximum retention volume per unit area. Supposing that the entire lost volume, S', at instant t, is equal to the infiltrated volume per unit area, F (mm), up to the same instant, and that in place of the whole rainfall volume up to instant t, or P, the expression $(P - I_a)$ is introduced, with I_a (mm) equal to the initial abstraction per unit area (interception by vegetation, accumulation in surface depressions and infiltration before the runoff) the equation can be rewritten as :

$$Q = \frac{F}{S}(P - I_a) \tag{8.7}$$

From the continuity equation we have:

$$F = P - I_a - Q \tag{8.8}$$

and thus:

$$\frac{P - I_a - Q}{S} = \frac{Q}{P - I_a} \tag{8.9}$$

from which the well-known equation :

$$Q = \frac{(P - I_a)^2}{P - I_a + S} \tag{8.10}$$

The evaluation of the soil potential maximum retention, S, is made with the use of the equation:

$$S = 10(100/CN - 1) * S_0 \tag{8.11}$$

with S_0 equal to 25.4, which represents a conversion factor. By studying results from many small experimental watersheds an empirical relation was developed :

$$I_a = \alpha S \tag{8.12}$$

The data provided by the SCS gives a value for α equal to 0.2, which, in soil conditions very similar to Italian ones, systematically leads to an underestimate of the runoff caused by an excessive evaluation of the initial abstraction. For which reason it was considered suitable to assume α values noticeably lower than 0.2, for the rainfall event under consideration (previous analyses conducted at a basin scale showed an α value equal to 0.003). The SCS-CN procedure is, therefore, typified by a single CN parameter (between 0 and 100) which increases with decreasing permeability. In particular, CN values equal to 100, correspond to completely impermeable surfaces, typified by a zero infiltration. CN values equal to 0 are, instead, relative to surfaces characterized by high permeability values. The information contained in the GIS data base related to soil nature and use, can be easily elaborated with the aim of obtaining the CN parameters at a local scale according to associative criteria observed in the foregoing section 8.2.2. The matrix of CN values can be elaborated inside GIS through simple map algebra operations. In this case, if the equations (8.11), (8.12) and (8.10) are utilized, estimates are obtained at a local scale, respectively of the soil potential maximum retention volume, of initial abstraction volume and finally, of surface runoff volume.

The SCS-CN method, even if greatly used for its simplicity, presents some shortcomings in its own basic theories. Having been chosen a priori without any physical justification, it leads to some incongruence in the relationship between infiltration capacity and rainfall intensity. From the equations (8.7) and (8.8) it is possible to obtain :

$$F/S = (P - I_a - F)/(P - I_a) \tag{8.13}$$

or:

$$\frac{F}{S} + \frac{F}{(P - I_a)} = \frac{P - I_a}{P - I_a} = 1 \tag{8.14}$$

from which :

$$F = \frac{S(P - I_a)}{(P - I_a + S)} \tag{8.15}$$

Differentiating, with I_a and S time independent, we obtain :

$$\frac{dF}{dt} = \frac{S^2 \frac{dP}{dt}}{(P - I_a + S)^2} = f(t) \tag{8.16}$$

in which dP/dt is rainfall intensity. The presence of rainfall intensity at the nu-
merator indicates that, as rainfall intensity increases, the infiltration rate tends to
increase. This property of the SCS-CN method is in disagreement both with multiple field and laboratory experiences, as well as with the theory which describes
the physical infiltration phenomenon. For saturated surface conditions, infiltration
rate is controlled by a monotonically decreasing infiltration capacity curve, independent of rainfall intensity, i. This circumstance is found in the SCS-CN method
only for a rainfall event typified by a uniform rainfall intensity. Obviously, if the
estimated infiltration rate with equation (8.16) does not satisfy the real physical
behavior of the phenomenon, the surface runoff rate which ensues, will also show
an unrealistic behavior. In fact, if one notes the following equation:

$$i - f = \frac{dP}{dt} - \frac{S^2 \frac{dP}{dt}}{(P - I_a + S)^2} = \frac{(P - I_a)(P - I_a + 2S)\frac{dP}{dt}}{(P - I_a + S)^2} \tag{8.17}$$

one senses that, once the initial conditions of surface saturation are satisfied, the
model allows an estimate of surface runoff non-zero values, for each rainfall
intensity non-zero value. This means that the ponding condition is maintained
throughout the event, something which generally is not found in real circumstances. Other uncertainties arise, finally, from the use of equation (8.12), since
it is well-known that the initial abstractions are not only a function of S but also
of rainfall intensity.

The use of physically based equations for the estimate of infiltration capacity seems indispensible for a better simulation of the surface runoff phenomenon.
There are different models which satisfy this requirement. They can vary from approximate solutions of the Richards' equation (1931), such as the model proposed
by Philip (1957), to exact analytical solutions of approximate physical models,
such as the model proposed by Green-Ampt (1911). This latter model is based on a
simplified schematization of the vertical profile of soil water content and on some
initial hypotheses which deal with, on the one hand, a soil surface covered by a
thin layer of water of dimensions h_0 and, on the other, the wetting front at instant,
t, from the start of the phenomenon, horizontal and to depth, L_w. Moreover the
same front divides a completely saturated upper area $\theta = \theta_s$, from a lower area at
moisture content $\theta = \theta_i$. If a vertical column of soil from a horizontal unit area
is considered, a control volume, V_t, can be assumed coinciding with that made up
of the soil comprised between the surface and the layer at depth, L_w. By means of

such a schematization, the soil presents an initial moisture content, θ_i, for all of its depth which, proceeding from bottom to top, undergoes an abrupt increase in its value with regard to the wetting front, moving from an initial value, θ_i, to that of saturation, θ_s. For this reason, the water stored within the control volume, as result of infiltration, is equal to:

$$(\theta_s - \theta_i)V_t = (\theta_s - \theta_i)L_w \tag{8.18}$$

of which the volume of infiltrated water up to instant, t, per unit area is:

$$F(t) = L_w(\theta_s - \theta_i) = L_w\Delta\theta \tag{8.19}$$

The continuity equation (8.19) is integrated with Darcy's Law, expressed with the equation:

$$f \approx K\frac{\Psi + L_w}{L_w} \qquad \text{with} \qquad h_0 \approx 0 \tag{8.20}$$

showing:

$$f = K\frac{\Psi\Delta\theta + F}{F} \tag{8.21}$$

If equation (8.21) is integrated, according to the equation $f = dF/dt$, then the Green-Ampt equation is obtained:

$$F(t) - \Psi\Delta\theta \ln\left(1 + \frac{F(t)}{\Psi\Delta\theta}\right) = Kt \tag{8.22}$$

The non-linear equation (8.22), can be solved with the successive substitution method, or with the iterative method of Newton-Raphson, using $F = Kt$ as an initial value. As the cumulative infiltration, F, is known, the infiltration rate, f, can be obtained using equation (8.21):

$$f(t) = K\left(\frac{\Psi\Delta\theta}{F(t)} + 1\right) \tag{8.23}$$

The Green-Ampt model requires a continuously higher rainfall intensity, i, than the infiltration rate, f. Whenever the infiltration rate is less than infiltration capacity, the potential storage capacity of the soil is not fully used. Therefore, the infiltration rate is not reduced with the same modalities of that corresponding to a rainfall intensity higher or equal to the infiltration capacity. Consequently, the use of this infiltration capacity law in situations typified by $i < f$, systematically leads to an underestimate of the loss volume, with a consequent overestimate of surface runoff. This circumstance is almost always found at the beginning of precipitation, where rainfall intensity is less than infiltration capacity, and persists up to when the soil surface is entirely saturated. The time interval between the start of the rain and reaching surface saturation, or ponding time t_p, therefore takes on an extremely important role in the infiltration process. Mein and Larson (1973) propose a variation of the Green-Ampt model aimed at an estimate of the ponding time, t_p, for a constant rainfall intensity, starting instantaneously and continuing indefinitely. If cumulative infiltration, F, corresponding to the ponding time, t_p, is equal to $F_p = it_p$ while the infiltration rate, f, is equal to i, the equation (8.23) rewritten in terms of t_p, gives:

$$t_p = K \frac{\Psi \Delta \theta}{i(i - K)} \tag{8.24}$$

To obtain the effective value of the infiltration rate, f, after ponding time, t_p, it is necessary to create an infiltration capacity curve starting from the instant t_0 so that t_p, in relation to the new time origin t_0, is found after a time interval equal to $t_p - t_0$. This is necessary so that the cumulative infiltration, F, and the infiltration rate, f, at time t_p, was equal to the same observed quantities with regard to a precipitation which begins at time $t = 0$. Therefore, substituting $t = t_p - t_0$ and $F = F_p$ in equation (8.22), we obtain:

$$F_p - \Psi \Delta \theta \ln \left(1 + \frac{F_p}{\Psi \Delta \theta} \right) = K(t_p - t_0) \tag{8.25}$$

with:

$$t_0 = t_p - \frac{1}{K} \left[F_p - \Psi \Delta \theta \ln \left(1 + \frac{F_p}{\Psi \Delta \theta} \right) \right] \tag{8.26}$$

which for $t > t_p$, gives:

$$F(t) - \Psi \Delta \theta \ln \left(1 + \frac{F(t)}{\Psi \Delta \theta} \right) = K(t - t_0) \tag{8.27}$$

subtracting equation (8.25) from equation (8.27) we obtain:

$$F - F_p - \Psi \Delta \theta \ln \left[\frac{\Psi \Delta \theta + F}{\Psi \Delta \theta + F_p} \right] = K(t - t_p) \tag{8.28}$$

which allows the determination of cumulative infiltration after surface saturation, and consequently, the value of f by means of equation (8.23). The problem is more difficult in the case in which a variable rainfall intensity event is considered. In fact, in relation to each time interval, Δt, three cases can be found:

  1) Ponding saturation occurs throughout interval Δt;
  2) Ponding absent during interval Δt;
  3) Ponding occurs during interval Δt.

A detailed description of the procedural steps for solving this problem is shown in Chow et al. (1988) and useful information regarding the corresponding computer code can be found in Mendicino (1993). These models however, require input parameter regarding hydraulic characteristics of the soil which, for most cases, is not available at a local scale. This, therefore, limits their use in the field of distributed models. An analysis of the Green-Ampt model underlines that the number of parameters, actually necessary for the characterization of the infiltration phenomenon, is equal to two. In the foregoing equations, in fact, the parameters θ_s, θ_i and Ψ are always expressed by the product $\Psi \Delta \theta$. This latter, defined as the storage suction factor, S_f, together with the hydraulic conductivity at natural saturation, K_s, represent the necessary parameters for estimate of surface runoff. Investigations carried out by Morel-Seytoux and Verdin (1982), into the relationships $CN - (K_s, S_f)$ enable a good approximation of the spatial distribution of some hydraulic parameters of the soil, once information related to its nature and use are known. These relationships, even if arguable for their empirical content, allow the local values of parameters K_s and S_f, for the whole basin to be fixed, whose average values, in many cases, can be compared with those estimated at a basin scale. Finally these spatial distribution, if compared to the theoretical ones used in distributed models, show a greater reliability owing to a correct representation of reality.

8.5. Saturation Excess Models

One of the most used models for the estimate of saturation excess is represented by TOPMODEL, proposed by Beven and Kirkby (1979). Different versions of this model were achieved for a distributed runoff estimate; some of them even capable of evaluating excess runoff, owing to the exceeding of the infiltration capacity (Beven, 1986; Sivapalan et al., 1987).

The model in its general outline offers some conceptual simplifications with regard to the impossibility of exactly defining the initial conditions, the initial water table profile and the soil moisture profile in the unsaturated zone above the water table (Fig. 8.18).

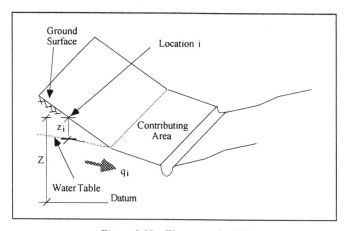

Figure 8.18. Elements of a hillslope

In this model some assumptions are made regarding:

- the recession discharge prior to the start of rain, $Q(0)$, derives from a steady rate of recharge, R, of the water table;
- the saturated hydraulic conductivity at a given point, i, of a hillslope decreases according to an exponential law such as:

$$K_{si}(z_i) = K_{0i}e^{-fz_i} \qquad (8.29)$$

in which z_i represents the depth of the soil profile, K_0 is the saturated conductivity at the soil surface and f is a parameter;
- the direction of the local hydraulic gradient is parallel to the local slope, consequently the downslope flow below the water table at a depth z_i, is represented for whatever point, i, by the following equation:

$$q_i = T(z_i) \tan \beta \qquad (8.30)$$

in which $T(z_i)$ represents a transmissivity that varies nonlinearly with depth to the water table. Hypothesizing negligible downslope flow through the capillary fringe, and integrating equation (8.29), we have:

$$q_i = \frac{K_{0i}}{f} \tan \beta e^{-fz_i} \qquad (8.31)$$

According to Beven (1986) can be assumed:

$$T_{0i} \approx \frac{K_{0i}}{f} \qquad (8.32)$$

and thus:

$$q_i = T_{0i} \tan \beta e^{-fz_i} \qquad (8.33)$$

in which T_{0i} represents the local value of the saturated transmissivity of the soil. Under quasi-steady state conditions, due to an assumed spatially uniform recharge rate, R, equation (8.33) can be locally rewritten:

$$aR = T_{0i} \tan \beta e^{-fz_i} \qquad (8.34)$$

in which a represents the upslope drainage area per unit contour length at point i. If equation (8.34) is rewritten in terms of z_i we obtain:

$$z_i = -\frac{1}{f} \ln \left(\frac{aR}{T_{0i} \tan \beta} \right) \qquad (8.35)$$

Multiplying the two terms of equation (8.35) by the moisture content deficit below the saturation value, $\Delta\theta$, and introducing the parameter $m = \Delta\theta/f$, the local initial soil moisture storage deficit, S_i, is obtained:

$$S_i = -m \ln \left(\frac{aR}{T_{0i} \tan \beta} \right) \qquad (8.36)$$

which integrated over the whole basin of surface, A, gives the areal average storage deficit:

$$\bar{S} = \frac{1}{A} \int_A -m \ln \left(\frac{aR}{T_{0i} \tan \beta} \right) dA \tag{8.37}$$

Since the value of R is constant over the area A, equation (8.37) can be rewritten as:

$$\bar{S} = m \left\{ \frac{1}{A} \int_A - \ln \left(\frac{aR}{T_{0i} \tan \beta} \right) dA - \ln R \right\} \tag{8.38}$$

in which, rearranging R according to the equation (8.36), as:

$$- \ln R = \frac{S_i}{m} + \ln \left[\frac{a}{T_{0i} \tan \beta} \right] \tag{8.39}$$

we obtain:

$$\frac{S_i - \bar{S}}{m} = [\ln(T_{0i} - \ln(\bar{T}_0)] - \left[\ln \left(\frac{a}{\tan \beta} \right) - \lambda \right] \tag{8.40}$$

with:

$$\lambda = \frac{1}{A} \int_A \ln \left(\frac{a}{\tan \beta} \right) dA \tag{8.41}$$

$$\ln(\bar{T}_0) = \frac{1}{A} \int_A \ln(T_{0i}) dA \tag{8.42}$$

The equation (8.40), if rewritten in terms of S_i, becomes:

$$S_i = \bar{S} + m\lambda - m \ln \left(\frac{a\bar{T}_0}{T_{0i} \tan \beta} \right) \tag{8.43}$$

in which the argument of the natural logarithm represents a combined form of the soil-topographic index. The equation (8.43), once noted parameters $\bar{S}$ and m allows an estimate of local values, S_i, of the initial soil moisture storage deficit. In particular, it enables the definition of the area which directly contributes to runoff, at start of rainfall, or all those areas typified by values of $S_i < 0$. The rainfall in these areas becomes saturation excess runoff.

After the beginning of rain all the areas typified by values of $S_i > 0$ are capable of storing water in the soil, by means of the infiltration phenomenon, until their soil moisture storage deficit is not filled. Obviously, those areas typified by values of $S_i > 0$ will contribute to that part of the runoff, defined excess infiltration runoff, due to an exceeding of the soil infiltration capacity.

Since the effect of redistribution of infiltrated water downslope during the rainfall event on the pattern of the soil moisture deficits is not taken into account, the deficit at any point is only reduced by local infiltration at that point.

It is useful to underline that the infiltration phenomenon can be sufficiently schematized by the Green-Ampt method described in the previous section. Finally, those areas which become saturated during rainfall contribute to an expansion of the runoff contributing area.

In the case where no information is available with regard to the values assumed from the soil transmissivity, reference can be made to the conceptual simplification proposed by Beven (1986), $T_{0i} \approx K_{0i}/f$, for which the local saturated hydraulic conductivity values (estimated by means of the procedure inside GIS, previously examined), in the identification process of the runoff contributing areas can be used.

8.6. Comparison Between the Models Observed

The models examined in the previous sections were applied to the Turbolo Creek basin with the aim of analyzing both the performances in terms of hydrological responses as well as the descriptive capability of the simulated phenomenon. In Fig. 8.19 the time variation of the rainfall event ($\Delta t = 20$ min) uniformly distributed over the basin, considered for the simulations, is represented. For this rainfall event, the observed surface runoff volume at the basin outlet was 177,977 m^3.

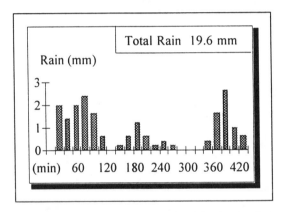

Figure 8.19. Rainfall event considered for the simulations

In Fig. 8.20 the good performances provided in the estimate of runoff volume by the SCS-CN method are shown. In fact, by means of the analysis carried out at a local scale with the aid of GIS, a percentage error of 3.57% is obtained with respect to the observed volume (Fig. 8.21).

This latter result is even more satisfactory when the estimate is carried out at a basin scale and without the aid of GIS. In this case, in fact, the corresponding percentage error was found to be equal to 24.91%. GIS, by means of the tools within it, on the one hand, allows those areas which most produce runoff to be selected and, on the other, their hydrological behavior during the rainfall event, to be studied. These areas are represented in Fig. 8.22, in which are emphasized the shortcomings typifying those models describing runoff production by means of a hortonian-type mechanism. In fact, these latter, because of their close dependence

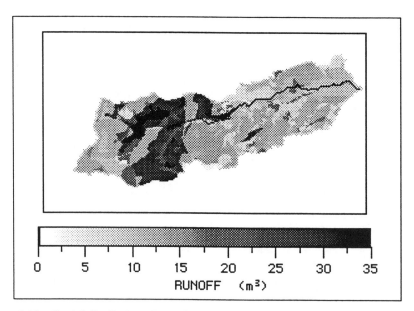

Figure 8.20. Spatial distribution of runoff volume computed with the SCS-CN method over the study area

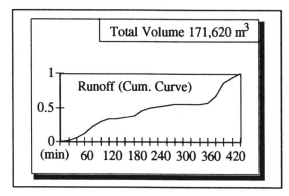

Figure 8.21. Cumulative runoff predicted by means of the SCS-CN method for the considered rainfall event

on spatial rainfall variation, on soil characteristics and on its initial moisture, are not capable of evaluating the effect of the topography and of the water table on the hydrological response of the basin. The models which describe runoff production by means of a hortonian-type mechanism, even if a distributed-type model is used, do not enable the highlighting of areas which directly contribute to runoff. Since they exclusively evaluate excess direct runoff, due to the exceeding of infiltration capacity they should be integrated with models which also take into account runoff production consequent to the exceeding of soil retention capacity. In this case, an automatic procedure was created inside GIS, capable of estimating both infiltration excess (Green-Ampt) and saturation excess runoff (TOPMODEL). The basin sur-

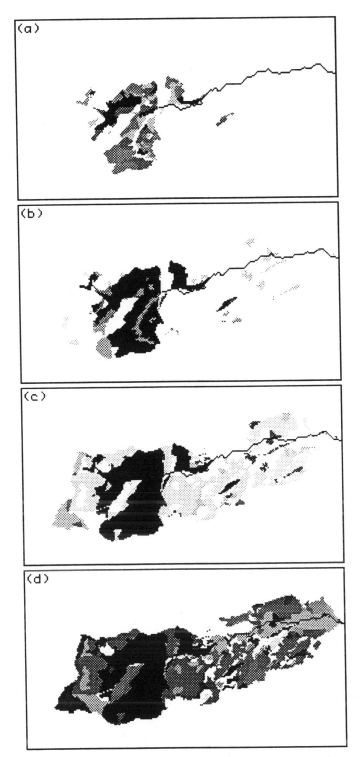

Figure 8.22. Runoff production at the beginning (a), during (b)-(c) and at the end (d) of the considered rainfall event

face (georeferenced to the hydrological model) which contributes directly to surface runoff for each time interval in a given rainfall event, was also created. This procedure, after an initial calibration phase, in which the parameters $\bar{S}$ and m were optimized comparing the calculated surface runoff with those recorded at the basin outlet (Fig. 8.23), allows a local estimate of the initial soil moisture storage deficit values, S_i, (Fig. 8.24). From these initial values for each successive rainfall interval, GIS is capable of showing infiltration excess runoff, saturation excess runoff, total runoff produced, and finally, the direct contributing areas to runoff. For the rainfall event considered, Figs. 8.25, 8.26, 8.27, show the results produced by GIS for some time instants (t = 0, t = 240 and t = 440 min).

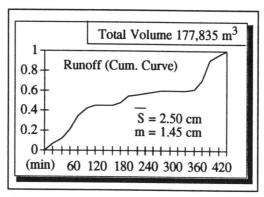

Figure 8.23. Cumulative infiltration and saturation excess runoff produced during the considered rainfall event

Finally, some consideration about the local estimate of the topographic index, $\ln(a/\tan\beta)$, should be given. The results were obtained using a FRho8 approach, which enabled an evaluation of the total runoff consequent to the considered rainfall event with a percentage error of 0.1% and a contributing area of 25% of the basin. Instead, using a Rho8 approach for the estimate of the topographic index, for the same runoff production, a value of contributing area equal to 24% was found.

8.7. Spatial Variability of the Parameters

Particular attention has been given in the last decade to problems related to the heterogeneity of the soil and to the effects produced on the infiltration phenomenon and to runoff. The heterogeneity and anisotropy of the parameters which typify the rainfall-runoff transformation imply that distributed analyses should be directed at the study of the relationships between the locally assumed values of the parameters when the distance between the different localities varies. Through an analysis of the degree of spatial persistence of the parameters it is intended to establish whether the theories of spatial independence assumed by distributed-type models appear to be more or less arbitrary. In the latter case, instead of a hydrological model based on a local scale analysis, an evaluation carried out on a sub-

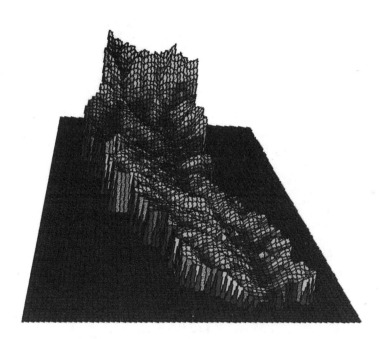

Figure 8.24. Spatial distribution of the initial soil moisture storage deficit superimposed on an iso-metric projection of the 50 m DTM of the Turbolo Creek basin

basin scale, taken as a representative elementary area, might prove more appropriate. In any case a change of scale is needed since a noticeable reduction in input data can be obtained as a result of the aggregation of cells, with a consequent time-saving in their processing. In consequence of the aggregation some conceptual forcing regarding the hypothesized homogeneous single cell can be eliminated (without taking the discretization step into account). The REA concept introduced by Sivapalan et al. (1987), Beven et al. (1988) and Wood et al. (1988) even if interesting in the identification of the representation scale of the hydrological process (typified by a minimum variance), does not allow the discrimination of some nonstationary spatial phenomena associated with long or complex scale correlations. Further studies carried out by Mancini and Rosso (1989) and Colosimo and Mendicino (1991, 1992), showed how spatial relationships between local CN parameter values can be analyzed by means of spatial autocorrelation. A fundamental contribution to the spatial distributed data analysis is that of Matheron (1971), who introduced the theory of regionalized variables to estimate area averages considered as realization of stochastic processes. This theory concerns the development of the technique better-known as "Kriging" which is a modified optimization technique

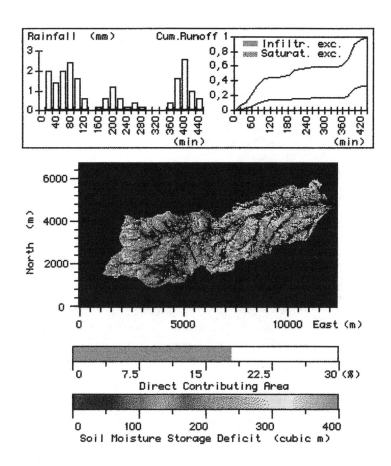

Figure 8.25. GIS results produced at the beginning of the considered rainfall event ($t = 0$)

of Russian origin, based on spatial correlation functions. In particular, Kriging is typified by the semivariogram, which represents the semivariance variation along a specific direction. As is well-known, the semivariance is a measure of the degree of spatial dependence between samples along a specific support. This dependence, for the most part, shows marked non-stationary characteristics, which can be analyzed by use of the Kriging techniques: Universal Kriging; Kriging with generalized co-variance; a modified Kriging to take into account an a priori estimate of the drift. (Delfiner and Delhomme, 1975; Delhomme, 1978; Gambolati and Volpi, 1979a, b; Gambolati and Galeati, 1990). The possibility of integrating the analysis tools contained in GIS with some geostatistical procedures aimed at the identification of the semivariogram related to a given direction, is of extreme importance for the direct identification of the representative dimensions of homogeneous areas. In the present case, by using a computer code making part of the GIS and which directly

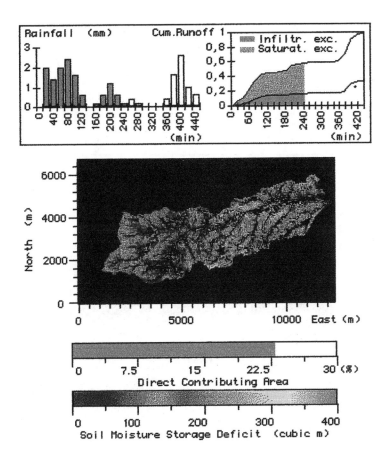

Figure 8.26. GIS results produced during the considered rainfall event ($t = 240$ min)

uses information contained in its data bank, N-S (latitude), W-E (longitude), NW-SE (diagonal 1) and NE-SW (diagonal 2) semivariograms were estimated for the Turbolo Creek basin, relating to the CN parameter, to the hydraulic conductivity, K_s, and to the topographic index, $\ln(a/\tan\beta)$. For example, a stationary behavior of regionalized variables, x, was supposed, which allowed the expression of the semivariogram, γ_d, through the following equation:

$$2\gamma_d(r) = \frac{1}{N_d(r)} \sum_i [z(x_i + r) - z(x_i)]^2 \qquad (8.44)$$

in which d represents the specific direction towards which the semivariance is estimated, r defines the extent of the spatial interval under consideration and, lastly, with $N_d(r)$ is indicated the number of pairs of points spatially separated by a distance equal to r. In the case of the spatial distribution of the CN parameter the

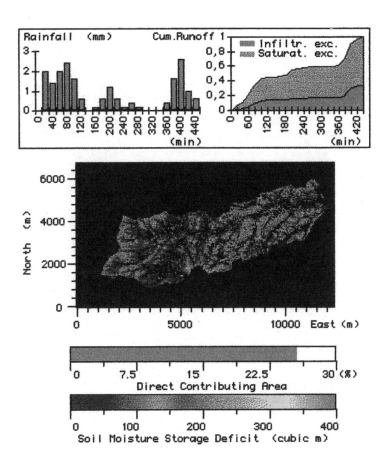

Figure 8.27. GIS results produced at the end of the considered rainfall event ($t = 440$ min)

estimated experimental semivariograms in the four directions showed an isotropic persistence for distances less than 350 m, which corresponds to a relevant area of about 0.4 Km^2 (Fig. 8.28).

Analogous results were carried out on the same basin by Colosimo and Mendicino (1991, 1992) using the autocorrelation function in the direction of W-E longitude, in the direction of N-S latitude and along the diagonals of the sample grid. This representative dimension was moreover, confirmed by analysis of the experimental semivariogram constructed by means of the values assumed by the hydraulic conductivity (Fig. 8.29). The representative dimension of the elementary area, obtained by the semivariogram analysis, was provided as input data for other procedures contained inside GIS with the aim of automatically extracting the set of sub-basins, which constitutes the Turbolo Creek basin, whose average areas are comparable to that of the representative elementary area. As is well-known, there are many procedures for the automatic extraction, starting from a DTM, of the corresponding river network and the sub-basins (Mendicino, 1990; Martz and Garbrecht, 1992). This procedures, moreover, through the introduction of threshold parameters defined in the input phase allow different generalization levels of the river network and the corresponding sub-basins to be obtained. For the estimated representative area (0.4 Km^2), the generalization level obtained is that shown in Fig. 8.30.

The rainfall-runoff transformation model used in the preceding section at a local scale can be applied to all the homogeneous areas made up of elementary sub-basins. The only differences to be considered regard the data format, no longer of matrix-type, but rather according to the lists which synthesize the parameter values relating to each elementary sub-basin. The parameter values to be introduced into these lists can be obtained directly using some functions contained in GIS which allow, for given homogeneous areas contained in the "sub-basin layer", the extraction of average values on the same areas of the parameters contained in other layers $(CN, K_s, a, \tan\beta, \text{etc})$. The analysis of spatial variability of transformation model parameters, using the procedure inside GIS, should not be understood only as aiming to identify the representation scale of the hydrological process. It also allows a deeper understanding of the hydrological behavior assumed by single parameters within the "basin" system. In this respect an examination of Figs. 8.28 and 8.29 shows a marked tendency on the part of the CN and Ks parameters to assume a homogeneous behavior in the N-S and NE-SW directions. The discrepancies shown by the semivariograms are confirmed by observing Figs. 8.10 and 8.12, which show substantial variations in the values assumed by the parameters of the same directions. Semivariograms offer the capability of discerning the main directions toward which a specific phenomenon evolves, therefore, appears to be of notable interest for the choice of differing procedures to use in rainfall-runoff models. The case of topographic index clearly supposes an anisotropic behavior due to a predominant runoff concentration in the W-E direction (Figs. 8.15 and 8.16). This is further confirmed by analysis of the semivariograms along the 4 directions N-S, W-E, NW-SE and NE-SW (Fig. 8.31).

The difference between the values assumed by the semivariograms of the topographic index in the N-S and W-E directions can be interpreted as the major or minor tendency to concentrate runoff along a single main direction. If these differences are analysed by the same estimated index, however, with two different approaches, either FRho8 and Rho8, the results which follow are noticeably different (Fig. 8.32).

The multiple FRho8 approach demonstrating smaller values with respect to that of the single Rho8 approach in fact, underlines a lesser tendency to concentrate runoff along a single main direction (in the specific W-E case) and a consequently, more realistic behavior of the simulated phenomenon. One can arrive at this result both by a study of the runoff contributing areas and by spatial analysis of the parameters, assumed to be regionalized variables.

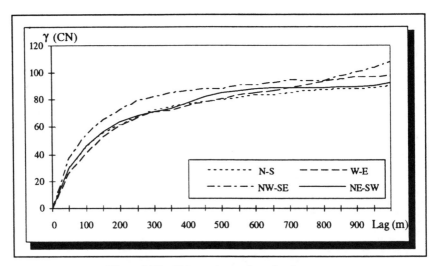

Figure 8.28. CN experimental semivariograms

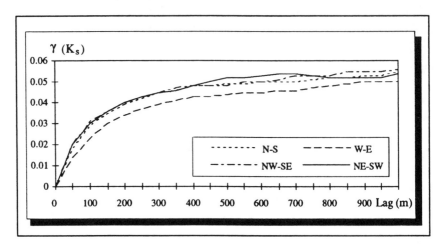

Figure 8.29. K_s experimental semivariograms

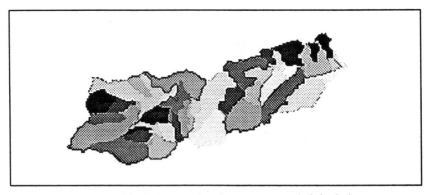

Figure 8.30. Representative elementary areas (sub-basins)

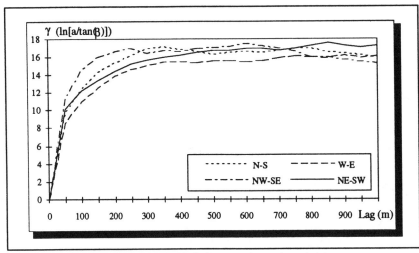

Figure 8.31. Topographic index experimental semivariograms

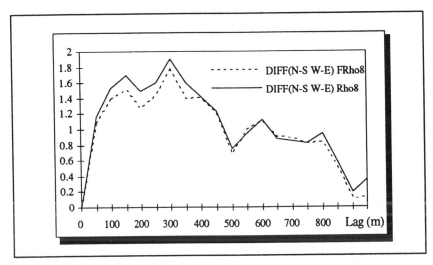

Figure 8.32. Difference between the values assumed by the semivariograms of the topographic index in the N-S and W-E directions

8.8. Conclusion

The use of GIS in hydrological practice assumes an important role, both for purely computational aspects and for a deeper understanding of the rainfall-runoff transformation phenomenon.

In computational terms, it can be observed how the aid of GIS allows the availability at a local scale of a remarkable amount of information with a high degree of accuracy and, for their elaboration in a greatly reduced timespan with appropriate changes of scale to perform the hydrological process. These scale changes can be achieved using several geostatistical procedures contained within GIS, capable of analyzing spatial parameter variations and corresponding persistence levels.

Some constraints which arise from the use of GIS, such as the management of the time component of the hydrological quantities, can be largely overcome by integrating with hydrological models. This integration enables the overcoming of the inherent limitations of hydrological models, regarding not only the capabilities offered by the models in characterizing hydrological processes, or the solution to corresponding equations, but above all the capability of accurately specifying the constituent parameters values of the models.

Moreover, further benefits can be derived from the possibility of allowing different transformation models to coexist within one GIS, which with any variation of the climatic and territorial conditions of the site to be analyzed, can use different infiltration excess mechanisms and/or saturation excess mechanisms, individually or combined. In the case of saturated excess models, it can be observed that a multiple FRho8 approach in the estimate of the topographic index should be preferred to the single Rho8 approach for a better evaluation of the areas which directly contribute to runoff.

Finally, some reflections should be made regarding the boundary which separates GIS from the rainfall-runoff transformation model. In the course of writing up it was observed that the GIS structure should not be understood as a single software package, but rather as an integrated and expandable system, made up of various procedures, which from time to time, are recalled and used for the solution of specific problems. This should orientate experts in the field toward developing different GIS packages, whose structure will vary according to the area of discipline and application.

References

Abbott, M.B., J.C. Bathurst, J.A. Cunge, P.E. O'Connell and J. Rasmussen, 1986, a. An Introduction to the European Hydrological System - Système Hydrologique Europeèn, SHE, 1: History and Philosophy of a Physically-Based, Distributed Modelling System. J. Hydrol., n. 87, pp.45-59.

Abbott, M.B., J.C. Bathurst, J.A. Cunge, P.E. O'Connell and J. Rasmussen, 1986, b. An Introduction to the European Hydrological System - Système Hydrologique Europeèn, SHE, 2: Structure of a Physically-Based, Distributed Modelling System. J. Hydrol., n. 87, pp.61-77.

Band, L.E., 1986. Topographic Partition of Watersheds with Digital Elevation Models. Water Resour. Res., Vol.22, n.1, pp.15-24.

Bathurst, J.C., 1986, a. Physically-Based Distributed Modelling of an Upland Catchment Using the Système Hydrologique Europeèn. J. Hydrol., n. 87, pp.79-102.

Bathurst, J.C., 1986, b. Sensitivity Analysis of the Système Hydrologique Europeèn for an Upland Catchment. J. Hydrol., n. 87, pp.103-123.

Betson, R.P., 1964. What is Watershed Runoff? J. Geophys. Res., Vol. 69, n.8, pp.1541- 1552.

Beven, K., 1986. Hillslope Runoff Processes and Flood Frequency Characteristics. In Hillslope Processes, edited by A.D. Abrahams, Allen and Unwin, Winchester, Mass..

Beven, K., 1989. Changing Ideas in Hydrology - The Case of Physically-Based Models. J. Hydrol., n. 105, pp.157-172.

Beven, K.J. and M.J. Kirkby, 1979. A Physically Based Variable Contributing Area Model of Basin Hydrology. Hydrol. Sci. Bull., Vol.24, n.1, pp.43-69.

Beven, K.J., E.F. Wood and M. Sivapalan, 1988. On Hydrological Heterogeneity - Catchment Morphology and Catchment Response. J. Hydrol., n. 100, pp.353-375.

Chow, V. T., D. R. Maidment and L. W. Mays, 1988. Applied Hydrology. McGraw Hill, New York.

Cline, T.J., Molinas, A. and P.Y. Julien, 1989. An Auto-CAD-Based Watershed Information System for the Hydrologic Model HEC-1. Water Resources Bulletin, Vol. 25, n. 3, pp.641-652.

Colosimo, C., and G. Mendicino, 1991. L'uso dei GIS per la Stima dei Parametri di Assorbimento di un Bacino Idrografico. Ingegneria Sanitaria Ambientale, n.6, pp.21-35.

Colosimo, C., and G. Mendicino, 1992. Application of GIS Techniques for Design Flood Hydrographs. International Conference on Interaction of Computational Methods and Measurements in Hydraulics and Hydrology, Hydrocomp'92, Budapest, Hungary, pp.401-409.

Colosimo, C., and G. Mendicino, 1993. L'Uso del Flood Hydrograph Package HEC-1 per il Calcolo delle Piogge Nette. Editoriale Bios, Cosenza, Italy.

Delfiner, P. and J. P. Delhomme, 1975. Optimum Interpolation by Kriging. In Display and Analysis of Spatial Data, J. C. David (Editor). NATO Advanced Study Institute, John Wiley and Sons, pp. 96-114.

Delhomme, J. P., 1978. Kriging in the Hydrosciences. Advances in Water Resources, vol. 1, n. 5, pp.251-266.

DeRoo, A.P.J., L. Hazelhoff, and P.A. Burrough, 1989. Soil Erosion Modeling Using 'ANSWERS' and Geographical Information Systems. Earth Surface Processes and Landforms 14, pp.517-532.

Dunne, T., 1978. Field Study of Hillslope Flow Processes. In Hillslope Hydrology, edited by M.J. Kirkby, pp. 227-293, John Wiley, New York.

Dunne, T., and R.D. Black, 1970. Partial Area Contributions to Storm Runoff in a Small New England Watershed. Water Resour. Res., Vol.6, n.5, pp.1296-1311.

Dunne, T., T.R. Moore and C.H. Taylor, 1975. Recognition and Prediction of Runoff-Producing Zones in Humid Regions. Hydrol. Sci. Bull., Vol.20, n.3, pp.305-327.

Engel, B.A., R. Srinivasan and C.C. Rewerts, 1991. A GIS Toolbox Approach to Hydrologic Modelling. In GRASS 1991 User's Conference, Berkely, California.

Engman, E.T., and A.S. Rogowski, 1974. A Partial Model for Storm Flow Synthesis. Water Resour. Res., Vol.10, n.3, pp.464-472.

Fairfield, J., and P. Leymarie, 1991. Drainage Networks from Grid Digital Elevation Models. Water Resour. Res., Vol.27, n.5, pp.709-717.

Fedra, K., 1993. GIS and Environmental Modeling. In M.F. Goodchild, B.O. Parks and L.T. Steyaert, eds, Environmental Modeling and GIS, Oxford University Press, Chapter 5.

Freeman, T.G., 1991. Calculating Catchment Area with Divergent Flow Based on a Regular Grid. Computers and Geosciences, Vol.17, n.3, pp.413-422.

Gambolati, G., and G. Volpi, 1979a. Groundwater Contour Mapping in Venice by Stochastic Interpolators. 1. Theory. Water Resources Research, Vol.15, n.2, pp.281-290.

Gambolati, G., and G. Volpi, 1979b. A Conceptual Deterministic Analysis of the Kriging Technique in Hydrology. Water Resources Research, Vol.15, n.3, pp.625-629.

Gambolati, G., and G. Galeati, 1990. Optimal Bathimetry of Tyrrhenian sea by Stochastic Interpolators. Water Resources Bulletin, Vol.26, n.4, pp.677-685.

Giorgini, A., 1990. Curve Number and Precipitation Losses. Technical Report CDHSE- 90-1, School of Civil Engineering, Pardue University, West Lafayette, IN 47907.

Green, W.H., and G. Ampt, 1911. Studies of Soil Physics. Part I-The Flow of Air and Water Through Soils. J. Agric. Sci. 4, pp.1-24.

Horton, R.E., 1933. The Role of Infiltration in the Hydrologic Cycle. Eos Trans. AGU, n. 14, pp.446-460.

Horton, R. E., 1940. An Approach Toward a Physical Interpretation of Infiltration Capacity. Soil Sci. Soc. Am. J., n. 5, pp. 399-417.

Hydrologic Engineers Center, 1990. HEC-1 Flood hydrograph package. User's manual, U.S. Corps of Engineers, Hydrologic Engineers Center, Davis, Calif..

Hodge, W.H., M. Larson and W. Goran, 1988. Linking the ARMSED Watershed Process Model with GRASS Geographic Information System. In Proceedings of the 1988 International Symposium on Modeling in Agricultural, Forest, and Rangeland Hydrology, Chicago, IL, ASAE Publication 07-88, ASAE, St. Joseph, MI., pp. 501-510.

Jenson, S.K., and J.Q. Domingue, 1988. Extracting Topographic Structures from Digital Elevation Data for Geographic Information Systems Analysis. Photogrammetric Engineering and Remote Sensing, Vol.54, n.11, pp.1593-1600.

Maidment, D.R., 1993. GIS in Hydrologic Modeling. In M.F. Goodchild, B.O. Parks and L.T. Steyaert, eds, Environmental Modeling and GIS, Oxford University Press, Chapter 14.

Mancini, M., and R. Rosso, 1989. Using GIS to Assess Spatial Variability of SCS Curve Number at the Basin Scale. In New Directions for Surface Water Modeling, Proceedings of the Baltimore Symposiumm, May 1989, IAHS Publ.no.181, pp.435-444.

Martz, L.W. and E. De Jong, 1988. CATCH: A Fortran Program for Measuring Catchment Area from Digital Elevation Models. Computers and Geosciences, Vol.14, n.5, pp.627-640.

Martz, L.W. and J. Garbrecht, 1992. Numerical Definition of Drainage Network and Subcatchment Areas from Digital Elevation Models. Computers and Geosciences, Vol.18, n.6, pp.747-761.

Matheron, G., 1971. The Theory of Regionalized Variables and Its Applications. Cahiers du Centre de Morphologie Mathematique, Ecole des Mines, Fountainbleau, France, pp.211.

Mein, R. G., and C.L. Larson, 1973. Modeling Infiltration During a Steady Rain. Water Resour. Res., Vol.9, n.2, pp. 384-394.

Mendicino, G., 1989. Generazione Automatica di Reticoli Idrografici. Università della Calabria, Memorie e Studi n. 193, Italy.

Mendicino, G., 1990. Banche Dati Georeferenziate per la Gestione delle Risorse Idriche di un Bacino Idrografico. Proc. of XXII Convegno di Idraulica e Costruzioni Idrauliche, Cosenza (Italy), Vol.4, pp.107-120.

Mendicino, G., 1993. Idrologia delle Perdite. Circolazione dell' Acqua nel Suolo: Teoria e Calcolo. Patron Editore, Bologna, Italy.

Moore, I.D., A.K. Turner, J.P. Wilson, S.K. Jenson and L.E. Band, 1993. GIS and Land Surface-Subsurface Modeling. In M.F. Goodchild, B.O. Parks and L.T. Steyaert, eds, Environmental Modeling and GIS, Oxford University Press, Chapter 19.

Morel-Seytoux H.J., and Verdin J.P., 1982. Correspondence between the SCS CN and Infiltration Parameters. Advance in Irrigation and Drainage, pp.308-319.

O'Callaghan, J.F and D.M. Mark, 1984. The extraction of Drainage Networks from Digital Elevation Data. Computer Vision, Graphics and Image Processing Vol.28, pp.323-344.

O'Loughlin, E.M., 1981. Saturation Regions in Catchments and their Relations to Soil and Topographic Properties. J. Hydrol., n.53, pp.229-246.

Panuska, J.C., I.D. Moore and L.A. Kramer, 1991. Terrain Analysis: Integration into the Agricultural Nonpoint Source (AGNPS) Pollution Model. J. Soil and Water Cons., vol. 46, n. 1, pp.59-64.

Philip, J. R., 1957. The Theory of Infiltration, 1: The Infiltration Equation and its Solution. Soil Science, n.83, pp. 345-357.

Quinn, P., K. Beven, P. Chevallier and O. Planchon, 1991. The Prediction of Hillslope Flow Paths for Distributed Hydrological Modelling Using Digital Terrain Models. Hydrological Processes, Vol. 5, n. 1, pp.59-79.

Ragan, R.M., 1968. An Experimental Investigation of Partial Area Contributions. Int. Ass. Sci. Hydrol. Publ. 76, pp.241-249.

Richards, L. A., 1931. Capillary Conduction of Liquids Through Porous Mediums. Physics, vol. I, pp. 318-333.

Rogers, C.C.M., K.J. Beven, E.M. Morris and M.G. Anderson, 1985. Sensitivity Analysis, Calibration and Predictive Uncertainty of Istitute of Hydrology Distributed Model. J. Hydrol., n. 81, pp.179-187.

Sasowsky, K.C. and T.W. Gardner, 1991. Watershed Configuration and Geographic Information System Parameterization for SPUR Model Hydrologic Simulations. Water Resources Bulletin,Vol. 27, n. 1, pp.7-18.

Sivapalan, M., K. Beven and E.F. Wood, 1987. On Hydrologic Similarity 2. A Scaled Model of Storm Runoff Production. Water Resour. Res., Vol.23, n.12, pp.2266-2278.

Soil Conservation Service, 1968. Hydrology. Supplement A to Section 4, National Engineering Handbook. Washington, D.C.: U.S. Department of Agriculture.

Spriet, J.A. and G.C. Vansteenkiste, 1988. Modelli Matematici e Simulazione. Gruppo Editoriale Jackson, Milano, Italy.

Stuebe, M.M. and D.M. Johnston, 1990. Runoff Volume Estimation Using GIS Techniques. Water

Resources Bulletin, Vol. 26, n. 4, pp.611-620.

Vieux, B.E., V.R. Bralts and L.J. Segerlind, 1988. Finite Element Analysis of Hydrologic Response Areas Using Geographic Information Systems. In Proceedings of the 1988 International Symposium on Modeling in Agricultural, Forest, and Rangeland Hydrology, Chicago, IL, ASAE Publication 07-88, ASAE, St. Joseph, MI., pp. 437-446.

Vieux, B.E. and Y. Kang, 1990. GRASS Waterworks: A GIS Toolbox for Watershed Hydrologic Modelling. In Proceedings of Application of Geographic Information Systems, Simulation Models, and Knowledge-based Systems for Landuse Management, November 12-14, 1990 Virginia Polytechnic Institute and State University, Backsburg, Virginia.

Vieux, B.E., 1991. Geographic Information Systems and Non-Point Source Water Quality and Quantity Modelling. Hydrological Processes, Vol. 5, n. 1, pp.101-113.

Wolfe, M.L. and C.M.U. Neale, 1988. Input Data Development for a Distributed Parameter Hydrologic Model (FESHM). In Proceedings of the 1988 International Symposium on Modeling in Agricultural, Forest, and Rangeland Hydrology, Chicago, IL, ASAE Publication 07-88, ASAE, St. Joseph, MI., pp. 462-469.

Wood, E.F., M. Sivapalan, K. Beven and L. Band, 1988. Effects of Spatial Variability and Scale with Implications to Hydrologic Modeling. J. Hydrol., n. 102, pp.29-47.

CHAPTER 9

GIS for Large-Scale Watershed Modelling

G.W. Kite, E. Ellehoj and A. Dalton

Abstract.   A hydrological model is an attempt to describe the natural processes which transpose precipitation into runoff. There are scores of models available but a general trend is seen from lumped conceptual models towards distributed physically-based models. As models become more complex and describe more physical processes and as they are used for larger watersheds and for interaction with atmospheric general circulation models, the data processing needs increase and the use of geographic information systems becomes more and more necessary. This chapter briefly describes the changes taking place in hydrological models, discusses macroscale modelling and describes, as an example, how a geographic information system is being used with the SLURP macroscale hydrological model. An appraisal of the present situation is given, limitations are noted and recommendations are made for future developments of GIS.

9.1. Introduction

9.1.1. LARGE-SCALE HYDROLOGICAL MODELLING

A hydrological model, like any other model, is intended to be a realistic representation of a physical system and the detail of its design will depend on the purpose of the modelling. Figure 9.1 shows a simple classification of hydrological models in which the complexity increases from the bottom-left hand corner towards the top-right hand corner as modelling capability advances.

The simplest form of hydrological model is probably a regression-type relationship between precipitation or snowpack and seasonal runoff and, for many purposes such as supply forecasting, this type of model works well (Warkentin, 1990). More advanced stochastic models such as the constrained linear system (CLS)

V. P. Singh and M. Fiorentino (eds.), Geographical Information Systems in Hydrology, 237–268.
© 1996 Kluwer Academic Publishers. Printed in the Netherlands.

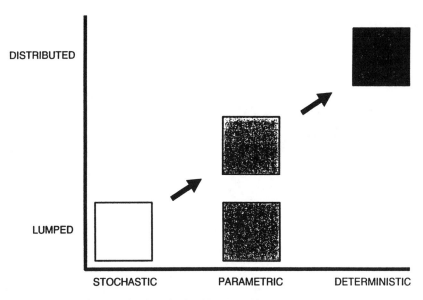

Figure 9.1. Simple classification of hydrological models

have also proved useful, having few parameters and being easy to calibrate (Natale and Todini, 1977). However, stochastic or 'black-box' models are only mathematical propositions and, having no physical basis, may only be applicable to the period and conditions of the original data.

The application of deterministic models, on the other hand, should not be restricted by space or time. In it's simplest form, a lumped parametric model simulates streamflow using basin average data and a single set of, often empirical, parameters. Such models are particularly applicable if the only modelling requirement is to simulate or forecast a streamflow hydrograph. For example, the early lumped Stanford Watershed Model (Crawford & Linsley, 1962) was applied successfully to many basins worldwide and spawned many modified and adapted versions. The lumped basin models were often developed into semi-distributed parametric models which divide the watershed into sub-elements based either on elevation, as for example in the SRM (Martinec at al., 1983), or on area, as for example in the PRMS model (Leavesley and Stannard, 1990). For accurate simulation of snowpack, soil moisture and other state variables at many points in a watershed it is necessary to use a physically-realistic deterministic model such as the Système Hydrologique Europeèn (SHE) (Abbott et al., 1986). Since the physical descriptions in such models include the lateral movements of surface and groundwater, physically-based models are almost inevitably also distributed models. Similarly, for scientific research into hydrological processes it is necessary to use a physically-based fully-distributed model within which alternate concepts may be tested. It should be noted, however, that there are many questions about the appropriateness of applying the physics of point processes to large areas. For example,

Morton (1990) has questioned the use of the Penman-Monteith evaporation equation in the SHE model.

As models are applied to larger watersheds, the lumped approach becomes unreasonable and the fully distributed models become impractical because of unrealistic data requirements. There are practical advantages to using a model somewhere in the middle ground between a lumped model and a full-blown physically-based distributed model. Such middle-ground models can incorporate the most important of the physical laws while retaining simplicity of operation. Such models are able to simulate the behaviour of a watershed at many points and in many variables but avoid the data demands and computation excesses of the fully-distributed models (Kite, 1993b). In practical applications, the users of both lumped and fully-distributed models may also tend towards this middle ground. For example, users of lumped models may simulate the use of a distributed model by successively applying the lumped model to sub-watersheds and, from the other end of the spectrum of models, Jain et al. (1992) describe how, in practice, the fully-distributed SHE model may use areally-averaged data and parameters.

This chapter describes, in more detail, the SLURP watershed model which is just such a middle-ground model, designed for use in macro-scale catchments and, eventually, in continental-scale applications. The ultimate aim of macro-scale hydrological modelling is to form a component of a global hydrological cycle model. The climatologists have atmospheric general circulation models, the oceanographers have ocean circulation models and now hydrologists are moving from catchment-scale models to continental-scale models.

9.1.2. DATA NEEDS OF HYDROLOGICAL MODELS

The data requirements of hydrological models vary with the geographical area, the type of model used and the output requirements. Table 9.1 presents a list of typical data types organised by dimension, i.e. time-invariant data and time series of data measured at a point, along a line, and over an area, which also vary with time. Again, depending on the model used, many of these data are used only as enhancements and are not vital to the operation of the hydrological model.

The biggest problem facing hydrological modellers is that of obtaining reliable distributed data. The most physically-sophisticated distributed hydrological model will give no better results than the simplest lumped model without accurately distributed input data. Many so-called distributed models are, in fact, simply lumped models run many times over with the same parameters and the same input.

9.1.3. REMOTELY SENSED DATA

The use of hydrological models for large watersheds and the shortage of good ground-based data sources have caused hydrologists to make increasing use of remotely sensed data. The use of aircraft is an early (and continuing) example of remote data collection. Most major field programs, e.g. the FIFE project (Sellers

TABLE 9.1. Typical data requirements for hydrological modelling

Category	"Invariant" data	Time series at a point	Time series for an area
Meteorological/climatological	snowmelt rate lapse rate rain/snow temperature	air temperature precipitation evaporation humidity dew point wind velocity/ distance hours of sunshine radiation snow water equivalent	air temperature precipitation evaporation cloud cover snow covered area snow water equivalent
Vegetation/land cover	classes of land cover roughness canopy properties		NDVI LAI
Physiography	elevation slope aspect stream pattern stream properties (dimensions, roughness) watershed boundary & area		
Soils	classes of soil roughness max. infiltration rate conductivity depression storage field capacity	albedo temperature	

et al., 1988) include a comprehensive range of aircraft for collection of data across the radiation spectrum. In the western US and Canada, areal snow cover, snow water equivalent and soil moisture are determined on a routine basis using airborne gamma radiation equipment (Carroll, 1987).

As ground-based radar coverage increases, the use of radar for estimation of precipitation distribution becomes more common, for example over the River Dee catchment in the U.K. (Cowther & Ryder, 1985) and over Canadian watersheds (Kouwen et al., 1986).

As the scale of hydrological models increases to provide interaction with general circulation models and regional climate models, hydrologists must turn to the world-wide coverage provided by satellites. Unfortunately, very few of the basic hydrological variables can be measured directly from space. This hasn't deterred the users, however, and many thousands of papers have been written on the subject. As only one example, the proceedings of the 1989 IGARSS conference in Vancouver contains 881 papers and over 3000 pages on the application of satellite data. The following paragraphs are necessarily therefore a very brief summary of some of the main areas of interest to hydrologists.

Precipitation is undoubtedly the most important input data for hydrological models and yet, at the moment, it can be estimated from satellite data only in special circumstances with limited accuracy. For example, Pietroniro et al. (1989) used thermal infrared data from Meteosat to estimate monthly precipitation over the Sahel region. Other techniques are available such as cloud indexing, thresholding and life-history methods using visible and infrared data from NOAA and Meteosat/GOES satellites and using regression algorithms with passive microwave data from the Nimbus, DMSP and MOS-1 satellites (Engman & Gurney, 1991). The most promising new technology is the series of recent active microwave satellites such as ERS-1, JERS-1 and RADARSAT which may allow hydrologists to use the same type of Z-R relationships that are presently used with ground-based radar.

In cold regions, the areal extent of snow cover is measured relatively easily at visible and infrared frequencies. Areal extent may be measured to within 30m using Landsat TM or SPOT images although this is at restricted return periods and in cloud-free conditions only (Rango, 1993). The geostationary satellites such as Meteosat and GOES with images every 30 minutes have inadequate resolution, particularly at high latitudes. NOAA satellites provide a good compromise at 1km resolution and 12-hour return period.

Snow water equivalent, as one measure of cumulative precipitation in cold regions, is now measured more or less routinely for flat homogeneous areas such as the Canadian prairies (Goodison, 1989) using passive microwave data from the SSM/I sensor on the DMSP satellite. Although the passive microwave data are available in all weather conditions, the DMSP resolution is only 25km. Rango (1990) and Slough and Kite (1992) have described the extension of passive microwave techniques into mountainous areas. Snowmelt timing and change in area can be detected using multi-temporal SAR data from the ERS-1 satellite (Maxfield, 1993).

Surface water extent, as an index of reservoir content and flood coverage, may be measured with both optical and microwave sensors. The latter has the advantage of all-weather coverage which is important during storm events. For example, Radarsat International (RSI,1993) was able to provide images of the 1993 Mississippi flooding during the extreme wet weather period by superimposing ERS-1 images of flood extent on to earlier clear-weather SPOT and Landsat images within a GIS. Lake and river ice is easy to detect when snowcovered but much more difficult when snow-free. Glacier extent and position of the snowline may be determined using optical and microwave sensors.

Hydrological processes such as interception, infiltration and runoff depend upon distributed catchment characteristics such as elevation, slope, aspect, vegetation cover and soil type. Land cover is a useful surrogate for many of these characteristics and can be accurately estimated by classification of data from optical sensors. Thus data from Landsat MSS and NOAA AVHRR are commonly used for watershed land cover measurements (Kite and Kouwen, 1992). Canopy interception forms an important component of total evapotranspiration from a watershed. The variation of leaf area index through the year for various species of vegetation can be modelled by computing vegetation indices from optical and near infrared frequencies; NOAA AVHRR data are commonly used (NOAA, 1992).

Evaporation has traditionally been one of the most difficult hydrological processes to estimate since direct measurements present many problems. Methods used in hydrological models include point-value methods such as Monteith (1981) and complementary relationship areal-value methods such as Morton (1983). Techniques are now becoming available to use satellite data to provide areal estimates of skin temperature, cloud cover/sunshine duration and surface albedo and to compute evapotranspiration from different land covers (Bussières and Louie, 1989).

In any hydrological model it is important to keep track of the temporal and spatial variation in soil moisture. Both passive and active microwave data have been used for this purpose (Engman, 1990).

The temptation in hydrological modelling, as elsewhere, is to graft modern technology onto traditional methods. For example, Ragan and Jackson (1980) made use of Landsat MSS images to estimate an SCS curve number for a simple lumped hydrological model. Such a use doesn't take full advantage of the satellite's high resolution distributed data. Similarly, satellite data have been used to improve empirical regression-type equations relating basin characteristics to seasonal water supply. For example, Chandra and Sharma (1978) used Landsat to determine basin characteristics such as slope, elevation, area and land cover type for input into water supply estimates. Such relationships are specific to the basin and the time period used.

9.1.4. GEOGRAPHIC INFORMATION SYSTEMS

A geographic information system (GIS) is a mechanism for capturing, storing, viewing and performing analyses of spatially distributed data. A good GIS will generally includes both raster (pixel) data and vector data and will include algorithms to transfer data between different projections. As the scale and complexity of hydrological models has increased, hydrologists have found the GIS to be a useful technique of coping with the vast numbers of data and the spatial relationships between data from disparate sources. There are three levels of linkage between a GIS and a hydrological model (Maidment, 1991):

i) Using the GIS only for determining parameters for existing hydrological models; that is, using the GIS to build up input data for a model without having to manually derive the data from paper. For example, Muzik (1993) used a GIS to determine the SCS curve number for a simple lumped watershed model. The SCS curve number method was developed to summarize basin conditions at a time when distributed data were not available and, as noted earlier, it seems rather inefficient to use distributed data to determine inputs to a lumped model. It would be better to use a distributed model that makes use of all the available data.

A second use for a GIS at this basic level is to act as a display device for the results of the hydrological model.

ii) Combining a GIS and a hydrological model to make full use of the GIS's spatial data analysis to derive distributed data. Leavesley and Stannard (1990) describe the USGS PRMS modelling system which uses a GIS to derive distributed parameters for Hydrological Response Units within a watershed. Other authors have used a GIS to derive distributed data sets for TOPMODEL (Romanowicz et al., 1993). The SLURP model, described later in this chapter, uses a GIS in this manner.

iii) Embedding the hydrological model within a GIS. At this linkage level the hydrological model is an integral part of the GIS and uses the programming language of the GIS to carry out the hydrological simulation. This option is severely restricted by the lack of an efficient temporal dimension in most GIS's. Models for estimation of mean annual variables or probable maximum variables can be envisaged. Batelaan et al. (1993) describe the integration of a one-layer regional groundwater model into a GIS to map seepage zones and Smith (1993) developed a GIS-based system for a distributed-parameter urban hydrology model. These models only operate for short events, however. Stuart and Stocks (1993) discuss the restrictions of the SPANS programming language for hydrological modelling.

There are many tested and proven GISs on the market, ranging from the public domain GRASS (US Army Corps of Engineers, 1991) through the non-profit IDRISI (Eastman, 1992) to the fully commercial systems such as Arc/Info and SPANS. GIS World (1990) lists over 160 commercially available GIS packages. This surfeit of software hasn't stopped hydrologists from writing their own systems. For example, a browse through the proceedings of a recent conference unearthed acronyms such as RAISON, ILWIS, Smallworld, TAPES, GSIS, EGIS,

REGIS, etc. In many applications the GIS is linked to relational databases, expert systems and decision support systems and terms such as "spatial fuzziness" appear in the literature.

9.2. The SLURP Hydrological Model

9.2.1. THE MODEL CONCEPT

SLURP is a distributed conceptual model which fits about two-thirds of the way along a line linking lumped basin models to fully-distributed physically-based models, see Figure 9.1. The name SLURP was originally an acronym for Simple LUmped Reservoir Parametric. The name has remained while the model has developed to a semi-distributed version and to the latest, distributed, version. SLURP was developed for meso-scale Canadian watersheds as an alternative to the use of larger and more complicated models (Kite, 1975 & 1978).

SLURP divides a watershed into Aggregated Simulation Areas (ASAs) and into areas of different land covers (Figure 9.2). An ASA is a sub-unit of the watershed and may be an individual grid square, a group of grid squares or a smaller watershed. The only constraint is that each ASA must have known land covers or land uses (Kouwen et al., 1990). Land cover data are obtained from satellite and are used in SLURP as an indicator of climatic zone, vegetation type, soil characteristics and physiography. This embedded use of land cover information makes the SLURP model particularly useful for studies in which land cover is expected to change; for example in climatic change studies. The model has been used in this way to investigate methods of application of an alternative climate scenario to a mountain watershed.

The model may use up to hundred ASAs and up to ten land covers, potentially dividing a watershed into one thousand sub-units. If the model is to be calibrated, then at least some of the ASAs should have recorded streamflow data at their outlets.

The SLURP model applies a water balance to each element of the matrix of ASAs and land covers using land cover roughnesses, infiltration rates and hydraulic conductivities. Details of the techniques used may be found in Kite (1993a) and in the manual for the model (Kite, 1993b). In brief, the vertical structure contains four non-linear reservoirs, a canopy, a snowpack, a rapid response (may be considered as a combined surface storage and top soil layer storage), and a slow response (may be considered as groundwater). The four tanks have specified initial contents; there is a maximum depression storage for the rapid response and a maximum field capacity for the slow storage. The vertical water balance operates at a daily time step. The resulting rapid and slow runoffs are routed within the ASA using physiographic data obtained from a GIS. The routed runoffs from each land cover are combined into a streamflow from the ASA and this is routed down the stream system to the next lower ASA.

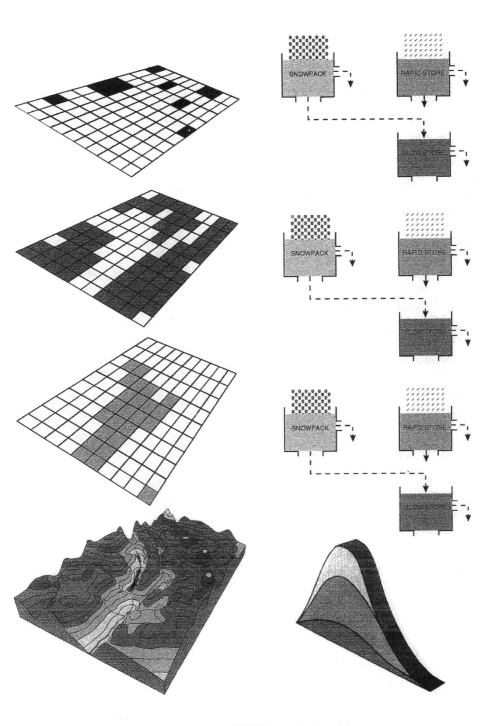

Figure 9.2. Concept of SLURP watershed model

In the ideal model, all parameters will be physically based and easily measured and calibration will be unnecessary. However, this ideal state is unlikely ever
to be reached given the extreme variability in hydrological conditions across a watershed. The SHE model is probably one of the most physically realistic of current
models, requiring vast numbers of distributed data; yet, in a practical application,
even SHE is calibrated (Refsgaard et al., 1992). SLURP uses parameters which
may be estimated directly for different land classes (e.g. Manning's n, infiltration
rate, hydraulic conductivity, soil depths, etc) but, because the model is distributed
and the land class parameters are applied over large areas, some form of calibration is necessary. Once parameters have been derived for a particular set of land
classes, the model may be applied to other watersheds containing the same land
classes without further calibration.

9.2.2. USE OF SATELLITE DATA IN SLURP

The SLURP model was designed to make maximum use of remotely-sensed data.
As well as using land cover information from Landsat or NOAA satellite (Figure
9.3), the SLURP model will also use NOAA AVHRR visible and infrared data to
augment the calculation of snow extent and DMSP satellite data to compute average snow water equivalent (SWE) over each ASA. This use of satellite data is
particularly beneficial when applying the model to macro-scale watersheds where
sufficient land-based data may not be available. Table 9.2 summarizes the use of
satellite data in SLURP.

TABLE 9.2. Satellite data used in the SLURP watershed model

Satellite & Sensor	Resolution	Frequency	Cost Can$/Image	Use in the model
Landsat MSS	80m	1 every 8 days	800	land cover
NOAA AVHRR	1km	2 per day	25	a) snow and cloud cover
				b) land cover
				c) NDVI
DMSP SSM/I	25km	1 per day	20	snow water equivalent

Snow and cloud cover for the SLURP model have been estimated (Kite, 1989)
using daily visible and near infrared data (bands 1 and 2) from the NOAA satellite
series. The snow cover may be used in the model to modify the volume of snowmelt while the cloud cover may be used to modify precipitation data from conventional sources.

SWE may be input directly to the model from snowcourse data or may be
computed from passive microwave data (Slough & Kite, 1992). It was found that
for a mountain watershed, data from individual snowcourses are of little or no

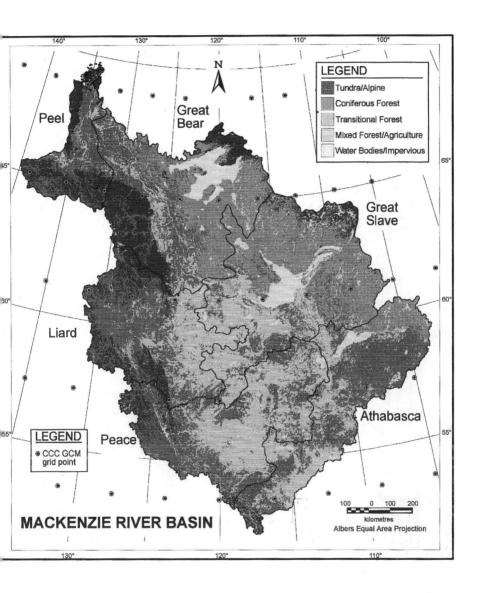

Figure 9.3. Land cover distribution for the Mackenzie Basin from NOAA AVHRR sensor, showing GCM grid points and sub-basins

use in a model; the point SWE measurements are not good estimates of ASA averages. SWE computed from passive microwave data in the 19 and 37 GHz bands, land cover, and daily maximum temperature provide better estimates because they themselves are average values from large areas (e.g. 25km × 25km for data from the DMSP satellite for south-eastern British Columbia). In addition, the microwave data are available in all weathers and at frequent sampling intervals.

The SLURP model uses the complementary relationship areal evapotranspiration (CRAE) model developed by Morton (1983). As described later, CRAE is very suitable for further development using measurements from satellite.

9.2.3. USE OF GIS IN SLURP

SLURP uses a GIS to relate data on topography, land cover and stream distribution to define the ASAs as well as for within-ASA and between-ASA flow routing. Table 9.3 summarizes the types of information from GIS used in SLURP. The following paragraphs summarize the uses of a GIS in SLURP; details of the methods used in the GIS are given in section 9.3.

TABLE 9.3. Information from GIS used in the SLURP watershed model

- ASA boundary and area
- Watershed boundary and area
- Land cover percentages
- Stream network
- Distances to the nearest stream
- Distances to the stream network
- Differences in elevation
- Weights for climatic station influence

Data from classified satellite images are used in the GIS to relate other variables to land cover (Figure 9.3).

A GIS is used to distribute the runoff from each land cover over time. An analysis of the land cover data combined with a ASA streamflow network yields a distribution of distances both to the nearest stream and then along the stream network to the ASA outlet. Figure 9.4 shows, as an example, the sum of the to-stream and the in-stream distances for the Great Slave ASA in the Mackenzie Basin. Figure 9.5 shows the distribution of these distances when associated with the five land classes used in the ASA. Table 9.4 shows some of the values derived from the GIS for the Great Slave ASA.

While other models (e.g. Naden,1993, and Wyss et al., 1990) have assumed a constant velocity, the SLURP model uses a GIS to calculate the changes in elevation between each pixel and the ASA outlet. Assuming that the hydraulic radius

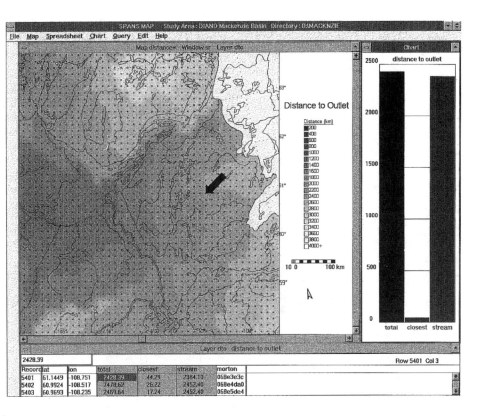

Figure 9.4. An example of a GIS used for a distance to outlet query for a highlighted point in the Great Slave area of the Mackenzie Basin, N.W.T.; the bar graph and the spreadsheet show the total, overland and in-stream distance; these displays update as the cursor is moved throughout the map

is unity for an in- finitely wide shallow channel and using a different roughness factor for each land cover, SLURP computes an average velocity for each land cover using Manning's equation. From the average velocity and the minimum and maximum distances to the ASA outlet, the minimum and maximum travel times for each land cover are computed. These travel times are used in a linear smoothing filter which distributes the runoff from each land cover over time. The runoffs are then weighted by the percentages of the ASA covered by each land cover and summed to give a total flow for the ASA.

Land cover, as used in the SLURP model, represents an index of vegetation, anthropogenic effects, climate and soil charac- teristics. The seasonal change in vegetation from spring greenup through summer peaks and fall senescence to winter

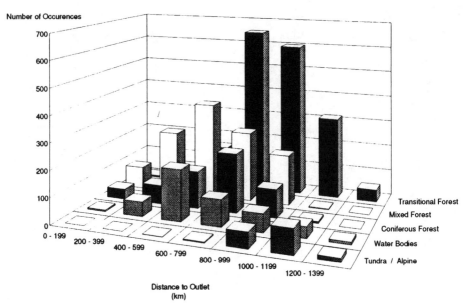

DISTANCE FROM NETWORK OUTLET
Great Slave Lake Sub-Basin

Figure 9.5. Distribution of the distances to the ASA outlet for each land cover for the Great Slave ASA, Mackenzie Basin, kilometres

TABLE 9.4. Physiographic data and travel times Great Slave ASA

Land cover	Water Bodies/ impervious	Tundra/ alpine	Coniferous forest	Transitional forest	Mixed agricu
Change in elevation, m.	117	292	246	228	24
Maximum distance, km.	1234	1250	1105	1256	101
Minimum distance, km.	10	529	23	7	1
Computed slope * 10^4	4.9	11.4	8.7	4.8	6.
Roughness, n	0.05	0.05	0.10	0.10	0.0
Velocity, m/s.	0.66	1.00	0.44	0.33	0.7
Minimum travel time, days.	1	8	1	3	0
Maximum travel time, days.	23	16	35	26	1

dormancy may be represented by the leaf area index (LAI), the projected vegetation leaf area per unit surface area. LAI has been identified as the single variable of greatest importance for quantifying energy and mass exchange by plant canopies at the landscape scale (Running et al., 1986). At field sites, LAI is computed from measurements of individual tree diameters; an extremely tedious procedure (Spanner et al., 1990). However, at watershed scales, many studies (e.g. Running et al., 1986) have shown that LAI may be estimated from remotely-sensed vegetation indices.

The most commonly used indicator of vegetation activity is the Normalized Difference Vegetation Index (NDVI) which is generally computed from AVHRR pixel intensities in band 2 (near infrared, 0.73-1.0μm) and band 1 (visible, 0.55-0.7μm). Using a normalized index partially compensates for changing illumination conditions, surface slope and viewing aspect. However, Spanner et al. (1990) note the importance of using LAC (Local Area Coverage) data and of using data with reasonably consistent solar zenith angle.

Kineman & Hastings (1992) and Gallo (in NOAA, 1992) have published bi-weekly and monthly generalized global vegetation indices (GVI) on CD-ROMs. The AVHRR data were screened to omit cloud covered scenes and those with low solar elevation ($< 15^0$) and were converted to a 10-minute grid. The maximum NDVI values in each two week period were converted to single byte format. Gallo (in NOAA, 1992) computed monthly GVI as the maximum values of the bi-weekly GVIs corrected using pre-launch calibration information and converted from mercator to latitude-longitude projection. Kineman & Hastings (1992) inverted the GVI values during processing so that high GVI values correspond to high vegetation signals and, instead of using the maximum values, used a root-mean-square average of the artifact-free values within each month.

NDVI is strongly related to the amount of chlorophyll in the vegetation cover and to the amount of absorbed photosynthetically-active radiation (APAR) absorbed by the plant canopy. NDVI can be used directly in hydrological models as an indicator of canopy evapotranspiration. For SLURP, the monthly NDVI values are tied to the underlying land cover data using a GIS so that changes in vegetative activity may be used in separate evapotranspiration calculations for each land cover. NDVI may also be converted to LAI for use in more detailed ecosystem models (Running et al., 1989).

9.2.4. EXAMPLES OF RECENT SLURP USES

a) Kootenay Basin. Because the SLURP model parameters are related to land cover in-formation, it can be used with data from general circulation models (GCMs) to study the effects of climatic and land cover changes on water resources.

The SLURP model was calibrated and verified for the Kootenay River Basin in the Rocky Mountains of British Columbia. A Landsat MSS image of the Kootenay River Basin above Skookumchuck was classified (Kite, 1991) into three land cover groups (bare ground & ice, coniferous forest, crop & grassland). The basin was divided into three grouped response units (GRU) consisting of groups of 12×12 km grid squares each containing land cover characteristics (Kite and Kouwen, 1992). The percentages of each of the three land covers were known for each grid square and were aggregated over each GRU. The changes in land cover associated with a doubled carbon dioxide scenario were estimated and the model was rerun incorporating the changed climate and the changed land cover (Kite, 1993a).

The $2 \times CO_2$ GCM results showed a considerable increase in precipitation and the hydrological model showed a corresponding increase in mean streamflow. Evapotranspiration was decreased by about 10% in the $2 \times CO_2$ run because of the changed temperatures and land covers. Although the maximum streamflow was virtually unchanged, the frequency of high flows increased significantly in the $2 \times CO_2$ run. For example, mean daily flows above 300 m^3/s were recorded only six times in the period 1986-1990 but were simulated 16 times under the $2 \times CO_2$ scenario. The inter-annual distribution of hydrological variables was also affected by the climatological changes. Despite the increased precipitation, the monthly snowpacks were reduced because of the higher temperatures. Accordingly, the mean monthly streamflows were increased and the annual peak was shifted from June to April. The GCM shows a much greater decrease in snowpack than does the hydrological model. This is because each land cover in the hydrological model has a different snowmelt rate so that, as the land covers change from one climatic scenario to the other, the overall melt-rate changes. In the GCM, on the other hand, no such changes take place. SLURP also shows a more reasonable annual cycle in the snowpack than does the GCM.

b) Mackenzie Basin. While the atmospheric components of GCMs are often very sophisticated and the land-phase components may be very detailed, most GCMs contain no lateral transfers of water within the land-phase. Any water excess or overflow is simply discarded and takes no further part in the model computations. Such GCMs are modelling an incomplete hydrological cycle (Kite et al., 1994). From a climatological viewpoint this loss of information may not be important. However, if the data from GCMs are to be used to investigate the possible effects of alternative climates then this shortcoming becomes important. The political impetus for climatic change studies is that such changes may affect our lifestyle and our lifestyle depends, to a large extent, on the lateral distribution of surface and groundwater. Also, measured data are needed to validate GCMs and while areal estimates of precipitation, evaporation and soil moisture are difficult to derive, river runoff data are accurate, readily available and represent the integration of all hydrological processes within the watershed.

A vertical water balance was carried out at 12-hourly intervals for a 10-year test period at each GCM grid point relevant to the 1.6 million km^2 Mackenzie Basin using the Canadian Climate Centre's GCM II (CCC, 1991). The 12-hourly water excesses from each grid point were aggregated to 10-year mean monthly figures for the whole Mackenzie Basin. No time delays or attenuation were included and no consideration was made of the geography or physiography of the basin since the GCM contained no such information. The resulting hydrograph was the wrong shape, had roughly twice the volume of the recorded hydrograph and a peak five times as large (Figure 9.6).

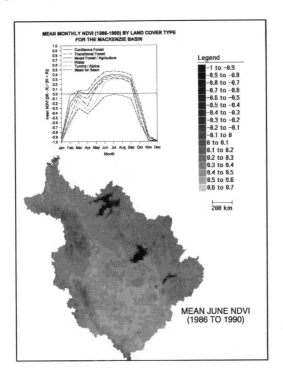

Figure 9.6. A comparison of mean monthly flows in the Mackenzie River as recorded at WSC gauging station no. 10LC014, 1972-1990, as calculated from the sum of 12-hourly water excesses at 39 grid points of the CCC GCM over a 10-year period, and as computed by the SLURP hydrological model using GCM-derived precipitation and temperature data for the same 10-year period

The gridded GCM outputs were then used as distributed climatological inputs to the calibrated SLURP watershed model. The resulting daily simulated flows were averaged to mean monthly values and compared to the recorded data and to the water excesses computed from the GCM. The improvement in the hydrograph due to the use of the hydrological model is dramatic, showing that hydrological models can be used in association with a GCM to replace the deficient land-phase component (Kite et al., 1994).

9.3. Use of the GIS

This section describes how a GIS was used to assemble, store, manipulate and display spatial information in the context of the large-scale hydrological modelling projects described earlier. Examples of specific methods will be presented. We have attempted to avoid the use of software-specific jargon and have tried to use more general terms to describe the various procedures used.

9.3.1. GIS SYSTEM DESCRIPTION

The software, hardware and data sources together form an integrated spatial analysis tool. The system currently used at the National Hydrology Research Centre includes several major software packages, several PC's with input, output and storage devices appropriate for spatial data as well as utility programs and network hardware and software.

The major spatial analysis systems used are SpansGIS and SpansMap from Intera Tydac and the Easi/Pace image analysis system from PCI Incorporated. Additional processing, data translation and presentation of results is accomplished with Lotus123G for OS2, CorelDraw, Picture Publisher, HiJack, IDRISI and MapInfo. For tape backup, software from Novastor, Sytos and Overland Data are used.

9.3.2. ASSEMBLY AND ORGANIZATION OF DATA

The data assembly phase is one of the most important aspects of any GIS-related project. Using existing sources can eliminate much of the traditional and time-consuming manual digitizing and data entry. CD-ROM compilations of hydrometeorological data for climate and flow gauging stations, digital vector map data with basin outlines and drainage vectors, 9-track tapes with general circulation model output, coarse-resolution elevation data and one- kilometre gridded land cover data are examples that are readily available for purchase prior to the start of a water resources project.

Required files may be downloaded from CD-ROM or 9-track tape or FTP'd over Internet, and arranged into binary or ASCII form as required by the SpansGIS or Easi/Pace software. Accompanying text containing such information as projection parameters or database history, should be assembled and kept on file for each spatial database to be used.

For two dimensional GIS systems the projection choice is important; whenever data is displayed on a computer monitor or printed out, it is being projected in some way and trade-offs between equal distance, equal area, or true direction must be considered. This is true even in the case of 3D systems since they display and print in 2D. An additional constraint in some 2D systems is that the internal calculations are carried out in the projected or 2D domain. If accurate area calculations are critical, an equal area projection must be used. In SpansGIS, for example, the projection and spatial extent of the study area must be declared at the outset; for the larger Mackenzie basin, an Albers equal area conic projection was chosen, while in the smaller Kootenay and Columbia basins the UTM projection was an acceptable compromise.

Once the various point, line, area and raster spatial data sets have been assembled and properly imported or translated into the GIS system, the analysis phase can begin.

9.3.3. DERIVATION OF NEW INFORMATION FOR MODELLING

Three examples are given where new information was added through the GIS analysis of assembled spatial data.

a) Area analysis

The SLURP model needs to know how much of each land cover type is found within each ASA. Using the Mackenzie Basin as an example, the basin was subdivided into five sub-basins or ASAs (see section 9.2.1 for a definition of a ASA). The generalized sub-basins contained in the National Atlas of Canada digital map were combined and adjusted to reflect the actual gauging station locations that were used. This process involved vector editing and map class recombination to arrive at a set of five ASAs covering the Mackenzie basin. Land cover data was purchased from the Manitoba Remote Sensing Centre as a set of one-kilometre resolution, classified and georeferenced raster images for various parts of Canada from NOAA satellite imagery. The Easi/Pace satellite image analysis system was used to combine the Northwest Territories and the Alberta images. This was complicated by the fact that each image had different (and undocumented) projection parameters. A standard polynomial or "rubber sheeting" method was used to warp a small portion of the Alberta image to the larger NWT image. The resulting raster image was imported into the GIS system, georeferenced and the original 12 land cover classes were recombined into 5.

  With the ASA and land cover information layers now assembled, an area cross-tabulation report was generated within SpansGIS using the two layers (or, in different GISs, coverages, themes, or maps). The report consists of a table indicating, in percent as well as square kilometers, the area occupied by each land cover type within each ASA.

b) Network analysis

The SLURP model routes runoff from the different land cover types both within and between ASAs; to accomplish this, the actual (as opposed to straight-line) distances from each point in the ASA to the outlet of the ASA must be calculated, and the results tabulated for each land cover type.

  A number of approaches are possible. If a sufficiently detailed, regular-grid digital elevation model (DEM) is available, then the procedure outlined by Jenson and Domingue (1988) can be used. This is a series of steps based solely on the relative elevations of each pixel or grid cell, whereby DEM depressions are filled, flow direction and accumulations calculated, watersheds and drainage networks delineated and, with additional raster procesing, an overland or flow path is calculated from the basin outlet. Similiar methods are available for TIN (triangulated irregular network) based DEMs (Tachikawa et al., 1993).

If a detailed DEM is unavailable, then an alternate approach is possible. Spans-GIS has a procedure which calculates the network shortest path, that is, the shortest distance from any point to other points along a vector network. For hydrology, a drainage network such as that from the National Atlas of Canada digital map forms a suitable vector network. By specifying the gauging station location (basin outlet) as one point and using a 15 km spacing interval, for example, the SpansGIS system will calculate the distance from the outlet to points spaced 15 km apart along the entire drainage network. For larger lakes, it was found that replacing the lake shorelines with a straight line beween the inlet and outlet produced more realistic results. In addition the network had to be 'cut', or vector edited at any lakes which outflow to two or more watersheds to prevent the flow path from 'bleeding' over into an adjacent watershed.

The step outlined above gives the along-stream or channel distances. However, a hydrologic model requires distances to points everywhere in the basin, not just along the streams. One solution is based on the assumption that, for areas between the stream channels, flow distance can be approximated by a simple straight line distance to the closest point on a stream. To accomplish this, the GIS was used to generate a dense regular grid of sample points throughout the basin so that the straight line distance from each sample point to the closest point along the stream network could be determined. No way to do this within the GIS could be found so the set of x/y grid-sample locations was exported to a spreadsheet (Lotus 123G) along with the previously created set of x/y/distance- to-outlet points. A macro was used which repeatedly calculated the minimum Pythagorean distance between sample points. The result was a table of 1) the straight-line distance to the closest stream point, 2) the along-stream distance associated with this closest point, and 3) the sum of distances 1 and 2 (the total distance from the outlet) for each x/y point.

The resulting table was re-imported into the GIS and, using an append class procedure, the land cover type associated with each point was added. A frequency distribution could then be calculated relating the distribution of land cover types by distance to outlet (see for example, Table 9.4 for the Great Slave ASA).

c) NDVI and land cover comparison
As discussed in section 9.2, one of the research studies currently underway involves introducing the Normalized Difference Vegetation Index (NDVI) and Leaf Area Index (LAI) into the SLURP model to investigate alternative evapotranspiration estimates.

The first task was to calculate the mean monthly NDVI for the Mackenzie basin for 1986 to 1990, compare these NDVI values to the classified land covers and produce a summary table of mean NDVI for each month for each land cover type. Monthly NDVI values have been calculated from the original NOAA imagery and are available on CD- ROM (NOAA, 1992) as georeferenced raster images with 10 minute pixel size. Software to subset specific rectangular blocks from the global files is contained on the CD-ROM and was used to download the sixty files required.

The PCI Easi/Pace system was used to average the images, creating twelve image files, one for each month.

SpansGIS was then used to import and georeference the twelve raster images into a previously created study area which contains the Mackenzie basin land cover information. Since SpansGIS carries out most of its spatial analysis tasks using a quadtree data structure (a special compressed form of raster), the images must be converted to quadtree format.

The average monthly NDVI layers and the land cover layer are subjected to the area cross-tabulation procedure and reports are produced which contain the statistical values required for tables and graphs summarizing the relationship between NDVI and land cover (Figure 9.7). In this instance, Lotus123G was used to organize the report files and produce the tables and graphs. SpansGis was also used to produce classified colour images of the monthly NDVI for the Mackenzie basin.

To summarize, while the procedures outlined above are conceptually straightforward, the 'real world' solutions are not. For example, when georeferencing raster images and moving them to different systems, the following must be considered:

— centroid-registered vs. corner registered pixel,
— spherical or rectangular grid,
— row/column numbering and ordering schemes,
— appropriate resampling methods, and,
— pixel format (8,16,32 bit; swapped or not; ASCII or binary).

Vector data sources can also be problematic. The lack of an accepted standard for exchanging topologically structured data is one particular difficulty encountered when dealing with vector datasets.

9.3.4. DISPLAY AND ANALYSIS OF MODEL RESULTS

In the last section, several examples were given illustrating the use of a GIS to derive new information for a hydrological model. A GIS can also be used to display and interpret model results; this is part of the rapidly-evolving field of scientific visualization. In terms of a GIS, we attempt to display complex spatio-temporal scientific information at a scale appropriate for humans so that patterns, anomalies, etc. can be picked up by the human eye/brain (which is still unmatched by even the most powerful computers for this kind of multi-dimensional analysis).

For example, the monthly NDVI images referred to in 9.3.3 above could be displayed sequentially on a high resolution computer monitor as an animation. Additional frames could be added using a form of interpolation called morphing, or by using weekly NDVI data. Additional variables could be brought into the animation in many ways: animated histograms or graphs showing other variables can be simultaneously displayed alongside the original image animation; or colour transformations such as intensity/hue/saturation could, in this case, make the five land cover classes basic colours and vary the intensity or brightness to show the NDVI.

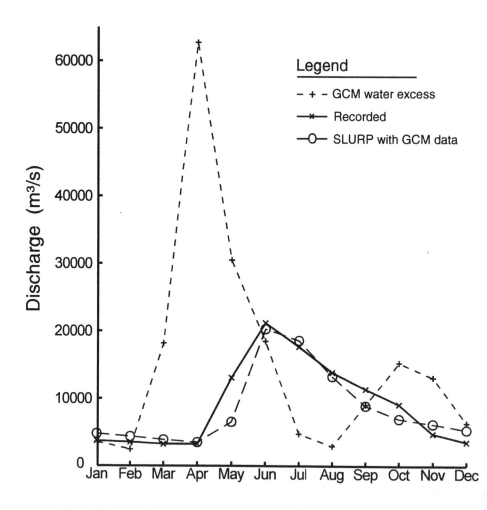

Figure 9.7. NDVI image for the Mackenzie Basin for the month of June with inset graph of mean NDVI for each land cover type for each month

New colour graphics Unix workstations are beginning to appear which will allow this kind of scientific visualization; the limiting factors are the high cost and the seamless integration with existing GISs.

9.4. Limitations of Present Geographic Information Systems for Large-Scale Watershed Modelling

In the practical use of a GIS for hydrological modelling, a number of limitations become obvious. This review limits itself to an overview of some of the major problems faced by researchers. A scientist developing a model is likely to find more difficulties than those listed here, but many of them would be system specific, and quickly solved with the help of the appropriate vendor. Some of these limitations are simple to understand, such as incompatible formats and limitations in memory or speed. Other limitations are not as obvious and include the geographic model used to store the data in a digital format, and the manner in which the software operates.

9.4.1. THE GEOGRAPHIC MODEL

In analyzing how GIS are useful in hydrological models we must first examine how geographic information systems store data. A theoretically infinite amount of information must be stored in a manner such that it is available and useful to any spatial scientist.

A GIS is a tool designed to model the spatial environment. As such it must be able to manipulate, analyze, enter, and store appropriate data. This information can be either geographic or attribute. Geographic information consists of all features that can be symbolized through sets of point, line and area coordinates. Often these coordinates are not representative of the exact location of the entire object, but are an approximation suitable to the scale of the study. Attributes are information attached to these geographic entities, and they provide further information about them. For example a set of x,y coordinates may indicate the presence of a sampling site, but the attribute information would contain useful information as to what type of sample may have been collected, when it was collected and by whom, and the values observed in the process.

Models of natural processes are created to gain an understanding of how these processes work. Geographic Models are created in order to represent the complex relationships of the objects that cover the surface of the planet according to their location. This representation can take place at any height (depth) or time. It is usually a snapshot; a representation of the geography in that location, depth and time. Geographic Models create a series of discrete objects which describe the processes occurring in a certain location at a scale appropriate to the scale of the map to be produced. According to the size of the object and the scale at which it is to be represented, these objects can be shown as points, lines, or areas. Rivers are areal features, but at small scales they are shown as lines, likewise cities are shown as points at certain scales.

Paper maps have served as a model for geographic data storage, and until recently they were the most efficient spatial data model. Two models stand out in providing a framework for storing and modelling geographic data in digital form: traditional and object models. Traditional models separate the information that would appear on maps according to rules similar to those used to assign appropriate symbology to all the elements of the map. Object models treat each element individually assigning rules and processes to each.

9.4.2. TRADITIONAL (LAYER) VS. OBJECT MODELS

The traditional model consists of treating each theme in a traditional map as a separate layer. For example, roads can be considered one or several layers (Principal Highways, Secondary, Gravel) and hydrography likewise. It is unlikely that both of these major themes would be stored as a single layer, as this would result in difficulties in assigning appropriate cartographic symbols to these entities. The information as stored in the system represents a snapshot in time. For example a different land use map would be generated each year, and this map would represent the acquisition date of the image used to create the map.

Object models identify each feature individually. Each object is entered as a discrete individual. Certain rules can be applied to this object, such as how it is to be represented at different scales (as a point or area) and what type of generalization will be used on it, how it changes over time; a formula(s) can define the object's behaviour over time. Even more interesting is that the object can change location over time following a certain set of rules. An object dropped in a stream can have a predicted location at any time; the system is able to generate the location of the object at any particular time desired. In a traditional, layer- based, system there would be a discrete number of layers of information representing the object's location at different time intervals. While a raster system assumes that all information within a single pixel is constant, a traditional system assumes that no changes occur in between the time slices being represented with maps.

Objects can have relationships with other objects. For example, islands usually are present only if they are surrounded by a body of water; where a road and river cross we usually find a bridge, and so on. If further information is known about these objects such as the width and flow of the river and the width, type of highway and daily traffic, it is possible to deduce what type of object would appear at their intersection (what type and size of bridge.)

There are many difficulties in using an object view over more traditional methods. The most important is that most data is already available or can be made available in the form of layers. For example map information is stored in most systems as a discrete set of layers of data. To perform any type of analysis, the user selects the appropriate layers and performs the task. A second problem is that many people find it difficult to grasp that geographic entities can have a set of properties associated with them. Defining these to the system is more difficult still. For example, a river may have many properties such as flow, width, lateral movement, depth, turbidity, etc., but this same line may also serve as a marine transportation route, as a political boundary, or as part of a vegetation or soil polygon. In each of these tasks, this line would assume different properties, and at some point might even split, such as when a large lateral movement changes the location of the river, but not that of the political boundary.

Many of the advantages available to the object model can be simulated in a layer-based system. For example, inheritance allows objects to retain certain characteristics of the objects that created them. As two rivers join, the new river channel will have some of the characteristics of the two rivers that formed it, and some unique to itself. This type of behaviour can be simulated in a layer-based system because a number of attributes are assigned to each entity. The attributes of the resulting river could be generated based on the attributes of the original two channels and relevant new information. Unfortunately, this type of inheritance has to be executed whenever current results are desired, while an object-based system would store these relationships in the database, and thus all information presented would remain current.

Modelling movement in layer-based systems is also difficult. This has been simulated by breaking up the landscape into a set of equal-sized land units. As an object is to move it can do so to an adjoining cell (or remain in the current cell) based on a set of probabilities. Difficulties lie in that these movements can only occur in given intervals of time which must be carefully chosen to accurately reflect the object's behaviour. Ecological models typically use this approach, but if an animal remains in a cell for too long or the cell is too small for that time period, the animal dies, while in reality it would move away more quickly than reflected by the model.

Programming a model using a GIS requires a thorough understanding of the assumptions made by the data storage model used; some of these are outlined above. The model can be written externally in such a manner that the GIS "feeds" the model the data it requires. If the model is continuous it would require that the GIS provide it the necessary information whenever it is needed. This is not always easy, but developments in integrated databases, and client-server systems have made it possible to proceed with such endeavours. In traditional systems, the data have to be prepared before the model is executed, and must be stored in such a manner that the program can read the information in a clear and consistent form.

Writing the model in the GIS programming language eases the process of creating, generating, and reading the information generated by the GIS. Unfortunately, these languages do not have the same depth as traditional programming languages used for these tasks. Also, most models are written using portions of existing models. In using the GIS programming language all programs must be written again. Object-oriented system anticipate the need for high level programming and provide means to link external models to the built-in programming language. 'Facets' is an example of such a tool.

The movement toward object-based programming has not taken place in isolation in GIS, but as a result of greater emphasis on objects within the computer industry as a whole. As more sophisticated object-based programming languages and databases become available, GIS vendors will be able to incorporate these into their products. The result will be a more seamless integration among all the compoments of a model.

9.4.3. OPEN VS. CLOSED SYSTEMS

Many operating system developers now promise open systems. In theory such systems are to provide the ability to link together a variety of tools for a common purpose. For example a GIS, a programming language, and a graphics package may be linked together by means of systems calls. These can be thought as subroutines where each passes on some information to the receptor program and in turn receives information from it. These types of systems greatly facilitate the process of developing complex models. Unfortunately the types of tools needed for the models are poorly standardized, and don't always work as expected. Some researchers (e.g. Chairat and Delleur, 1993) use GRASS, a public domain source code GIS, which allows all routines to be called from any C program.

9.4.4. MEMORY/SPEED/FUTURE DEVELOPMENTS

Hydrological modelling and GIS make great demands on existing computer systems. Fortunately, there is a consistent trend toward systems which offer greater speeds. There is also greater emphasis on systems which are able to perform floating point math. operations quickly and effectively. These systems are based on Reduced Instruction Set Chips (RISC) and are commonly called workstations. As Intel-chip based computers have increased in performance, they have approached performance levels recently expected only from a workstation. At the same time, workstation manufacturers have designed new systems with lower price tags for entry systems. IBM and DEC have introduced new chips that promise enhanced performance levels at affordable costs. Operating systems are also undergoing significant changes. Unix is becoming more standardized and easier to use thanks to improved user interfaces. OS/2 and Windows NT provide many of the advantages of Unix while still having the ability to run DOS and Windows applications without change. These types of developments are critical in order to be able to create more sophisticated hydrological models.

9.4.5. FORMATS

Data that are entered from hardcopy sources can be formatted in a manner suitable to the system being used; however, a considerable number of digital data may already be available for a particular area. The format of the data is usually determined by the originating system. Data are available in raster and vector formats. Raster is better for terrain information, especially natural resources data, while vectors are better for linear features such as rivers or roads or property boundaries.

The vector model assumes that all geographic information can be represented with points, lines, and areas. In turn, these are composed of a finite set of coordinates which define their location or extent. Within this model it is possible to imbed a certain amount of intelligence regarding the behaviour of some of this information. This is known as the Arc Topology Model; it provides a certain amount of object behaviour even in a layer-based system. Discussion about this model appear at length in this and other books (Burrough, 1986, Aronoff, 1989.)

The raster model assumes that all the features present on the surface of the earth can be represented by laying a fine grid over the landscape, and indicating which cells are occupied by those features. Point features occupy a single cell, while line and area features occupy many cells. In these systems it is difficult to represent features that occupy the same cell; in such situations the information is divided into a set of layers. It is difficult to embed information about the behaviour of objects into these systems.

Some GIS software now offer the ability to use both raster and vector data, each for the tasks that it is better suited to. At the same time, these systems provide the ability to translate between vector and raster formats. The higher end systems provide the most flexibility in accepting different formats of data. There are a number of common data formats available in GIS. Almost every system is able to produce and read .DXF files (Autocad Data eXchange Format.) These files are easy to create because the DXF standard is commonly published. Problems with the format include its inability to represent geographic coordinates beyond those of plane coordinate systems, such as UTM. Furthermore the format does not offer the possibility of encoding topology and assigning attributes is cumbersome. Any relationships among the objects are lost as data are exported in this format. Equally troublesome is the number of versions of this format, each version of Autocad introduces a new version of the DXF format. Many vendors do not upgrade their translator at the same rate as Autodesk introduces new versions of their software. For many, the solution is to purchase a copy of Autocad or Hijaak to create older versions of DXF files.

Another CAD-based format is that of Intergraph Standard Interchange Format (SIF.) This format was created by Intergraph in order to move data among a variety of platforms running Intergraph software. As many mapping agencies used Intergraph products, demand for translators to and from this format grew naturally. This format suffers from the same problems as the DXF format. As Intergraph embraces a policy of Open Systems it has also embraced more common interchange formats such as DXF.

ESRI's ArcExport format has proven popular mostly because of the number of system operating that software. The ArcExport format was designed to be a true GIS format, but details about its format are not widely available. All translators outside those produced by ESRI are reverse-engineered, thereby causing some problems, especially as new versions of the format are delivered. The format appears in compressed and uncompressed format. Uncompressed ArcExport (E00) files are preferred because they avoid the common problems involved in moving binary data between workstations and PC's (flipped bits.)

Intera-Tydac's VEC/VEH file format was also designed for GIS. It has not gained the popularity of ESRI's format, but its format is widely published, and it is relatively simple to write translators to and from the format for systems that cannot accept data in such a form. A variety of Canadian Government Agencies distribute data in this format. Translators are available from Intera-Tydac or can be used as-is with the SPANS line of GIS products.

The Data Line Graph (DLG) format was an attempt at creating a format that was suitable for GIS data, and at the same time was vendor-independent. Despite extensive testing, early versions of the format proved to be less than satisfactory with certain systems. At least three versions of the format have been created, and therefore some incompatibilities still exist. Some agencies and vendors still argue that even the current format falls short of its goal. The CCOGIF format is another attempt by Canadian Federal and Provincial agencies to produce a vendor-independent GIS data format, suitable for data distribution. This format has not been well received by vendors yet.

A number of other formats also exist. Almost every GIS software has its own ASCII format which allows the user to move data among systems running the same software. The format used in these packages is suitable only for the product which it supports, and it is sometimes difficult (and a violation of copyright) to import such data into other systems.

A variety of raster formats is also present in the market. Typically, however, raster data is much easier to translate than vector data as rasters are nothing more than a large grid of data values. The number of data bits used to represent the data can change from raster to raster, and it is usually documented in an ASCII header as are the raster dimensions (number of rows and columns.) The manner in which the system reads the raster can also vary, and includes such alternatives as row order, row prime order, and Peano. The header should also document the size of the pixels, the point(s) of reference and its coordinates, and any data compression techniques used. The most common formats supported by GIS are those used by popular image analysis software products, such as PCI.

9.5. Conclusions

As hydrological models are applied to larger basins, the use of remotely-sensed data becomes more important and the need for interaction with GISs becomes more and more necessary.

A GIS has been used to effectively organize and analyze physiographic, land cover, climatological and hydrological data within large Canadian watersheds.

Near-future work in hydrological modelling will include developing new components, better methods of applying small-scale physically-based techniques at larger scales, more useful algorithms for using remotely sensed data and ways of combining atmospheric, oceanographic and hydrological models into a global model of the hydrological cycle.

The problem of spatial and temporal scale becomes important as hydrological models interact with atmospheric models. At the moment such interaction is one way; output from the atmospheric model is used as input to the hydrological model but the outputs from the hydrological model are not fed back to the atmospheric model. One solution may be to develop a hierarchical hydrological model in which appropriately scaled data are passed between different types of hydrological models and to and from the GCM.

Despite the number of publications describing the "integration" of GIS and hydrological models, GISs are still mainly used in hydrology for processing spatially-distributed data before input to the model and for displaying the model results. This is unfortunate because there is an obvious affinity between the distributed hydrological model with its layers of groundwater flow, interflow and surface flow and the ability of the GIS to relate layers of physiographic data, soils data and vegetation cover.

The reasons why hydrological models have not been fully integrated within a GIS include restrictions in the GIS programming languages and the lack of an efficient mechanism for dealing with time series within a GIS. A GIS is usually only two dimensional with the third dimension achieved by using overlays. When the three-dimensional watershed is also changing with time, as in a distributed hydrological model, then the storage requirement within the GIS becomes too large. New methods of incorporating both the third spatial dimension and time are needed.

Also, there is little standardization in GIS data formats; this results in much wasted time translating and 'fixing' the required files. Attempts have been made to create such standards, but so far with mixed results. Vendor-based standards are also unreliable as they tend to change with each new version of the software.

Object-oriented GIS offer hope in that the programming language will become better integrated with the geographic data. Developments in object-oriented databases and programming languages should provide better, more integrated, tools for researchers. These systems should be better able to handle the larger volumes of data required by temporal information in traditional GIS-based models.

The change to a more object-oriented GIS as well as incorporation of three spatial dimensions plus seamless temporal data should result in increasingly sophisticated and accurate large-scale hydrological models.

References

Abbott, M.B., J.C. Bathurst, J.A. Cunge, P.E. O'Connell & J. Ras mussen, 1986. 'An introduction to the European Hydrological system - Système Hydrologique Europeèn, "SHE", 1: History and philosophy of a physically-based distributed modelling system'. J. Hydrol. 87,45-59.

Aronoff, S., 1989. 'Geographic information systems: A management perspective', WDL Press, Ottawa.

Batelaan, O., F. de Smedt, M.N. Otero Valle and W. Huybrechts, 1993. 'Development and application of a groundwater model integrated in the GIS GRASS', in K. Kovar and H.P. Nachtnebel (eds.) Application of Geographic Information Systems in Hydrology and Water Resources, Publ. 211, IAHS, Wallingford, 581-589.

Burrough, P.A., 1986. 'Geographical information systems for land resources assessment', Clarendon Press, Oxford.

Bussières, N. & P.Y.T. Louie, 1989. 'Implementation of an algorithm to estimate regional evapotranspiration using satellite data'. Canadian Climate Centre, Downsview, Ontario.

Carroll, T.R., 1987. 'Operational airborne measurements of snow water equivalent and soil moisture using terrestrial gamma radiation in the United States' in: Large Scale Effects of Seasonal Snow Cover (Proc. of the Vancouver Symposium), Publ. 166, IAHS, Wallingford.

CCC, 1991. 'Application of the Canadian Climate Centre general circulation model output for regional climate impact studies; Guidelines for users'. Canadian Climate Centre, Downsview, Ontario, 25pp.

Chairat, S. and J. W. Delleur, 1993. 'Integrating a physically based hydrological model with GRASS' in: HydroGIS 93: Applications of Geographic Information Systems in Hydrology and Water Resources 143-150. IAHS Publication No. 211.

Chandra, S. and K.P. Sharma, 1978. 'Applications of remote sensing to hydrology', Proc. Symp. on Hydrology of Rivers with Small and Medium Catchments, Roorkee, India, 2, 1-13.

Crawford, N.H. and R.K. Linsley, 1962. 'The synthesis of continuous streamflow hydrographs on a digital computer', Technical Report 12, Stanford University, Dept. of Civil Engineering, Stanford, California.

Cowther, L. & P. Ryder, 1985. 'North West weather RADAR project'. Report by UK Meteorological Office and North West Water Authority, Bracknell.

Eastman, J.R., 1992. 'IDRISI users guide', Clark University, Worcester, Massachusetts.

Engman, E.T., 1990. 'Use of microwave remotely sensed soil moisture in hydrological modeling', in: G.W. Kite and A. Wankiewicz (eds.), Applications of Remote Sensing in Hydrology, Proc. Symposium No. 5, NHRC, Saskatoon, 279-292.

Engman, E.T. and R.J. Gurney, 1991. 'Remote sensing in hydrology', Chapman and Hall, London, 225pp.

Gehegan, M.N and S. Roberts, 1988. 'An intelligent, object- oriented geographical information system', International Journal of Geographical Information Systems, Vol 2:101-110. bibitem Goodison, B.E., 1989. Determination of areal snow water equivalent on the Canadian Prairies using passive microwave satellite data. IGARSS'89 Proceedings, 3:1243-1246.

Jain, S.K., B. Storm, J.C. Bathurst, J.C. Refsgaard and R.D. Singh, 1992. 'Application of the SHE to catchments in India, Part 2. Field experiments and simulation studies with the SHE on the Kolar subcatchment of the Narmada River', J. Hydrol., 140, 25-47.

Jenson, S.K. and J.O. Domingue, 'Extracting Topographic Structure from Digital Elevation Data for Geographic Information System Analysis', Photographic Engineering and Remote Sensing, Vol 54, Mo. 11, November 1988, 1593-1600.

Kineman, J.J. & D.A. Hastings, 1992. 'Monthly generalized global vegetation index (April 1985-December 1988)', in: NOAA-EPA, 1992. 'Global ecosystem database version 1.0 (on CD-ROM), documentation manual, disc-A', National Oceanic and Atmos- pheric Administration, National Geophysical Data Center, Boulder, Colorado, A01-15 - A01-22.

Kite, G.W., 1975. 'Performance of two deterministic models', in: Application of mathematical Models in Hydrology and Water Resources Systems (Proc. of the Bratislava Symposium, September 1975) IASH Publ. no. 115, 136-142.

Kite, G.W., 1978. 'Development of a Hydrologic Model for a Canadian Watershed', Canadian Journal of Civil Engineering, 5, 1, 126-134.

Kite, G.W., 1989. 'Using NOAA data for hydrological modeling'. In: Quantitative Remote Sensing: An Economic Tool for the Nineties, Proc. IGARSS'89, IEEE, 2, 553-558.

Kite, G.W., 1991. 'A watershed model using satellite data applied to a mountain basin in Canada',

J. Hydrol., 128, 157-169.

Kite, G.W., 1993a. 'Application of a land class hydrological model to climatic change', Wat. Resour. Res., 29(7), 2377-2384.

Kite, G.W., 1993b. 'Manual for the SLURP hydrological model, v7.0', NHRI, Saskatoon, 67pp.

Kite, G.W. and N. Kouwen, 1992. 'Watershed modeling using land classifications', Wat. Resour. Res., 28(12), 3193-3200.

Kite, G.W., Dalton, A. and K. Dion, 1994. 'Simulation of streamflow in a macro-scale watershed using GCM data', Wat. Resour. Res., 30, 5, 1547-1559.

Kouwen, N., G. Rosseel and G. Garland, 1986. 'Hydrologic applications of weather radar data', Proc. 10th Canadian Symposium on Remote Sensing, Edmonton, Canadian Aeronautics and Space Institute, Ottawa, 749-756.

Kouwen, N., E.D. Soulis and A. Pietroniro, 1990. 'Enhancing rainfall-runoff modelling of mixed land-use/land-cover areas with remote sensing', in: G.W. Kite and A. Wankiewicz (eds.), Applications of Remote Sensing in Hydrology, Proc. Symposium No. 5, NHRC, Saskatoon, 94-108.

Leavesley, G.H. and L.G. Stannard, 1990. 'Application of remotely-sensed data in a distributed-parameter watershed model', in: G.W. Kite and A. Wankiewicz (eds.), Applications of Remote Sensing in Hydrology, Proc. Symposium No. 5, NHRI, Saskatoon, 47-57.

Maidment, D.R., 1991. 'GIS and hydrological modeling', in M.F. Goodchild, B.O. Parks and L.T. Steyaert (eds.) Proc. First Int. Symp./Workshop on GIS and Environmental Modeling, Oxford University Press, New York.

Martinec, J., A. Rango and E. Major, 1983. 'The Snowmelt-Runoff Model (SRM) user's manual', NASA Reference Publication 1100, NASA, Washington, 18pp.

Maxfield, A., 1993. Personal communication.

onMonteith, J.L., 1981. 'Evaporation and surface temperature', Q. J. Roy. Meteor. Soc., 107, 1-27.

Morton, F.I, 1983. 'Operational estimates of areal evapotranspiration and their significance to the science and practice of hydrology', J. Hydrol., 66, 77-100.

Morton, F.I, 1990. 'Studies in evaporation and their lessons for the environmental sciences', Can. Wat. Resour. J., 15,3,261-286.

Muzik, I., 1993. 'GIS supported flood prediction model', in: Program and Abstracts, Scientific Meeting of the Canadian Geophysical Union, Banff, Alberta, CGU, Calgary, 100.

Naden, P.S., 1993. 'A routing model for continental-scale hydrology', in: Macroscale modelling of the hydrosphere (Proc. Yokahama Symposium, July 1993), IASH Publ. no. 214, 67-79.

Natale, L. and E. Todini, 1977. 'A constrained parameter estimation technique for linear models in hydrology', in: T.A. Ciriani, U. Maione and J.R. Wallis (eds.), Mathematical Models for Surface Water Hydrology, John Wiley, Chichester, U.K., 109-147.

NOAA, 1992. 'Earth systems data, global change data base volume 2: Experimental calibrated global vegetation index from NOAA's Advanced Very High Resolution Radiometer (AVHRR)'. National Oceanic and Atmospheric Administration, National Environmental Satellite, Data and Information Service, National Geophysical Data Center, Boulder, Colorado.

Peuquet, D.J., 1984. "A conceptual framework and comparison of spatial data models", Cartographica, 21,4,66-113.

Pietroniro, A., W. Wishart and S.I. Solomon, 1989. 'Use of remote sensing satellite data for investigating water resources in Africa',IGARSS'89, Vancouver, 2169-2172.

Ragan, M.R. and T.J. Jackson, 1980. 'Runoff synthesis using Landsat and SCS model', J. Hydr. Div., ASCE, 106, 667-678.

Rango, A., 1990. 'Average areal snow water equivalent determination in a mountain basin using passive microwave satellite data', in: G.W. Kite and A. Wankiewicz (eds.), Applications of Remote Sensing in Hydrology, Proc. Symposium No. 5, NHRI, Saskatoon, 317-319.

Rango, A., 1993. 'Snow hydrology processes and remote sensing', Hydrol. Proc., 7, 121-138.

Refsgaard, J.C., S.M. Seth, J.C. Bathurst, M. Erlich, B. Storm, G.H. Jorgenson and S. Chandra, 1992. 'Application of the SHE to catchments in India, Part 1. General results'. J. Hydrol., 140, 1-23.

Romanowicz, R., K. Bevan, J. Freer and R. Moore, 1993. 'TOPMODEL as an application module within WIS', in: K. Kovar and H.P. Nachtnebel (eds.) Application of Geographic Information Sys-tems in Hydrology and Water Resources, Publ. 211, IAHS, Wal-lingford, 581-589.

RSI, 1993. 'Radar satellite tracks Mississippi River flood', Reflections, Radarsat International, Vancouver, 2,2,2-3.

Running, S.W., D.L. Peterson, M.A. Spanner and K.B. Teuber, 1986. 'Remote sensing of coniferous forest leaf area', Ecology, 67,1,273-276.

Running, S.W., R.R. Nemani, D.L. Peterson, L.E. Band, D.F. Potts, L.L. Pierce & M.A. Spanner, 1989. 'Mapping regional forest evapotranspiration and photosynthesis by coupling satellite data with ecosystem simulation', Ecology, 70,4,1090-1101.

Sellers. P.J., F.G. Hall, G. Asrar, D.E. Strebel and R.E. Murphy, 1988. 'The first ISLSCP field experiment (FIFE)', Bull. Amer. Met. Soc., 69,1,22-27.

Slough, K. and G.W. Kite, 1992. 'Remote sensing estimates of snow water equivalent for hydrological modelling applications', Can. J. Wat. Res., 17(4), 1-8

Smith, M.B., 1993. 'A GIS-based distributed parameter hydrologic model for urban areas', Hydr. Proc., 7, 45-61.

Spanner, M.A., L.L. Pierce, S.W. Running and D.L. Peterson, 1990. 'The seasonality of AVHRR data of temperate coniferous forests: Relationship with leaf area index', Remote Sens. Environ., 33, 97-112.

Stuart, N. and C. Stocks, 1993. 'Hydrological modelling within GIS: An integrated approach', in: K. Kovar and H.P. Nachtnebel (eds.) Application of Geographic Information Systems in Hydrology and Water Resources, Publ. 211, IAHS, Wallingford, 319-329.

Tachikawa, Y., M. Shiiba, and T. Takaso, 1993 'Development of a basin geomorphic system using a TIN-DEM data structure', in: J. M. Harlin and K.J. Lanfear, (eds.), Proceedings of the Symposium on Geographic Information Systems and Water Resources, AWRA TPS-93-1, 163-172.

US Army Corps of Engineers, 1991. 'Geographic Resources Analysis Support System, version 4.0, user's reference manual', Con- struction Engineering Research Laboratory, Champaign, Illinois.

Warkentin, A.A., 1990. 'A qualitative evaluation of data networks for streamflow simulation in Manitoba', in: G.W. Kite and A. Wankiewicz (eds.), Applications of Remote Sensing in Hydrology, Proc. Symposium No. 5, NHRI, Saskatoon, 215-226.

Worboys. M.F., H.M. Hearnshaw, and D.J. Maguire, 1990. 'Object-oriented data modelling for spatial databases. International Journal of Geographical Information Systems, Vol 4:90-105.

Wyss, J., Williams, E.R. & R.L. Bras, 1990. 'Hydrologic modelling of New England river basins using radar rainfall data', J. Geophys. Res., 95,D3,2143-2152.

CHAPTER 10

Lumped Modeling and GIS in Flood Prediction

I. Muzik

Abstract.   Hydrology is a geoscience linked to processes occurring at the earth's surface. Application of geographic information systems (GIS), a computer based methodology of capture, storage, manipulation and display of spatially distributed data, is a predictable step in the evolution of hydrology. The ability of GISs to map spatial attributes such as land use and cover, soil types, rainfall, etc., to any degree of resolution is not matched by the present technology of hydrologic data collection for majority of watersheds. It is suggested that lumped hydrology models, specifically the unit hydrograph method, will continue to play a significant role in prediction of design flood hydrographs. Application of GISs improves estimation of unit hydrographs by new techniques, making the use of whatever spatial information is available for the watershed. Three approaches to unit hydrograph derivation and application using a GIS are discussed; the geomorphoclimatic, distributed, and dimensionless unit hydrographs, respectively. It is shown that the GIS derived unit hydrographs are in fact distributed rainfall-runoff models. The application of a GIS also facilitates the mergence of deterministic and stochastic models into one unified modeling approach. Regional flood analysis are greatly improved through GIS support. Examples of GIS supported flood hydrograph and flood frequency curve prediction are given. It is a widely held view that hydrology as a science has failed to make a significant progress in recent years. It is suggested that the future of hydrology may largely depend on our ability to successfully develop and integrate new generation of hydrologic models with the emerging technologies of GIS and remote sensing.

V. P. Singh and M. Fiorentino (eds.), Geographical Information Systems in Hydrology, 269–301.
© 1996 Kluwer Academic Publishers. Printed in the Netherlands.

10.1. Introduction

A flood, loosely defined as an abnormally high discharge or water level, results from an unusually high meterological input in the form of rainfall or snowmelt or from a rapid loss of storage such as a failure of a dam or a levee. The subject of this chapter is the prediction of flood runoff generated principally by rainfall. The objective is to describe the application of geographic information systems (GIS) in flood hydrograph simulation by means of practical methods based on lumped modeling approach. It is shown that the GIS technology enables estimation of the watershed unit hydrograph from spatially distributed excess rainfall, thus improving greatly its prediction capability and making the distinction between distributed and lumped models rather arbitrary. It is also the aim of this chapter to show that the division between deterministic and stochastic modeling is mostly artificial, and that through the power of data capture, storage and analyses afforded by GISs, the two aspects of hydrologic processes, deterministic and stochastic respectively, can be successfully brought together.

At present, the two main routes of flood runoff estimation are through deterministic and probabilistic modeling. For calculation of design floods at sites where observed flood data are available, a choice can be made between probabilistic approach, usually involving some form of flood frequency analysis, or a deterministic approach based on a model converting design rainfall into flood runoff. Very little quantitative guidance is available on choice of flood estimation methods or when it is better to change from flood frequency analysis to an estimate computed from a design rainfall. Pilgrim and Cordery (1993) provide some general recommendations on the choice of a flood estimation method but note that rather arbitrary rules are recommended for most regions. For example, Bulletin 17B of the Interagency Advisory Committee on Water Data (1982) recommends that flood estimates from precipitation should be used only as an alternative method for estimating floods with exceedance probabilities of 1 percent or less if the length of available stream flow record is less than 25 years. The U.K. Flood Studies Report (1975) simply states that if only the peak discharge of a finite return period is required, and a detailed hydrograph is not needed, the statistical approach provides the simpler technique. The report recommends that a flood frequency curve should be extrapolated to a period of 2 N years only, where N years is the length of record. Beyond a return period of 4 N years, a regional frequency curve is recommended up to a return period of 200 years. Where there is a need to specify the shape of the flood or to estimate floods for return periods over 500 years then the report recommends that the unit hydrograph approach is appropriate. Evidently, a new approach to calculation of design floods is needed, which would reflect the prevailing view (Hawking, 1993; Klemes, 1978; Yevjevich, 1972) that the best approach to modeling of geophysical processes is to consider them as being both deterministic and stochastic at the same time, because such is the nature of these processes. This chapter is aimed at encouraging the use of GISs and other new technologies, such as remote sensing toward realizing this goal.

10.2. Modelling Approaches

Mathematical catchment models can be classified in a number of various ways. In general, a mathematical model can be regarded as theoretical, conceptual, or empirical. In addition, a mathematical model can be either deterministic or probabilistic, linear or nonlinear, time-invariant or time-variant, lumped or distributed, continuous or discrete, analytical or numerical, and event-driven or continuous-process. Detailed discussion regarding classification of catchment models is presented by Ponce (1989). This chapter is specifically concerned with lumped modeling of flood hydrographs in a GIS environment, but the subject is treated within a broader context of, what might be called, the mergence of lumped and distributed, deterministic and stochastic modeling approaches.

10.2.1. LUMPED MODELS

Many practical hydrologic models are deterministic lumped models. For example, the rational formula for peak discharge, the unit hydrograph method for prediction of storm hydrographs, the ϕ-index method for estimation of excess rainfall depth, the Thiesen polygon method for estimating the average depth of rainfall, etc.

The common feature of all lumped models is that they ignore spatial variations of precipitation, water flow and other related processes, and deal only with spatially averaged inputs, outputs and parameter values. The advantage of lumped models is their relative simplicity. Thus lumped models are often used in flood prediction, but their usefulness has been limited by the inability to account for internal variations of hydrologic systems and processes. Lumped models have therefore been restricted for simulation of such hydrologic events in which spatial variability of catchment and process parameters does not dominate the outcome of the event. It might be argued that the spatial variability is broadly related to the size of a catchment and therefore a maximum catchment size could be established, up to which in general, lumped models could be safely applied. Such generalization, however, may be possible only in a regional context, since hydrology is the product of complex climate-catchment interaction, with local variations superimposed on considerable global variations of both climate and catchment characteristics. The geomorphoclimatic approach to runoff studies tries to encompass such interactions and it may lead to the establishment of guidelines regarding the applicability of lumped models to specific catchments.

10.2.2. DISTRIBUTED MODELS

Distributed hydrologic models are models that undertake the task of accounting for spatial variations of hydrologic processes and parameters. A truly distributed model of a process is possible only if the process can be described by an equation having an analytical solution. An example of such a model is the kinematic wave model of overland flow, given by the following momentum and continuity equations

$$q = \alpha y^m \qquad (10.1)$$
$$\partial y/\partial t + \partial q/\partial x = i_e \qquad (10.2)$$

in which q is the discharge per unit width, y is the depth of flow, t is the time, x is the distance in the direction of flow, i_e is the excess rainfall intensity, and m and α are coefficients.

The kinematic wave equations can be solved analytically for the case of rainfall having a constant intensity uniformly distributed over the runoff plane of constant slope (Henderson and Wooding, 1964). For example, the peak discharge at any distance x from the upstream end of the plane, resulting from a rainfall of duration t_0 which is shorter than the time to equilibrium, is given by the simple equation (Muzik and Beersing, 1989)

$$q_p = S_0^{\frac{1}{2}}(i_e t_0)^{\frac{5}{3}}/(nx) \qquad (10.3)$$

where S_0 and n are the slope and Manning's roughness coefficient of the plane.

The vast majority of equations used in hydrologic modeling cannot, however, be solved analytically for the varied watershed conditions encountered in practice, and recourse must necessarily be made to numerical methods. In our preceding example, if S_0, i_e or n vary with x, the kinematic wave equations (10.1) and (10.2) do not have analytical solutions and must be solved numerically. Otherwise, using average values in place of variable S_0, i_e and n in the analytical solution would turn the truly distributed model of an idealized runoff into a lumped model of a real basin.

Numerical solutions require assigning of average values over finite distances or elements due to their discretization scheme. Thus, most distributed models involve some degree of lumping and it becomes our decision as to the scale at which the lumping of parameters will take place. The scale of lumping can vary widely, from distances measured in centimeters to areas of a subwatershed size. The choice of scale of distributed elements in our models, or for that matter the choice between the distributed and lumped modelling approach, will generally depend on the purpose of simulation and level of accuracy required, knowledge, skill and experience of the modeller, and financial resources, time and technology available. Intuition suggests that lumped models may be suitable for prediction of average or typical, even if extreme events, while distributed models may be expected to do well simulating even individual events with all their particular variation, which lumped models cannot differentiate. Yet, the increased detail and accuracy to be gained by distributed modeling has its price and perhaps also its limits. Distributed models, by their nature, require a vast amount of data, compared to lumped models, and consequently they are also computationally very demanding and intensive. The computational aspects of distributed modeling are becoming of a lesser problem nowadays, thanks to the advances in computer hardware and software technology.

The real limitation on distributed modeling at present is the non-availability of distributed hydrologic data. While digital terrain models can describe surface features of a watershed to almost any degree of detail, there is no technology available at present, which would supply data of comparable spatial detail on soil moisture and rainfall distribution, to name just two of many significant hydrologic parameters. In the light of this limitation it is imperative to carefully consider the useful degree of resolution of a distributed model, if it is to be used in place of a lumped model of a particular watershed.

10.2.3. PROBABILISTIC MODELS

Outcomes of hydrologic events, flood discharges included, are affected by a very large number of causative factors. Inclusion of all causative factors in a deterministic model is not realistically possible. A probabilistic approach to hydrologic modeling thus evolved, which considers hydrologic variables as random, and strives to describe hydrologic processes by means of laws of chance. Probabilistic models can be classified as statistical and stochastic. Statistical models are postulated mathematical models for describing hydrologic random variables, most often in terms of their probability density functions. Inferential statistics is used to estimate the parameters of the models by extracting the maximum information about the populations of random variables from available hydrologic data, and to test the validity of the postulated models. Typical example of statistical modeling is the flood frequency modeling procedure.

Stochastic hydrologic models are mathematical models of sequences of hydrologic variables governed by probability laws. In stochastic modeling a mathematical dependence model for a given sequence is postulated, model parameters are estimated from available hydrologic data, and the validity of the model tested. Usually, it is assumed that the modeled variable has the same probability distribution at all positions along the sequence. A typical example of stochastic modeling is the time series analyses, although stochastic models of sequences associated with other dimensions, such as a line or an area, are also used in hydrology.

Probabilistic models, by the virtue of being developed and tested on a limited sample data, and by having no explicit physical bases, are prone to fail if the physical conditions of the process they simulate undergo changes. This means, for example, that extrapolation of probability distributions fitted to a sample of observed floods may produce very misleading results, since the hydraulics of extreme floods may be quite different from the hydraulics of observed floods included in the sample and used for the model development.

Deterministic models could, in theory, simulate all conditions by correct application of physical, chemical, and biological laws. However, as mentioned earlier, it is unlikely that all causative factors affecting the simulated process can ever be included in a model. Thus, the world's leading hydrologists maintain that a simultaneous use of both deterministic and stochastic (or statistical) methods of analysis and description of hydrologic processes in nature is necessary for producing the best scientific and practical information for hydrology, in general, and for water resources development, conservation and control, in particular (Yevjevich, 1972).

To illustrate and to explore the idea that most hydrologic processes can be considered as a combination of deterministic and stochastic processes, let us consider as we did before, the case of rainfall generated runoff from a uniformly sloping impervious plane. Specifically, we would like to determine the probability density function of the peak discharge q_p, given by eq. (10.3), resulting from rainfalls of various intensities but having the same duration t_0. Each runoff event in this experiment is as close to a deterministic process as can ever be expected. The intensity $i = i_e$ has a known probability density function $f(i)$. The rainfall pulses of duration t_0 with randomly determined intensity i are applied to the runoff plane under the same initial conditions. The probability density function of the peak discharge $f(q_p)$ can be, in this idealized case, obtained analytically (Muzik and Beersing, 1989). According to the theory of derived distributions it is given by

$$f(q_p) = f(i)|di/dq_p| \tag{10.4}$$

Since the derivative di/dq_p can be expressed from eq. (3) exactly as

$$di/dq_p = \frac{3xn}{5S_0^{\frac{1}{2}} t_0^{\frac{5}{3}} i^{\frac{2}{3}}} \tag{10.5}$$

the relationship between the two densities is then given by

$$f(q_p) = f(i) \left[\frac{3xn}{5S_0^{\frac{1}{2}} t_0^{\frac{5}{3}} i^{\frac{2}{3}}} \right] \tag{10.6}$$

Eq. (10.6) is a probabilistic model of an idealized peak discharge. Two points are noteworthy. First, the relationship between the two densities is influenced (deterministically) by the physical parameters of the process, namely by S_0, x and n. Any change in these physical parameters will change the probability distribution of q_p. Second, eq. (10.6) may be viewed as a geomorphoclimatic relationship, thus lending support to the geomorphoclimatic approach in flood runoff prediction.

In nature, runoff conditions are far too complex to be described by an analytical model. A recourse is then made to an experimental technique known as Monte Carlo simulation. In this modeling approach we can use deterministic models of major processes in combination with stochastic models of inputs and model parameters. Conducting thousands of simulations in sequence and performing frequency analysis of predicted peak discharges, provides another way of how to derive the probability density function of peak discharge.

10.3. GIS Modelling Environment

Geographic information systems are distinguished from other modern data processing systems by their ability to manipulate spatial data. This capability makes GISs extremely useful in hydrologic modeling. Unlike many kinds of other data (e.g. banking or medical records), geographical and hydrologic data are complicated by the fact that they must include information about position and possible topological connections of recorded objects, in addition to objects' attributes. For example, a stream may be stored in a GIS as a number of connected reaches represented by starting and ending XY coordinates and reach attributes, such as the average slope, cross-sectional area and Manning's roughness coefficient. Another example is a subwatershed represented by an area delineated by a set of XY coordinates plus attributes, such as the average runoff curve number, time of concentration, unit hydrograph, etc.

In essence, GISs store geographical objects as a set of points, lines and areas that are defined both by their location on earth's surface using a standard system of coordinates and by their non-spatial attributes. The coordinate system may be purely local, as in the case of a study of a limited area, or it may be that of a national grid or an internationally accepted projection such as The Universal Transverse Mercator Coordination System (UTM). There are basically two distinctly different approaches to storing spatial data and thus, modeling geographical objects. These are the raster and vector methods, respectively. Both are valid methods for presenting spatial data, each having certain advantages or disadvantages in specific applications (Burrough, 1986). This does not necessarily create a problem in deciding between the two, since both structures are interconvertible.

10.3.1. RASTER DATA STRUCTURE

The raster data structure consists of an array of grid cells made up of regularly spaced lines. Each grid cell is referenced by a row and column number and it contains a number representing the type or value of the attribute being mapped. Because each cell in a two-dimensional array can only hold one number, different geographical attributes must be represented by separate sets of arrays, known as overlays.

The raster data structure is easy to handle in the computer, however, usually at the expense of some loss of precision in modeling geographical objects associated with the cell size. This is because in the raster method a point is represented by a single grid cell, a line by a string of neighboring cells and an area by an agglomeration of neighboring cells. Thus, estimates of lengths and areas may involve errors when grid cell sizes are large with respect to the features being represented.

Raster systems are storage intensive but offer strong spatial analysis operators for distance, overlay, connectivity, and neighborhood computations (Meyer, Salem and Labadie, 1993). Two moderately priced raster-based GIS packages available for PC use are pMAP (Spatial Information Systems, Inc.) and IDRISI (Eastman,

1990). Another popular system is GRASS(CERL, U.S. Army Corps of Engineers), which is in the public domain and runs under the UNIX operating system.

10.3.2. VECTOR DATA STRUCTURE

Vector systems store information as points, lines and polygons, and associated attributes with these factors. The coordinate space is assumed to be continuous, not quantized as with the raster method, allowing all positions, length and areas to be defined more precisely. Simple lines or polygons do not carry information about connectivity such as might be required for drainage network analysis. The ways in which line and polygon networks are structured are discussed by Burrough (1986). Arc/Info (ESRI, Inc.), Intergraph (Intergraph Corp., Huntsville, Ala.) and ATLAS*GIS (Strategic Mapping Inc., San Jos, Calif.) are leading commercial vector- based systems (Mayer, Salem and Labadie, 1993). MOSS (U.S. Bureau of Land Management) is a public domain vector system.

10.3.3. DIGITAL ELEVATION MODELS

One of the capabilities of GISs, very useful to distributed hydrologic modeling, is the description of the topography of a region, that is, the attribute being mapped is the surface elevation. Any digital representation of the variation of relief over space is known as a digital elevation model (DEM). DEMs help determine the structure of a distributed rainfall-runoff model by being able to detect ridges and stream lines as well as the boundary of a catchment.

Three principal methods for structuring a network of elevation data in DEMs are square-grid networks (altitude matrix), contour-based network, and triangulated irregular network (Tachikawa et al., 1994). Square-grid networks are the most common form of DEMs used for topographic analysis of river basins and rainfall-runoff modeling (Wyss et al., 1990, Quinn et al., 1991). Zhang and Montgomery (1994) examined the effect of grid size (2-,4-,10-, 30-, and 90- m scale) on the portrayal of the land surface and hydrologic simulation using TOPMODEL (Bevan and Kirby, 1979). They suggest that for many landscapes, a 10-m grid size presents a rational compromise between increasing resolution and data volume for simulating geomorphic and hydrologic processes.

Grid-based DEMs have advantages for their ease of computational implementation, efficiency, and availability of topographic databases. However, when considering the directions of water flow, these methods may not be appropriate for hydrological applications because those trajectories represent only crudely the movements of water from one grid to one of the eight neighboring grids. A more applicable approach for hydrological modeling is the Triangulated Irregular Networks (TINs), (Tachikawa et al,. 1994). Palacios and Cuevas (1986, 1991) made it possible to delineate river-course and ridge of a watershed basin automatically with this method and to simulate surface runoff production. Jett et al. (1979), Jones et al. (1990), and Vieux (1991) also used TINs for representation of a watershed basin.

10.3.4. DATA INPUT

Data input is the operation of encoding the data and writing them to the database. The creation of a good digital database is a most important and complex task upon which the usefulness of the GIS depends. Data input to a geographical information system consists of (a) entering the spatial data to define where the geographic or cartographic features occur, (b) entering the associated attributes that record what the features represent, and (c) linking the spatial data to the non-spatial data.

The sources of both spatial data and attributes are varied, including topographical maps, soil and ecological maps, field notes, aerial photographs, remotely sensed data, point-sample data (e.g. raingage data), and data obtained by prior analysis (e.g. lag- time, unit hydrograph parameters) and surveys (e.g. Manning's roughness coefficient, land use and cover data).

There are several methods which can be used singly or in combination to enter the data into a GIS. Basically, data can be entered manually, by digitizing, or by means of automatic scanners. Some data which already exist in digital form such as remotely sensed satellite data or discharge records may be directly transferred into a GIS. Details of data input methods are discussed by Burrough (1986).

10.4. Lumped Modelling with GIS

A lumped parameter model treats a drainage basin as a single hydrologic response unit, so that all attributes must be averaged over the basin area. The unit hydrograph typifies the lumped modeling approach. The distinction between lumped and distributed models, however, is not as clear as might be desired. For example, if the total watershed being considered is divided into a number of arbitrarily small subbasins, each represented by a unit hydrograph, the total watershed model becomes essentially distributed (Quimpo, 1993), but some researchers (De Vautier and Feldman, 1993) may still classify it as being a lumped model. Putting the fine points of model classification aside, the following sections are concerned with the application of GISs in both derivation of unit hydrographs and the use of unit hydrographs in flood prediction studies.

10.4.1. UNIT HYDROGRAPH DERIVATION

The unit hydrograph of a watershed is defined as a direct runoff hydrograph resulting from a unit depth of excess rainfall generated uniformly over the drainage area at a constant rate for an effective duration (Chow et al., 1988). The unit hydrograph is a lumped linear model of a watershed. The assumption of a linear watershed behavior allows the use of two basic principles in hydrograph calculations: (1) the proportionality of outflow to inflow, and (2) the superposition of outputs from successive individual inputs. However, many watersheds may display nonlinear behavior over a wider range of discharges. To minimize errors resulting from the assumption of linearity, Pilgrim and Cordery (1993) suggest that unit hydrographs should be derived from floods of magnitudes as close as possible to those that will be calculated using the derived unit hydrograph. In other words, it is advisable to derive a family of unit hydrographs for a considered watershed, each unit hydrograph being applicable within a certain range of excess rainfall. Thus, the strict assumption of input-output proportionality is no longer present, but the principle of superposition is considered to apply. This approach is supported by the finding of Kundzewicz and Napiorkowski (1986) that some nonlinear systems fulfil the temporal superposition principle.

Three methods of unit hydrograph derivation, which are suitable for GIS application and which can accommodate the concept of a variable watershed response, will be briefly discussed. The three methods are: the geomorphoclimatic instantaneous unit hydrograph , the distributed unit hydrograph, and the regional dimensionless unit hydrograph. Application of the last two methods will be illustrated by examples.

10.4.1.1. *Geomorphoclimatic instantaneous unit hydrograph*

. Rodriguez-Iturbe et al. (1982) proposed a geomorphoclimatic theory of the instantaneous unit hydrograph (IUH) as a link between climate, the geomorphologic structure, and the hydrologic response of a basin. The probability density functions of the peak and time to peak of the IUH can be derived, according to this theory, as functions of the rainfall characteristics and the basin geomorphological parameters. A similar concept was originally introduced by Eagleson (1972) who derived a catchment model based on the kinematic wave equations. Muzik and Beersing (1989) presented results of experimental and theoretical studies of the dependency of the probability density function of overland flow peak discharge on the rainfall and runoff plane characteristics.

According to the geomorphoclimatic theory the peak discharge q_p and time to peak t_p of a triangular IUH are given as

$$q_p = \frac{0.871}{\pi_i^{0.4}} \tag{10.7}$$

$$t_p = 0.585\pi_i^{0.4} \tag{10.8}$$

where

$$\pi_i = \frac{L_\Omega^{2.5}}{i_e A_\Omega R_L \alpha_\Omega^{1.5}} \tag{10.9}$$

and

$$\alpha_\Omega = \frac{S_\Omega^{\frac{1}{2}}}{n_\Omega b_\Omega^{\frac{2}{3}}} \tag{10.10}$$

in which L_Ω and A_Ω are the length and drainage basin area of the highest-order stream, i_e is the mean excess rainfall intensity, R_L is the Horton's length ratio, b_Ω and S_Ω are the mean width and slope of the highest-order stream in the basin and n_Ω is the corresponding Manning roughness coefficient.

Convolution of the triangular geomorphoclimatic IUH with a pulse of excess rainfall of intensity i_e and duration t_r leads to the following expression for the peak of the direct runoff hydrograph (Brass, 1990).

$$Q_p = 2.42 \frac{i_e A_\Omega t_r}{\pi_i^{0.4}} \left(1 - \frac{0.218 t_r}{\pi_i^{0.4}} \right) \tag{10.11}$$

The time to this peak discharge is given by Rodriguez-Iturbe et al. (1982) as

$$T_p = 0.585 \pi_i^{0.4} + 0.75 t_r \tag{10.12}$$

The geomorphoclimatic theory of IUH is attractive to both the theoretician, because it establishes the links between climate, geomorphology and hydrology, and to the practitioner, because it offers a simple method of deriving a watershed's unit hydrograph without the need for observed rainfall and runoff data. Furthermore, the method can be easily incorporated within a GIS on a regional basis. Required parameters, eqs. (10.9) and (10.10), for any watershed within the region, could be quickly extracted by the GIS.

The geomorphoclimatic IUH in its present form, however, is not without problems. One of the more questionable assumptions underlying the theory presented by Rodriguez-Iurbe et al. (1982) is "the belief that hydrologic response at the basin scale depends only on some of the gross features of the basin and not on the details of the network geometry". Thus, according to eqs. (10.9) and (10.10) the geomorphoclimatic IUH depends only on the characteristics of the highest-order stream, and is independent of hillslope and lower-order stream processes. This assumption clearly cannot apply over a wide range of geomorphological features. Other comments on the geomorphoclimatic IUH can be found in Brass (1990). The concept of the geomorphoclimatic IUH has been discussed here to introduce it as a possible candidate for GIS application in future. For the present, however, it is recommended to regard the concept to be in a developmental research stage.

10.4.1.2. Distributed Unit Hydrograph

. The concept of a spatially distributed unit hydrograph, proposed by Maidment (1993), is based on the fact that the unit hydrograph ordinate at time t is given by the slope of the watershed time-area diagram over the interval $[t - \Delta t, t]$. The time-area diagram is a graph of cumulative drainage area contributing to discharge at the watershed outlet within a specified time of travel. The validity of the above can be shown as follows:

An S-hydrograph, defined as the runoff at the outlet of a watershed resulting from a continuous excess rainfall occurring at rate i_e over the watershed, is given by the equation

$$Q(t) = i_e A(t) \tag{10.13}$$

where $A(t)$ is the watershed area contributing to flow $Q(t)$ at the outlet at time t. The direct runoff hydrograph discharge at time t, resulting from a pulse of excess rainfall $P_e = i_e \Delta t$, is equal to the difference between S-hydrograph value at time t and its value lagged by Δt time, that is

$$Q(t) = i_e A(t) - i_e A(t - \Delta t) \qquad (10.14)$$

The unit hydrograph ordinates are $U(t) = Q(t)/P_e$, and thus

$$U(t) = \frac{A(t) - A(t - \Delta t)}{\Delta t} \qquad (10.15)$$

Determination of the time-area diagram for a watershed is greatly facilitated by a GIS. The GIS is used to describe the connectivity of the links in the watershed flow network and to calculate distances and travel times to the outlet for various points within the watershed. Also, with GIS's capability for rainfall mapping, the assumption of a uniform spatial rainfall distribution is no longer necessary. Hence the term spatially distributed unit hydrograph. The conceptual procedure of time-area diagram derivation using a raster-based GIS has been summarized by Maidment (1993) as follows: "Given a digital elevation model for a watershed, a grid of flow direction is defined from each cell to one of its eight neighboring cells in the direction of maximum downhill slope. From this, a grid of flow distance can be compiled for the watershed by tracing from the lowest cell to all cells upstream and storing in each cell its flow distance to the outlet. By assigning a velocity of flow to each cell, a grid of flow times to the outlet can similarly be computed from which isochrones of flow time at intervals Δt are laid out on a grid. The incremental areas $A_i, i = 1, 2, \ldots$, of cells between isochrones can be determined and by accumulating these areas upstream from the watershed outlet, the time-area diagram for the watershed is determined."

The spatially distributed unit hydrograph can in fact be classified as a type of the geomorphoclimatic unit hydrograph, since its derivation depends on watershed geomorphology, rainfall and hydraulics.

Example of application.   The Waiparous Creek watershed, with its major channel network, shown in Fig. 10.1, is used to illustrate the derivation of a distributed unit hydrograph and its application in simulation of an observed flood hydrograph. The watershed has an area of 229 km^2 and is located on the eastern slopes of the Rocky Mountains in Alberta, Canada. The discretized watershed boundary shown in Fig. 10.1 resulted by using a 1 km × 1 km grid in the GIS (IDRISI) operations.

Derivation of the time-area diagram requires the knowledge of the distribution of flow directions and velocities over the watershed. To obtain this information a simple digital elevation model of the watershed was constructed by assigning an average elevation to each 1 km × 1 km grid cell. Water on grid cell is permitted to flow to one of its eight nearest neighbor cells, as shown in Fig. 10.2. A grid of flow directions was then created by choosing for each cell the direction of steepest descent among the eight permitted choices. This grid is shown in Fig. 10.3 as a set of arrows. Connecting cell centres along the arrows will create an equivalent channel network as shown in Fig. 10.4. Comparison of Figs. 10.1 and 10.4 indicates that even when using relatively low resolution of 1 km × 1 km grid cells the main channel network has been reproduced reasonably well.

The generated channel network is used to calculate flow velocities and travel times between each grid cell and the watershed outlet. Using aerial photographs and field survey data channel reaches were assigned average width B and Manning roughness coefficient n values.

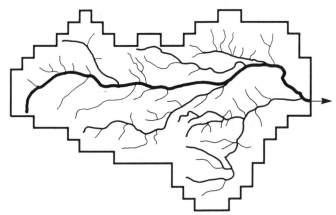

Figure 10.1. Waiparous Creek watershed

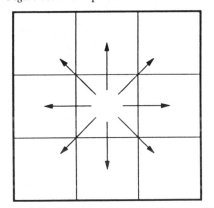

Figure 10.2. The eight-direction cell flow model

The average flow velocity V through a reach was computed by combining the Manning and continuity equations for a wide channel, written respectively as

$$V = \frac{S_0^{\frac{1}{2}}}{n} y^{\frac{2}{3}} \tag{10.16}$$

and

$$Q = VBy \tag{10.17}$$

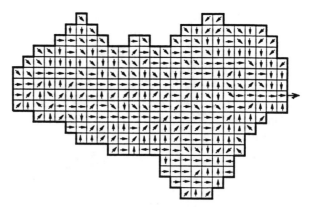

Figure 10.3. The grid of flow directions

Figure 10.4. The equivalent channel network

into one equation for velocity

$$V = \left[\frac{S_0^{\frac{1}{2}}}{n} \left(\frac{Q}{B} \right)^{\frac{2}{3}} \right]^{\frac{3}{5}} \tag{10.18}$$

where S_0 is the grid cell slope determined by the digital elevation model.

The discharge Q in eq. (10.18) is taken as the average of the inflow and outflow for the cell. For example, Fig. 10.5 shows a cell with two inflows, Q_1 and Q_2, from two upstream cells, plus locally generated overland flow q_3. The outflow from the cell is $Q_3 = Q_1 + Q_2 + q_3$, and the Q in eq. (10.18) would be $Q = (Q_2 + Q_3)/2$. The estimated average flow velocity is used to calculate the travel time through the cell along the main channel under equilibrium conditions.

The equilibrium overland flow discharge generated from the cell area A is $q = Ai_e$, where i_e is the average excess rainfall intensity computed by the Soil Conservation Service (SCS) runoff curve method (National Engineering Handbook, 1972), discussed in section 10.4.1.3. This requires the creation of curve num-

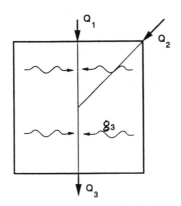

Figure 10.5.   Routing model of cell flow

ber (CN) data base in the GIS, followed by computation of excess rainfall intensity i_e for each 1 km × 1 km cell. The rainfall input used in this example consists of a rainfall of infinite duration, uniformly distributed over the watershed. Each cell, however, generates a different excess rainfall depth, resulting in a spatially distributed excess rainfall input.

For cells which do not have a defined channel the travel time was computed as the time to steady flow for overland flow given by the kinematic wave equation (Chow et al., 1988).

$$t_c = \frac{L^{0.6} n^{0.6}}{i_e^{0.4} S_0^{0.3}} \tag{10.19}$$

where L is the length of overland flow.

The general approach to determining the time-area diagram was to calculate the cumulative time of travel and contributing areas for various points along the main channel, starting from the watershed outlet. The key idea is to determine the time of travel from a given location to the outlet by starting the calculation from time zero, that is, initially there is no inflow from upstream, and as the flow progresses downstream it encounters equilibrium flow conditions only. When the flow arrives at a confluence, the contributing area upstream of the confluence along the minor stream must be found by trial and error for the known travel time found for the main channel up to the confluence. In essence, the approach is analogous to the method of characteristics for flow over a runoff plane due to rainfall (Henderson, 1966), where one is trying to trace the movement of characteristics emanating from various locations within the watershed at time zero.

Because equilibrium flow was assumed throughout the basin, the procedure neglects the time required to establish the equilibrium via channel storage. To account for the storage effect, the computed equilibrium travel times through each cell were increased by a certain percentage, ranging from zero to 60 percent, estimated on the basis of channel geometry and discharge within each cell. A simplified isochrone map computed for the study watershed for a one-hour pulse of uniform

rainfall with intensity of 3.7 mm/h is shown in Fig. 10.6. The figure shows isochrones with the time increment of 3 hours determined by interpolation between
travel times to the watershed outlet calculated for each cell. IDRISI was then used
to determine cumulative areas for travel times increased by one hour increments.
The result is shown in Fig. 10.7 as the time-area diagram. Lagging the time-area
diagram by one hour and subtracting it from the original one resulted in a one-hour
unit hydrograph shown in Fig. 10.8.

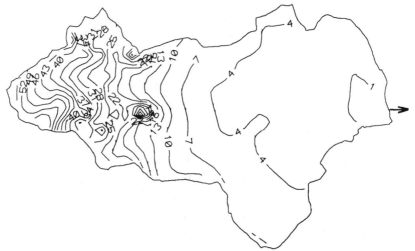

Figure 10.6. Simplified isochrone map

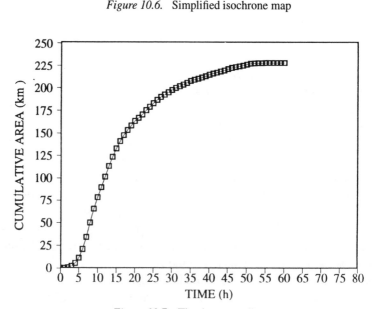

Figure 10.7. The time-area diagram

The derived distributed one-hour unit hydrograph was used to simulate an
observed hydrograph produced by a 21-hour storm which was recorded at two

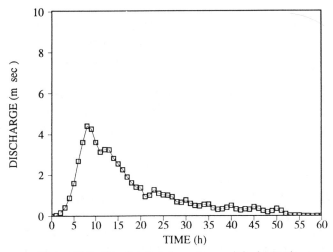

Figure 10.8. The distributed one-hour unit hydrograph

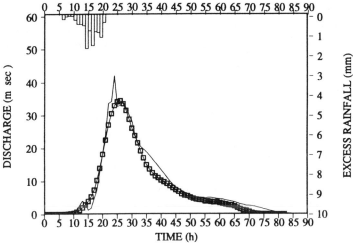

Figure 10.9. Observed and simulated (squares) hydrographs

neighboring raingage stations. The storm hyetograph, discretized into 1-hour incremental rainfalls, was applied uniformly over the watershed's 229 grid cells. Each cell generated an excess rainfall hyetograph according to its assigned runoff curve number. The incremental one-hour excess rainfalls were spatially averaged to obtain a representative excess rainfall hyetograph for the watershed. This average hyetograph was convoluted with the one-hour distributed unit hydrograph and resulted in the predicted hydrograph shown in Fig. 10.9. The agreement between simulated and observed hydrographs is very good, considering that practically no parameter optimization was performed. It can readily be seen that the outlined procedure would be very difficult to perform without the support of a GIS.

10.4.1.3. *Dimensionless Unit Hydrograph*

Dimensionless unit hydrograph is a unit hydrograph whose discharge and time co-ordinates have been expressed in terms of dimensionless ratios. The purpose of such conversion is to try to eliminate the effect of excess rainfall duration (and thus indirectly the effect of possible nonlinearities involved) and the effect of the size and other geomorphological watershed parameters on the shape of the unit hydrograph. Although a number of generalized dimensionless unit hydrographs have been proposed, such as the SCS dimensionless unit hydrograph (National Engineering Handbook, 1972), their universal applicability cannot be taken for granted, because there might be local or regional variations due to climatic and strong geomorphological differences. It is therefore preferable, if practicable, to develop, on the basis of observed rainfall and runoff data, a regional dimensionless unit hydrograph applicable to the so called hydrologically homogeneous region. The concept and delineation of a hydrologically homogeneous region, however, is not that well understood. The hydrologically homogeneous region is usually considered to be delineated by those watersheds having a sufficiently long record of observed flows which pass a statistical homogeneity test (Chow, 1964) based on 10-year flood discharges estimated by flood frequency analysis. The discriminative power of the test, however, is not very high to be totally relied on. The test results should be supplemented by more or less subjective evaluation of hydrograph analysis, that is to say, whether or not the variation in peak and shape of dimensionless unit hydrographs derived for various stations in the region, is deemed to be within acceptable limits, and whether or not a reasonably good relationship between lag-time and watershed geomorphologic parameters can be derived for the region.

Derivation of a dimensionless unit hydrograph described here is based on the procedure developed by the Bureau of Reclamation (Design of Gravity Dams, 1976). In this method, a carefully derived direct runoff hydrograph is converted into dimensionless graph as follows. The elapsed time from the beginning of a hydrograph to its centroid is computed; this is called the lag-plus-semiduration value for the hydrograph. The lag-time, LG, is defined as the elapsed time between centroids of excess rainfall and the direct runoff hydrograph, respectively. For a pulse of excess rainfall with duration D the lag-plus-semiduration is equal to LG + D/2, as illustrated in Fig. 10.10(a). The abscissae of the direct runoff hydrograph is converted from actual hours into percent of the lag-plus-semiduration value. Each ordinate of the hydrograph is multiplied by the lag-plus-semiduration value, and the product is divided by the total direct runoff hydrograph volume. The converted ordinates and abscissae are dimensionless and may be plotted as a dimensionless unit hydrograph as shown in Fig. 10.10(b).

To determine the average shape of a group of dimensionless hydrographs, to be representative of a watershed or a region, the Bureau of Reclamation (Design of Gravity Dams, 1976) recommends to first determine the average of the peak ordinates and the average of the corresponding abscissae. These two values become the coordinates of the peak of the average graph. Points on the lower portions of

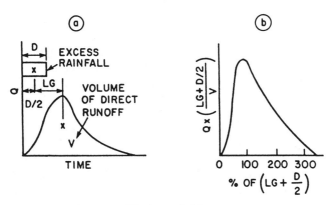

Figure 10.10. Conversion of a direct runoff hydrograph (a)
into a dimensionless unit hydrograph (b)

the accession and recession are averaged on the horizontal, that is, an ordinate is
assumed and the average of the abscissae corresponding to that ordinate is determined.

The objective of developing a dimensionless unit hydrograph for a watershed
or a whole region is, of course, to facilitate the determination of a synthetic unit hydrograph at an ungaged site anywhere within the watershed or the region. Lag-time
is a key parameter for estimating synthetic unit hydrographs. An average lag-time
value for a watershed is obtained by averaging the results obtained during hydrograph analyses. Such average values for different gages on a stream and/or different streams of similar runoff characteristics can be correlated empirically with
certain measurable watershed features. The correlation equation suggested by the
Bureau of Reclamation (Design of Gravity Dams, 1976) is of the form

$$LG = C \left[\frac{LL_{ca}}{S_0^{\frac{1}{2}}} \right]^x \qquad (10.20)$$

where C and x are regional constants, L is the length of longest watercourse from
outlet to watershed divide, ca is the centroid of the watershed, L_{ca} is the length
of longest watercourse from outlet to intersection with a perpendicular from ca to
stream alinement, and S_0 is the average slope of longest watercourse from outlet
to divide.

Examples of Application. It is often the case that suitable records of observed
discharge are unavailable at the exact stream point for which a unit hydrograph
is needed. The lag-time-dimensionless unit hydrograph method is one of the few
methods which allow the derivation of unit hydrographs for ungaged watersheds.
When the method is implemented in combination with a GIS on a regional basis
it becomes a very expedient tool for prediction, even forecasting of flood hydrographs anywhere within the region. The role of the GIS is to provide data to the

hydrologic model for any watershed in the region. The data would include rainfall depth, soil type, land use and land cover data, to enable calculation of excess rainfall, a proper lag-time curve and dimensionless unit hydrograph for a given watershed, and all required geomorphological parameters.

Two examples will illustrate GIS application in a flood prediction procedure based on the lag-time-dimensionless unit hydrograph method. The procedure was developed for a study region comprising about 11,000 km². The first example is a single event simulation of a flood hydrograph, the second example demonstrates the derivation of a synthetic flood frequency curve by Monte Carlo technique. The GIS used is called HYDROGGISS (Hydrograph Generating GIS Software) and was developed as a dedicated GIS to support the lag-time-dimensionless unit hydrograph method.

Study area.   The study region, covering about 11,000 km², runs generally in the north-south direction along the eastern slopes of the Rocky Mountains in Alberta, Canada, as shown in Fig. 10.11. Thirty-one streamflow and 38 rainfall gaging stations were used in hydrologic analyses. Squares shown in Fig. 10.11 are the GIS blocks, 100 km × 100 km each, used to store various attributes on a 1 km × 1 km raster grid.

The topography of the region varies from rolling, forested foothills in the east to rugged, bare mountain ridges in the west. The elevation ranges from 1100 m above mean sea level in the low foothills portion to 3000 m above sea level in the mountain areas. Limestones, dolomites, sandstones and shales dominate the bedrock of the study region. Surficial geology is mainly formed by glacial deposits (such as ground moraine) and colluviums. Soil in the area generally falls into the B or C hydrologic soil group, according to the SCS classification. Runoff curve numbers determined from standard SCS tables and for antecedent moisture condition II, range from 65 to 85. The area is mostly forested, predominantly by spruce and pine mixed with occasional stands of poplar. In lower elevations the forest gives way to meadows and pastures. The average summer rainfall is between 300 and 400 mm.

Regional dimensionless unit hydrograph and lag-time relationship.   Sixty-one dimensionless unit hydrographs were derived in the region from selected observed events. Details of screening for suitable events, baseflow separation, and determination of the duration of excess rainfall are given by Muzik and Chang (1993) and Chang (1992). To simplify all calculations the derived unit hydrographs were replaced by triangular unit hydrographs with the same peak discharge as the original graphs. Although there was some variation in the peak and time to peak values, it was decided to average all sixty-one hydrographs to compute one regional dimensionless unit hydrograph, shown in Fig. 10.12, applicable to all four GIS blocks in Fig. 10.11. In comparison with the SCS triangular dimensionless unit hydrograph,

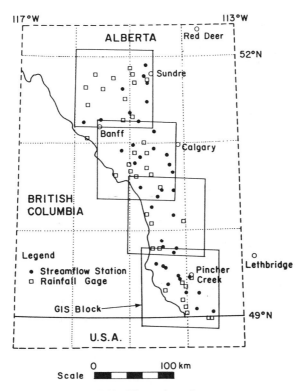

Figure 10.11. The study region

(National Engineering Handbook, 1976) the derived regional hydrograph has the recession times 13 percent longer, which may not be a significant difference.

The lag-time values determined during hydrograph analyses were used in an adaptive nonlinear least-squares algorithm to find the following best correlation equation for lag- time in hours

$$LG = C \left[\frac{L L_{ca} D_s}{\sqrt{S_0} I_e^{1.0574}} \right]^x \qquad (10.21)$$

where C and x are regional constants having the values of 8.53 and 0.0977, respectively; L is the length of the main watercourse from the point of interest to the watershed divide; L_{ca} is the length of watercourse from the point of interest to the intersection of the perpendicular from the centroid of the watershed to stream alignment, in km; S_0 is the mean slope of the main watercourse, in m/km; D_s is the duration of the storm, in hours; and I_e is the average intensity of excess rainfall, in mm/h.

The lag-time relationship plotted in Fig. 10.13, shows relatively large scatter of points about the regression line. This may be caused by anyone of the following four factors, or their combination: (1) too much noise in data, (2) nonhomogenuity

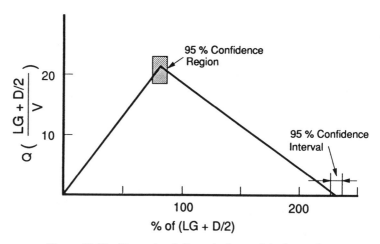

Figure 10.12. The regional dimensionless unit hydrograph

of the region, (3) pronounced nonlinearity in response of some watersheds, and (4) improper choice of variables. However, further exhaustive analyses and checking of data could not achieve any significant improvement of the relationship.

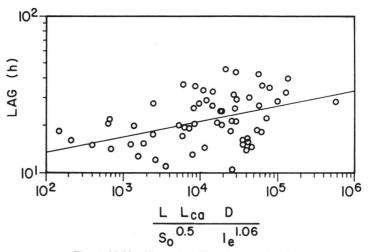

Figure 10.13. The regional lag-time relationship

Rainfall input and abstractions. Two types of rainfall inputs, in terms of total rainfall hyetographs, were considered. First, for the purpose of simulating observed flood hydrographs, the actual rainfall depths, observed at nearby existing raingage stations, would be used. The GIS would calculate the average incremental depth of precipitation for the watershed using the Thiessen polygon method.

Second, for determination of a design flood hydrograph or a synthetic flood frequency curve, a hypothetical storm hyetograph would be used. For this purpose, the GIS database stores, on a 10 km × 10 km grid, the extreme rainfall statistics extracted from the Rainfall Frequency Atlas of Canada (1985). The stored statistics are the means and standard deviations of the Gumbel distributions fitted to rainfall depths corresponding to the following rainfall durations: 1, 2, 3, 4, 5, 6, 12, 24, 48 and 72 hours. Thus, the GIS can calculate, for user specified return period and storm duration, the depth of rainfall for each 10 km × 10 km cell using the frequency factor method (Chow et al., 1988), and subsequently the weighted average depth of rainfall for the watershed. The total depth is then partitioned into, say, hourly increments, according to one of the nine time distribution curves derived by Huff (1967) for heavy storm. These curves are stored in GIS data base, and the selection is made either by the user or by random draw.

The total storm hyetograph is further processed for abstractions, to determine the excess rainfall hyetograph. This is achieved by means of the SCS runoff curve method (National Engineering Handbook, 1972). A brief summary of the method follows.

The Soil Conservation Service procedure for computing abstractions from storm rainfall relates the depth of excess rainfall or direct runoff P_e to the total rainfall depth P, initial abstraction I_a, and potential maximum retention S by the following equation

$$P_e = \frac{(P - I_a)^2}{P - I_a + S} \tag{10.22}$$

An empirical relation was developed by the Soil Conservation Service

$$I_a = 0.2S \tag{10.23}$$

which, after substitution into eq. (10.22), yields

$$P_e = \frac{(P - 0.2\,S)^2}{P + 0.85\,S} \tag{10.24}$$

The potential maximum retention S is assumed to depend on the soil type, land use and land cover. It is high for deep sandy soils and low for shallow clayish soils. For convenience, the potential maximum retention S, in mm, is converted into a parameter CN, having a range of values between 0 and 100, by the following equation

$$CN = \frac{25,400}{S + 254} \tag{10.25}$$

The parameter CN, called the curve number, is equal to zero when there is no direct runoff generated. The value of 100, on the other hand, means no abstractions, and the depth of excess rainfall equals the depth of total rainfall. Curve numbers have been tabulated by the Soil Conservation Service on the basis of soil type and land use generalized classes for three antecedent moisture conditions.

The study area is relatively homogeneous regarding the land use and cover, as well as the predominant soil types. It was found that better flood simulation results were obtained when using a regional relationship for S in the eq. (10.22) for excess rainfall depth. This relationship, shown in Fig. 10.14, in which S depends on the five- day antecedent precipitation P5, was developed using observed rainfall and runoff data in the region. Details of the method are given by Muzik and Chang (1993) and Chang (1992). The method allows the estimation of the S-P5 relationship, shown as a solid line enveloping the plotted points in Fig. 10.14, and the initial abstraction values, Ia. The initial abstraction appears to be a random variable which can be fitted by a log- Pearson III distribution. For the study region it was found that the best fit was obtained when the Ia values were divided into two groups, according to whether P5 was less than 30 mm, or greater and equal to 30 mm. The fitted distributions are shown in Fig. 10.15. Two additional regional distributions were derived. These are the probability distributions fitted to observed rainfall durations D, and to five-day antecedent precipitation values P5. These distributions are required in Monte Carlo simulation. Both variables were fitted by log-Pearson III distributions. Good fit was obtained in both cases, as shown in Fig. 10.16 for D, and in Fig. 10.17 for P5.

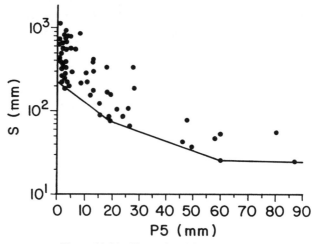

Figure 10.14. The regional S-P5 relationship

HYDROGGISS. Hydroggiss stands for "Hydrograph Generating GIS Software" (Muzik and Chang, 1993). The software is an example of a dedicated GIS developed for the sole purpose of supporting a lumped, unit hydrograph-based flood prediction model. The raster data structure is used in the GIS, with a fixed grid of 1 km × 1 km. Four types of attributes were stored for each cell in the data base: (1) land use/cover class (2) soil type, (3) runoff curve number CN, and (4) means and standard deviations of extreme rainfalls for durations of 1, 2, 3, 4, 5, 6, 12, 24, 48 and 72 hours. In addition, the regional dimensionless unit hydrograph, lag-time curve, S-P5 curve, probability distributions of storm duration D, 5-day antecedent precipitation P5, and initial abstraction I_a were also stored in digital form in the system.

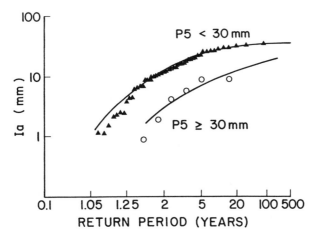

Figure 10.15. Regional frequency distributions of I_a fitted by log-Pearson III distributions

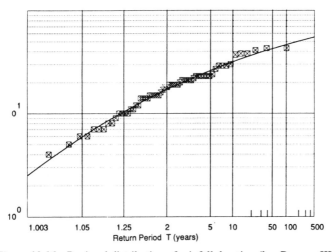

Figure 10.16. Regional distribution of rainfall duration (log-Pearson III)

The software consists of 15 major menu-driven programs, as illustrated in Fig. 10.18. The software can simulate single flood events, and by Monte Carlo simulation can derive synthetic flood frequency curves for both gaged and ungaged watersheds. The software also contains hydrograph analyses programs to facilitate the creation of a regional data base.

When using the software for simulation of flood hydrographs, the procedure begins with placing a topographic map containing the watershed of interest on the digitizer and digitizing two points to orientate the X- and Y-axes. Next, the watershed boundary and the main watercourse are digitized, and the elevations of the end points of the main stream are input through the keyboard. From this information the software computes the watershed area A, the main stream length L, the length along the main stream from outlet to the point nearest the watershed centroid

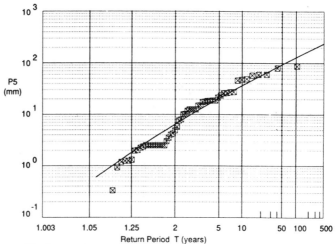

Figure 10.17. Regional distribution of five-day antecedent rainfall (log-Pearson III)

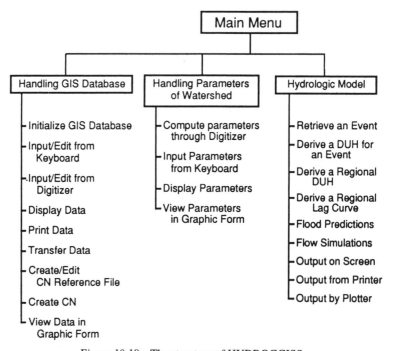

Figure 10.18. The structure of HYDROGGISS

L_{ca}, the mean slope of the main stream S_0, the average runoff curve number CN for the watershed, and the areal averages of means and standard deviations of the rainfall depth (Gumbel distribution) for various durations. At this point, the software is ready to proceed with the simulation of single flood hydrographs and/or synthetic flood frequency curve at the watershed outlet.

Example 1: hydrograph prediction. Simulation of observed events simply requires that the observed rainfall be input through the keyboard as a hyetograph discretized into one-hour increments. In design flood hydrograph studies however, the fundamental problem is that the relation between the probabilities of the design rainfall and the flood predicted from it is not easily established. There is no rigorous solution to the problem of how to achieve the selected probability of a design flood hydrograph, but four general approaches have been discussed by Pilgrim and Cordery (1993). The present example simply illustrates the prediction of a flood hydrograph from rainfall depth- duration-frequency data, without considering the exceedance probability of the estimated flood. The procedure can be outlined by the following steps:

(1) User specifies the design storm duration D_s, its return period T, and one of the nine temporal patterns according to Huff (1967).

(2) The software calculates the total rainfall depth P of the selected design storm, using the frequency factor method (Chow et al., 1989) given by the equation

$$P = \bar{x} + K_T\, s_x \qquad (10.26)$$

where $\bar{x}$ and s_x are the watershed average mean and standard deviation of the Gumbel distribution of P for the selected duration D_s, and K_T is the frequency factor for the return period T (stored in the data base).

(3) The software determines the design storm hyetograph according to the user selected temporal pattern.

(4) The software computes the excess rainfall hyetograph. If the standard SCS runoff curve number method is used the watershed average CN value, calculated previously, is used in eqs. (10.25) and (10.24) to obtain the depth of excess rainfall P_e. Alternatively, the S value can be obtained from the regional S-P5 relationship (Fig. 10.14) for the user specified value of the five- day antecedent precipitation P5. The initial abstraction I_a is either selected by the user or drawn randomly from the regional frequency curves (Fig. 10.15). The excess rainfall depth is then calculated by eq. (10.22).

(5) The software calculates the watershed lag-time LG by eq. (10.19) and converts the regional dimensionless unit hydrograph into the watershed synthetic unit hydrograph.

(6) The software convolutes the design excess rainfall hyetograph with the synthetic unit hydrograph to calculate and plot the design direct runoff hydrograph.

Illustration of the procedure is shown in Fig. 10.19, in which an observed hydrograph was simulated using observed rainfall data.

The total time required for the simulation, starting with taping the map to the digitizer, is less than 10 minutes. The procedure can be implemented for any location included in the data base.

To gain some insight on the accuracy of the procedure, the sixty-one observed events (previously used in deriving the regional dimensionless unit hydrograph)

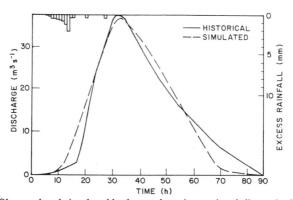

Figure 10.19. Observed and simulated hydrographs using regional dimensionless unit hydrograph

were simulated. The regional S-P5 and I_a curves were used. The relative error E of the simulated peak discharge Q_{ps} was computed by the equation

$$E = 100\frac{Q_{ps} - Q_{po}}{Q_{po}} \tag{10.27}$$

where Q_{po} is the observed peak discharge. The histogram of the errors is plotted in Fig. 10.20.

Example 2:   synthetic flood frequency curves.   One possible approach to the problem of how to estimate the exceedance probability of a predicted hydrograph is to preform frequency analysis of a synthetic annual flood series, that is, to generate a synthetic flood frequency curve. Of course, in many hydrologic studies, estimation of the flood frequency curve at a site is the major objective of the study in itself. A synthetic flood frequency curve can be obtained by Monte Carlo simulation. In general, this method is used to estimate the probability distribution of a system's output from a given probability distribution of the input, when the relation between the two probability distributions is too obscured to be determined analytically.

The analytical solution was obtained in section 10.2.3 for the case of an idealized runoff from a uniformly sloping plane. Watershed runoff processes introduce joint probabilities which cannot be readily expressed analytically. Thus, Monte Carlo simulation is presently the only practical method to derive synthetic flood frequency curves.

The technique, illustrated in the following example, basically involves three steps. Namely, determining the inputs, transforming the inputs into flood hydrographs, and then statistically analyzing the generated flood peaks to estimate their probability distribution. A flow-chart of the simulation procedure is shown in Fig. 10.21. Basically the prediction of a flood hydrograph, as described in Example 1, was repeated N times, except that all parameters needed for calculation of design

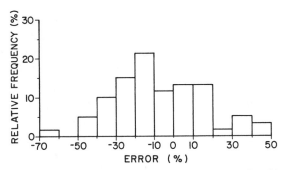

Figure 10.20. Distribution of relative errors in peak discharge predicted by regional DUH

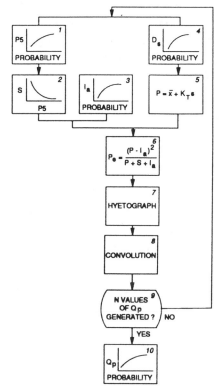

Figure 10.21. Flow chart illustrating Monte Carlo simulation

storms (D_s, temporal distribution) and abstractions (P5, I_a) were drawn randomly from their respective regional probability distributions (Figs. 10.16, 10.17, and 10.15). When the number of simulations N reached the predetermined value of 5000, the peak discharges of simulated hydrographs were used in flood frequency analyses. The derived synthetic flood frequency curve is shown in Fig. 10.22 together with the frequency curve of observed annual floods and the fitted log-Pearson III distribution. Several comments are warranted at this point. First, the synthetic flood discharge series was generated by combining stochastic inputs with a deterministic runoff model. Such combination is possibly the best approach to hydrologic modeling, as discussed in section 10.2.3. Second, the synthetic flood frequency curve is derived through modeling of the runoff process, not by a statistical fitting of a hypothetical probability distribution to a limited sample. When using the Monte Carlo simulation method, the modeller has the means of extrapolating the frequency curve beyond the range of observed values on the bases of his judgement with respect to the selection of the simulation model, model parameters and model inputs. Process modeling may be the only justifiable means of extrapolating empirical flood frequency curves. Lastly, application of the GIS greatly reduces the time necessary for the completion of hydrograph simulation, and makes it a very simple task to rerun the simulation for a different set of model parameters and inputs calculated on the basis of various attributes stored in the GIS. But perhaps the greatest advantage provided by the GIS in applications such as the one illustrated in the present example, is the ability to perform flood analyses anywhere in the region with great expediency.

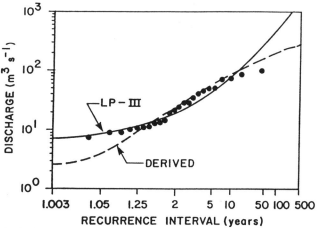

Figure 10.22. The derived synthetic flood frequency curve compared to observed (dots) and fitted log-Pearson III distributions

10.5. Conclusions

The aim of this chapter has been to illustrate the application of GISs in lumped modeling of flood hydrographs. Undoubtedly, GISs are an ideal tool for distributed modeling, however, the GIS technology allows new and more physically based methods to be used in deriving lumped model parameters from spatially distributed information, so that the distinction between lumped and distributed models is becoming rather arbitrary. For example, the distributed unit hydrograph method, described in this chapter, is in essence a hybrid between lumped and distributed model. This may be the best compromise for modeling watersheds with the typical amount of available data, which would not support a fully distributed model. In other words, the GIS technology of storage and manipulation of spatial data is not matched, at present, by the hydrologic data-collection technology, which still, in the main, relies on point-data measuring techniques. Thus, a significant amount of spatial averaging, based on a few measuring points, is the rule rather than an exception. Consequently, lumped and semi-distributed models can benefit greatly from the GIS technology.

Besides the lack of adequate data, there is a host of other reasons, discussed by Klemes (1978), why a fully-distributed model, such as a hydrodynamical model based on Saint- Venant and groundwater hydraulics differential equations, may not be the ideal and final goal of hydrologic modeling. The two most compelling reasons may be restated as follows. First, the level of inquiry at which the movement of water can be described is arbitrary. It may begin at the atomic or molecular level on one end of the scale, or with a control volume comprising thousands of cubic meters of water, on the other side of the scale. The level of inquiry, however, does not make a method necessarily superior to methods applied at lower or higher levels.

Second, "it is the fact that as the number of interacting elements increases, their individual behavior is still less and less reflected in the behavior of the whole, until eventually the later is best described by a set of properties based on averages; i.e. an adequate description of the whole is achieved at the expense of sacrificing the detailed description of its components".

Simply put, the above ideas can be expressed by saying that the nature of the problem at hand should be matched by the level of inquiry applied to solve the problem.

Examples used in this chapter were selected with the intention to emphasize the strong geomorphoclimatic component present in flood prediction methods supported by a GIS. Spatial mapping of attributes, including elevation and rainfall, is of course the essence of every GIS. Digital elevation models make it even possible to delineate watershed boundaries and channel networks automatically. Thus, GISs greatly enhance our ability to represent physical features of watersheds and the spatial distribution of hydrologic parameters, including rainfall inputs. This means, we now have a tool to analyze geomorphoclimatic relations pertinent to flood hydrology on a regional, national and perhaps the global scale, within one computer environment. The geomorphoclimatic approach is essential to our understanding of hydrology. Of the three methods, illustrated by examples in this chapter, the distributed unit hydrograph, although still in a research stage, appears to have the greatest potential for application with GIS, and at the same time significantly improving the accuracy of simulation. The dimensionless unit hydrograph is a standard technique eminently suited to GIS application on a regional basis. The lag-time relationship, however, requires much refinement. The geomorphoclimatic instantaneous unit hydrograph proposed by Rodriguez Iturbe et al. (1982), although conceptually interesting, seems to be flawed in its present form by the negligence of the effects of hillslope and lower-order streams runoff on the watershed hydrograph.

It has been shown, by way of an example of runoff from an impervious uniformly sloping plane subjected to rainfall inputs, that the deterministic and stochastic features of a given process, such as runoff, are interrelated through its physical mechanism. It is suggested further, that extrapolation of flood frequency curves should be based principally on Monte Carlo simulation, in which dominant physical processes are simulated by deterministic models, and model parameters and inputs are stochastic variables having specific probability distributions. This type of simulation is again significantly improved if supported by a GIS. The ability of the GIS to capture, store and manipulate spatial data improves model calibration procedures, and allows the exploration of numerous runoff scenarios during the simulation.

The cost of implementing a GIS can be significant, especially in terms of data collection and data base development. Then there is an additional problem of interfacing the GIS with a hydrologic model. For these reasons, a simple and user friendly GIS is preferable at present for lumped hydrologic modeling. Nevertheless, the future of hydrologic modeling appears to be linked to GIS and remote sensing applications. The focus of future advancements should be improved data collection techniques, expanded data bases, and the development of a new generation of hydrologic models capable of being integrated with the emerging GIS and remote sensing technologies.

References

Bevan, K., Kirby, M.J. 1979. "A physically based, variable contributing area model of basin hydrology", Hydrol. Sci. Bull., 24, pp. 43-69.

Brass. R.L. 1990. Hydrology, An Introduction to Hydrologic Science, Addison- Wesley Publishing Comp., New York, p. 643.

Bureau of Reclamation. 1976. Design of Gravity Dams, United States Department of the Interior, United States Government Printing Office, Washington, D.C.

Burrough, P.A. 1986. Principles of Geographical Information Systems for Land Resources Assessment, Oxford University Press, New York, p. 194.

Chang, C. 1992. A Physically-based Flood Prediction Model Aided by a Geographic Information System, Ph.D. thesis, Department of Civil Engineering, The University of Calgary, Calgary, Alberta, Canada, T2N 1N4, p. 377.

Chow, V.T., Maidment, D.R. and Mays, L.W. 1988. Applied Hydrology, McGraw- Hill, New York, p. 572.

Chow, V.T. 1964. "Statistical and Probability Analysis of Hydrologic Data", Handbook of Applied Hydrology, V.T. Chow editor, Mc-Graw Hill, New York, pp. 8.1-8.97.

De Vautier, B.A., Feldman, A.D. 1993. "Review of GIS applications in hydrologic modeling", Journal of Water Resources Planning and Management, ASCE, Vol. 119, No. 2, pp. 246-261.

Eagleson, P.S. 1972. "Dynamics of flood frequency", Water Resources Res., Vol. 8, No. 4, pp. 878-897.

Eastman, J. 1990. IDRISI: A grid-based geographic analysis system, version 3.2, Grad. School of Geography, Clark University, Worcester, Mass.

Flood Studies Report. 1975. Hydrological Studies, Vol. I, Whitefriars Press Ltd., London.

Hawking, S. 1993. Black Holes and Baby Universes and Other Essays, Bantam Books, New York, p. 182.

Henderson, F.M. 1966. Open Channel Flow, The MacMillan Comp., New York, p. 522.

Henderson, F.M., Wooding, R.A. 1964. "Overland flow and groundwater flow from a steady rainfall of finite duration", Journal of Geophysical Research, Vol. 69, No. 8, pp. 1531-1546.

Huff, F.A. 1967. "Time distribution of rainfall in heavy storms", Water Resources Research, Vol. 3,

No. 4, pp. 1007-1019.

Interagency Advisory Committee on Water Data. 1982. "Guidelines for determining flood flow frequency", Bull. 17B of the Hydrology Subcommittee, Office of Water Data Coordination, Geological Survey, U.S. Department of the Interior, Washington, D.C.

Jett, S.C., Weeks, A.A., Grayman, U.M. and Gates, W.E. 1979. "Geographic information systems in hydrologic modeling", Proceeding, Hydrologic Transport Modeling Symposium, A.S.A.E., New Orleans, Louisiana, pp. 127- 137.

Jones, N.L., Wright, S.G. and Maidment, D.R. 1990. "Watershed delineation with triangle-based terrain models", Journal of Hydraulic Engineering, Vol. 116, No. 10, pp. 1232-1251.

Klemes, V. 1972. Physically Based Stochastic Hydrologic Analysis. Advances in Hydroscience, Chow, V.T. editor, Academic Press, New York, pp. 285-356.

Kundzewicz, Z.W., Napiorkowski, J.J. 1986. "Nonlinear Models of dynamic hydrology", Hydrolog. Sci. Journal, Vol. 31, pp. 163-183.

Maidment, D.R. 1993. "Developing a spatially distributed unit hydrograph by using GIS", Application of Geographic Information Systems in Hydrology and Water Resources Management, IAHS Publication No. 211, IAHS Press, Institute of Hydrology, Wallingford, U.K., pp. 181-192.

Meyer, S.P., Salem, T.H. and Labadie, J.W. 1993. "Geographic information systems in urban stormwater management", Journal of Water Resources Planning and Management, ASCE, Vol. 119, No. 2, pp. 206-228.

Muzik, I., Chang, C. 1993. "A microcomputer-based geographic information system for hydrologic simulation", Microcomputers in Civil Engineering, 8, pp. 355- 365.

Muzik, I., Beersing, A.K. 1989. "Stochastic-deterministic nature of an elemental rainfall runoff process, Water Resources Research, Vol. 25, No. 8, pp. 1805- 1814.

National Engineering Handbook. 1972. Section 4, Hydrology, United States Department of Agriculture, Soil Conservation Service, United States Government Printing Office, Washington, D.C.

Palacios-Velez, O.L., Cuevas-Renaud, B. 1991. "SHIFT: A distributed runoff model using irregular triangular facets", Journal of Hydrology, 134, pp. 35- 55.

Pilgrim, D.H., Cordery, I. 1993. Flood Runoff, Handbook of Hydrology, Maidment, D.R. editor, McGraw-Hill, New York, pp. 9.1-9.42.

Ponce, V.M. Engineering Hydrology Principles and Practice, Prentice Hall, New Jersey, p. 640.

Quimpo, R.G. 1993. "Distributed models of the watershed", Advances in Hydro- Science an Engineering, Wang, S.S.Y. editor, Center for Computational Hydroscience and Engineering, School of Engineering, The University of Mississippi, University, MS 38677, p. 548-555.

Quinn, P., Bevan, K., Chevallier, P. and Planchon, O. 1991. "The prediction of hillslope flow paths for distributed hydrological modelling using digital terrain models", Hydrological Process, 5, pp. 59-79.

Rainfall Frequency Atlas of Canada. 1985. Canadian Climate Program, Environment Canada, Canadian Government Publishing Centre, Ottawa, Ontario, Canada.

Rodriques-Iturbe, I., Gonzalez-Sanabria, M. 1982. "A geomorphoclimatic theory of the instantaneous unit hydrograph", Water Resources Research, Vol. 18, No. 4, pp. 877-886.

Tachikawa, Y., Shinba, M., and Takasao, T. 1994. "Development of a basin geomorphic information system using a TIN-DEM data structure", Water Resources Bulletin, Vol. 30, No. 1, pp. 9-17.

Vieux, B.E. 1991. "Geographic information systems and non-point source water quality and quantity modeling", Hydrological Processes, 5, pp. 101-113.

Wyss, J., Williams, E.R. and Bras, R.L. 1990. "Hydrologic modeling of England River basins using radar rainfall data", Journal of Geophysical Research, 95 (D3), pp. 2143-2152.

Yevjevich, V. 1972. Probability and Statistics in Hydrology, Water Resources Publications, Fort Collins, Colorado, U.S.A., p. 302.

Zhang, W., Montgomery, D.R. 1994. "Digital elevation model grid size, landscape representation, and hydrologic simulations", Water Resources Research, Vol. 30, No. 4, pp. 1019-1028.

CHAPTER 11

GIS in Groundwater Hydrology

S. Gupta, G. Woodside, N. Raykhman and Jim Connolly

11.1. Introduction

Groundwater basins are commonly interconnected over large areas and exhibit a high degree of variability. Many groundwater systems are multilayered, and require three-dimensional evaluation of the subsurface hydrogeologic system. Discontinuities in geologic layers, major breaks in thickness or slope of hydrogeologic units, and spatial variations in hydraulic properties impact groundwater flow and contaminant transport. Groundwater flow analysis also requires quantitative data on the temporal variations of recharge, discharge and boundary conditions.

Information to describe the complexities of a hydrogeologic system are compiled in a groundwater basin data base. A groundwater basin data base includes lithologic logs, well construction information, water levels, recharge/discharge rates, and chemical concentrations, and other associated information, which normally adds up to several megabytes of electronic data. Available information in the form of maps and tables are also usually converted to an electronic form.

Up to the early 1980's, most groundwater system evaluations and data displays were performed manually with limited pre- and post-processors. With the availability of affordable computers, efficient software packages, more plentiful digital data, and advanced groundwater computer codes, hydrogeologists are developing the means to couple data processing, data display, and groundwater modeling. **Geologic Information System (GIS) provides an efficient data integrator**. Recently, data base tools and GIS software packages have been developed that include the capability to import and export information from diversified sources. The United States federal government's efforts are also assisting in standardization of data formats. Acceptance of a Spatial Data Transfer Standard (SDTS) into Federal Information Processing Standard (FIPS) 173 (DCDSTF,1988) is an illustration of these efforts.

V. P. Singh and M. Fiorentino (eds.), Geographical Information Systems in Hydrology, 303–321.
© 1996 *Kluwer Academic Publishers. Printed in the Netherlands.*

In this chapter, role of "loosely coupled" (rather than "fully coupled") GIS-groundwater models is presented with field application examples. The illustrations included in this chapter provide a general feel of the procedures and potential outcome of interfacing GIS with groundwater system analyses.

11.2. Role of GIS for Data Integration

In addition to efficient data integration, GIS provides means to develop large and complex groundwater basin data. The data base for groundwater basins are normally based on diverse data sources in varying coordinate systems and scales. GIS capabilities provide an easy means to convert available information into a consistent format. Useful GIS capabilities also include the ability to input or extract information in a user-defined map extent, and scale and projection change, including transformation between coordinate systems such as Universal Transverse Mercator and State Plane Coordinates.

Regional groundwater data are usually compiled from available information and supplemented with local-scale hydrogeologic investigation data. Hydrogeologists usually define alternative hydrogeologic interpretations of the available geologic information. Due to the limitation of available information, considerable professional judgement is used to develop hydrogeologic cross-sections and contour maps of key hydrogeologic units of the groundwater system. GIS generated displays of interpreted hydrogeologic layers, available groundwater levels, and chemical data of the given layers provide means for critical reviews of the interpretations to increase the "confidence level" in the hydrogeologic system definition. With increased awareness for long term environmental protection, thorough evaluations of all the available data, both spatially and temporally, requires a well-organized data base and efficient tools for accessing and displaying the data. The GIS is being increasingly used for importing, converting, editing, and displaying the data with the relevant background information.

11.3. "Loosely Coupled" GIS for Groundwater System Evaluations

The large size groundwater system data base and also the complexities associated with conceptual model development, simulations, and interpretation of results currently restrict groundwater simulations under the GIS platform. Listed in this section are the key explantations for use of "loosely coupled" approach rather than "fully coupled" GIS-groundwater models. The "loosely coupled" approach assists in efficient use of the technological advancements for solving the field problems. With increase in computation speed and availability of the additional cheap memory, the fully coupled GIS and groundwater models are likely to evolve in future. At present most of the large groundwater project use a team approach. Field hydrogeologists characterize the given basin, GIS analsists develop data structure and data base, and the groundwater system analyst performs the simulations. The simulation results are displayed using GIS capabilities. In this chapter the term "loosely coupled" also supports a team effort, rather than one single professional performing the entire groundwater system evaluations. Following identifies some of the steps and complexities which are involved in groundwater system evaluations.

Conceptual model development is a necessary precursor to groundwater model development. A well-organized data base in GIS assists in use of available data, and fosters development of a well-founded conceptual model. The hydrogeologist's understanding of the groundwater system is summarized in the conceptual model through cross-sections, areal maps, and time-series plots of hydrogeologic data. These steps involve several interations of data compilation, data processing, data displays, indentificaion to data gaps, field investigation. GIS analysts and geohydrologist work as a team to develop a mathematical representation of the groundwater system, which involves following efforts:

- Selection of the model dimensions (two-dimensional or three-dimensional).
- Identification of major hydrogeologic units and their properties (e.g top elevation, bottom elevation, hydraulic properties).
- Discretization of the model area (into rectangular or irregular elements)
- Selection of the model code
- Selection of the time frame for model simulation (steady state or transient; if transient, then how long).
- Assignment of model parameters to elements and/or nodes
- Execution of the model code
- Calibration and verification of the model
- Analysis of the model sensitivity

Because of data gaps and uncertainties in the conceptual model, calibration and verification of the model typically involves multiple model simulations for matching the simulation results with the historical data.

Evaluating projected future scenarios using the calibrated model also involves multiple simulations. Groundwater model computer codes usually require extensive memory and integration of several computer files to repeatedly solve a system of thousands of simultaneous equations. In general the groundwater simulations are perfomed outside the GIS platform and GIS capabilities are routinely used to modify the grid, hydraulic property distributions, and generate data tables and visual representation.

GIS and groundwater model, both require large memory. A coupled GIS-groundwater model may not be feasible with in the memory available in most of the workstations. The groundwater models are generally large with sophisticated numerical simulation methodology for the system of simultaneous equations. On the other hand, GIS may be characterized by efficient display and query capabilities. Current practices are to advocate "hybrid" or "loosely coupled" modes of GIS and groundwater computer codes. By seamlessly integrating the functionality of each, the superior qualities of GIS and a sophisticated groundwater model code are married to provide a powerful analysis tool.

GIS capabilities are efficiently used to process spatial data, generate input data files for simulations, and make maps of inputs and model results. Processing of spatial data may consist, for example, of compliation of hydraulic conductivity data at wells, and estimating areally-averaged hydraulic conductivity values across the model area from the point measurements at wells. Preprocessors integrate the GIS-generated information, such as the elevations for each unit, groundwater levels, and hydraulic properties, with the groundwater simulation model.

GIS capabilities also provide an efficient means to assist with model calibration. GIS is used to easily redefine the spatial parameters such as hydraulic conductivity or specific yield, and to provide the redefined parameters to the model. Calibration of large models is rendered more efficient through integrating the display and query abilities of GIS with the model simulation code. GIS capabilities are also routinely used for elegant display of groundwater simulation results (e.g. simulated heads, concentration distributions, and particle tracking results.) GIS-generated maps of simulation results with relevant background information are an integral part of the key project decisions made with the groundwater model.

11.4. Proposed GIS-Groundwater Modeling Coupling Approach

For the past seven years, the authors of this chapter have integrated ("loosely coupled") GIS (ARC/INFO) and CFEST (Coupled Flow, Energy, and Solute Transport), a three-dimensional finite element groundwater code (Gupta et. al, 1987) for several large groundwater projects. Since GIS capabilities to display temporal data and three-dimensional information are commonly lacking, we have developed supportive packages to supplement GIS capabilities. In this chapter we have included appropriate illustrations from integrated GIS and CFEST models to solve various groundwater flow problems. Use of the following steps assist with efficient integration of GIS and CFEST:
GIS for data integration:

- Groundwater basin data upload
- Conceptual model development
- GIS-CFEST interface

CFEST for model simulation:

- Calibration and scenario simulations

GIS for model presentation:

- Display of analysis results.

The following sections describe each of the above items with illustrations from groundwater simulation of large complex basins.

11.5. Groundwater Basin Data Upload

Groundwater basin data include definition of the areal extent of the basin, lithologic, hydrologic, chemical, and groundwater system management historical data. For the last several years, federal and state agencies have been centralizing storage of available information in a specific format. These centralized sources provide information on well logs, water levels, and concentration distributions. However, additional local-scale information are invariably compiled on a continuing basis from all possible sources. Below is a brief discussion on major categories of the groundwater system data.

External Extent of Groundwater Basin or Study Area
The first step in groundwater system evaluation is to identify the external extent of the study area. Groundwater system evaluations are normally extended to a natural boundary (e.g. large surface water bodies, groundwater divides). Available information on surface topography, surface geology, land uses, major roads, surface water bodies, and other associated information important for groundwater flow or for display of results are normally compiled in the data base. For some of groundwater basins, the above information may be in electronic format. Otherwise, hard copy maps are transformed to digital format by digitizing or scanning.

Lithologic Information
For each groundwater basin, the top elevations and thickness of each layer, location of identified faults, and variations in lithologic properties are defined using information from earlier studies or from well log data. Driller's logs, geologist's logs, and geophysical data are the basic information. New well log information are usually overlaid on earlier interpreted contour maps of each layer to verify or improve earlier interpretations. GIS provides an efficient means to import available information and redefine the properties of each geologic layer (such as top elevation, thickness, aerial extent). Raw geologic data, and interpreted information, such as top elevations and thicknesses of geologic units are maintained in the GIS data base. The GIS data base also includes zones of hydraulic properties interpretted from geologic data.

Water Level Measurements
Groundwater level data have historically been compiled in a tabular format. These data can be stored in a GIS or a relational data base. Groundwater level information are often stored in a spread sheet file and each record has well identification. Separate relational data file stores information associated with the geographic locations and construction details of the wells. Extraction of data for a given time period at selected wells normally requires development of computer programs to extract the user-specified data. Groundwater level data extracted from the data base are used for definition and interpretation of flow conditions, specification of initial conditions for model simulation, and model calibration.

Chemical Concentration Data

Chemical concentration data in a groundwater basin or Superfund site may add up to millions of data records. Because of the large size of some study areas, and the enormous amounts of data, all of these information cannot be displayed directly in GIS. In these cases, the GIS may serve primarily as a data base. For large sites, plotting with GIS allows display of concentrations at wells for a specific compound, or display of the maximum exceedance of drinking water standards for all compounds. GIS-generated point data displays such as these are used to define the plume extent and concentration variations. Figure 11.1 illustrates the use of GIS capabilities to display chemical concentration variations of volatile organic compounds. The manually interpreted contour lines are digitized and uploaded in the GIS data base. Concentration zones in Figure 11.1 are efficiently interpolated by a GIS package such as ARC/INFO into the groundwater model for contaminant transport simulation.

Water Budget Data

The locations of recharge and discharge rates are important components of groundwater system evaluations. These data include the locations and size of spreading grounds, river recharge areas, recharge from rainfall and septic systems, extraction and recharge wells, and flow across basin boundaries. Information regarding septic systems may include housing density data, and delimiting areas that are served by a sewage collection system versus area that have septic tanks. Data regarding pumping wells includes the locations and construction details of each extraction/injection well.

Spatial data describing the location and physical characteristics of recharge and discharge features must be coupled with temporal data defining the recharge and discharge rates. These data may include, for example, monthly pumpage quantities for each well and monthly recharge rates at spreading grounds. These data are normally stored in the GIS data base and extracted in an appropriate form for groundwater system evaluations.

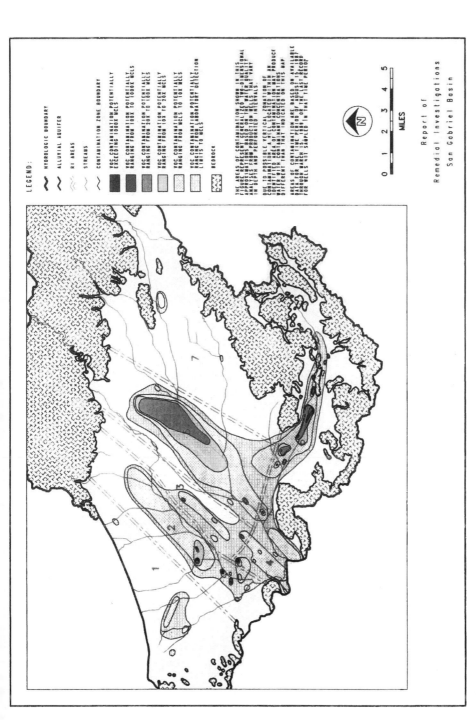

Figure 11.1. A GIS generated volatile organic compound distributions used to define initial concentrations for a 170 square mile groundwater basin analysis

11.6. Conceptual Model Development

The authors use GIS capabilities to display all the key groundwater system information on a given base map. Figure 11.2 illustrates one of the base maps which was generated to evaluate hydraulic performance of effluent disposal facilities. Included in this map are surface water bodies, boundaries of water disposal facilities, surface elevations, swamps, and other key background information. Base maps include injection/extraction well locations, recharge basins, chemical plumes, and all other key information which may impact groundwater flow or contaminant movement. Base maps that effectively present all the relevant information assist with efficient conceptual model development.

Large scale maps (usually 36 by 36 inches) are used to develop the finite element grid for the given basin. Figure 11.3 represents a conceptual finite element representation of area shown in Figure 11.2. Large elements on top and right side of the basin represent extension of the model boundary extent to reduce the impact of the boundary conditions on scenario evaluations. Figure 11.4 represent an additional example where GIS capabilities are used to represent all key features of an abandoned mine underneath a mountain. Using this GIS generated base map, a finite element representation of the system was developed. GIS-generated tabular data identifying top and bottom elevations of each key feature that are used to develop input simulation files.

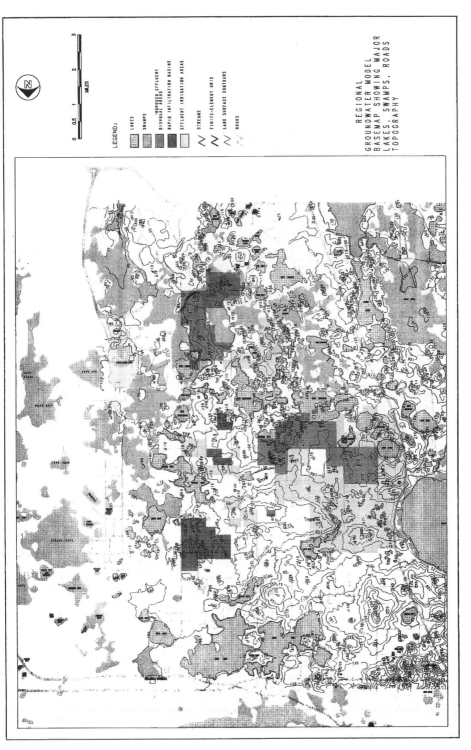

Figure 11.2. A GIS generated base map for groundwater model development

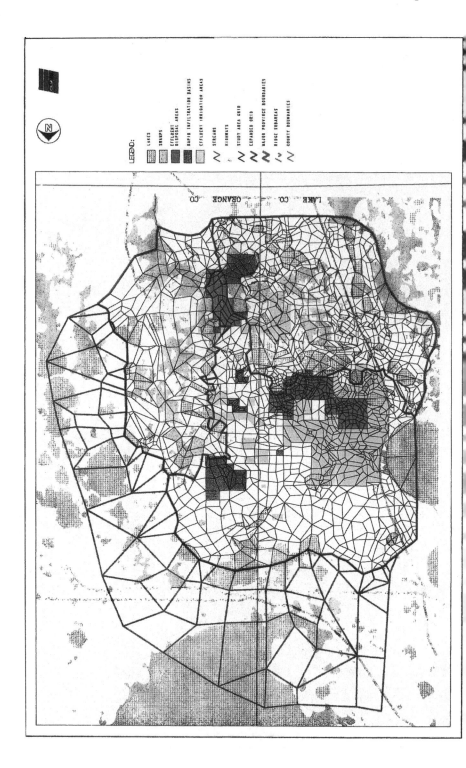

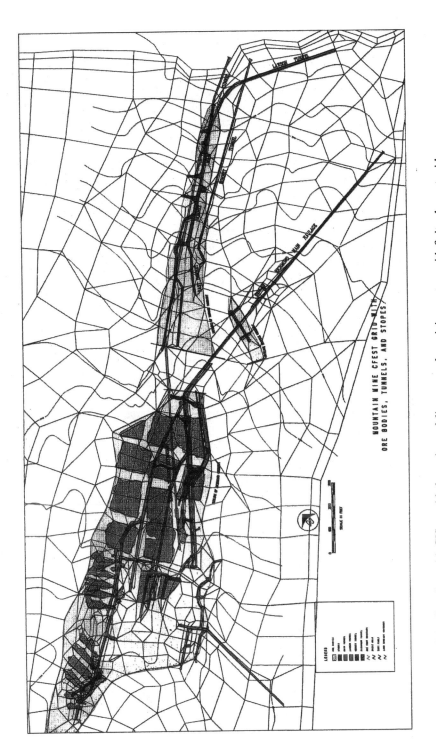

Figure 11.4. Coupling with GIS provided ease in modeling a complex mining system with finite element grid

11.7. GIS-CFEST Interface

The authors have developed interface processors that generate CFEST input or output in a format that is easily uploaded in GIS, or transferred from GIS to CFEST. Figure 11.5 illustrates a finite element grid generated using both manual and interactive GIS modifications. One of the key constraints in groundwater modeling is complexities or manual labor involved in each change in the finite element grid. GIS-integrated groundwater simulation, provides relatively simple means to redefine the grid and generate new descriptions of hydraulic properties, initial conditions, and layering details.

The GIS-CFEST interface consists of a series of computer programs that transfer data from the data base to CFEST, and from CFEST to the data base. Development of a seamless interface between GIS and CFEST allows the interface between the two softwares to be essentially transparent. This methodology allows the combination of the most powerful features of each software package into one working tool. The interface has been developed on a project-by- project basis, and updated as project-driven needs to demonstrate the necessity of additional features.

11.8. Calibration and Scenario Simulations

The data base describing the groundwater basin in normally limited. Hydrogeologic system definition typically have considerable associated uncertainties. Many groundwater basins have been historically over-extracted, and these basins are in a continual changing stage, requiring transient calibration. The transient calibration requires historical information of recharge rates, injection, and pumping rates. GIS capabilities provide efficient means to "tune" each spatial parameter and resimulate the basin. Historical water level measurements for a given time are compared with the simulation results using GIS map generation capabilities.

Flexible temporal data display in GIS is normally not provided in many software packages. Display of temporal data is commonly facilitated using software routines outside of the GIS. Figure 11.6 and 11.7 represent three-dimensional grid and calibration results. These graphical capabilities are incorporated in the CFEST software package to supplement the GIS limitation in visual display of historical and simulated results.

Figure 11.6 demonstrates the mathematical representation of a complex groundwater system in Ventura Coutny, California. Review of Figure 11.7 shows that groundwater levels on the left of fault are declining, but that on the right side of the fault there has been a continual rise in groundwater levels. This type of complex aquifer behavior is difficult to investigate and model without integrated GIS capabilities. Figure 11.7 also demonstrates that a reasonable matching of historical and simulated results are feasible for the basin using integration of GIS and CFEST.

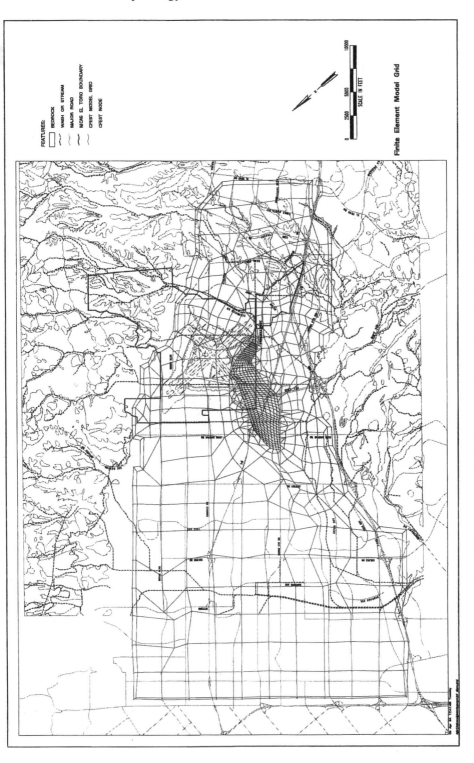

Figure 11.5. A GIS interactive grid (in plume area) for simulation of a Superfund site

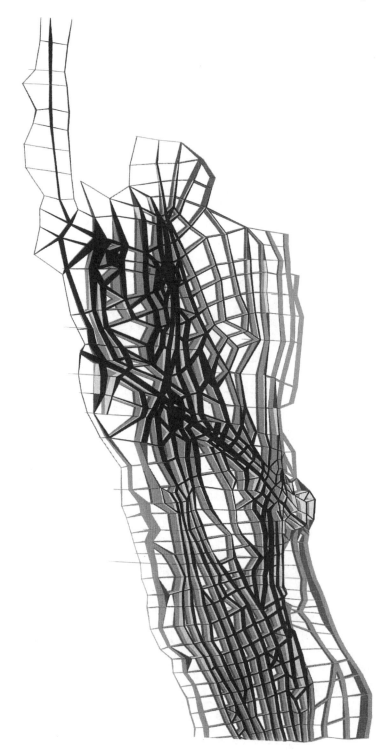

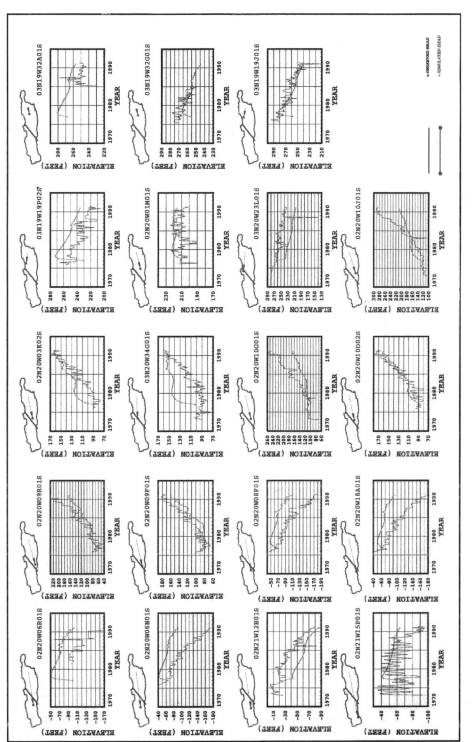

Figure 11.7. Transient calibration results of basin model shown in Figure 11.6 - temporal data display need other than GIS packages

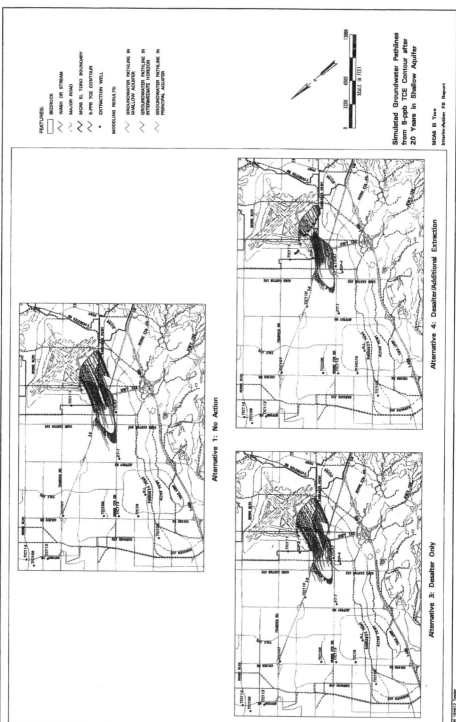

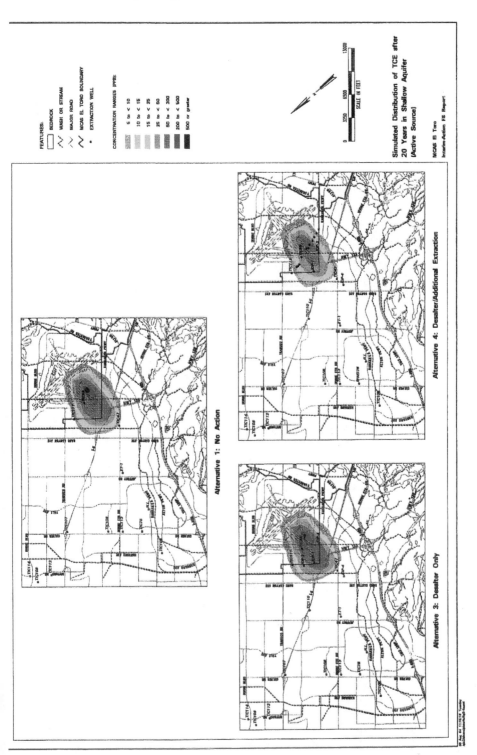

Figure 11.9. The GIS capabilities provide elegant means to display transport results from simulation grid shown in Figure 11.5

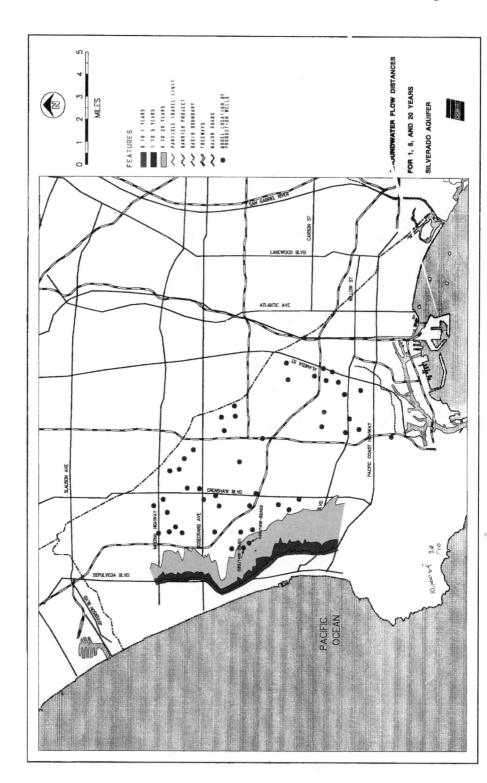

11.9. Display of Analysis Results

Figure 11.8 illustrates the use of GIS capabilities to display particle tracking results for several scenarios. This groundwater model corresponds to the conceptual model shown in Figure 11.5. Groundwater pathlines, or "particle tracks", are assigned different colors to demonstrate which aquifer the pathlines are traveling through. Figure 11.9 is an elegant display of contaminant transport results with background information (e.g. roads and other key features of the study area). Similar integrated use of GIS and CFEST were made for simulation of a seawater barrier project. Figure 11.10 illustrates the GIS-interfaced grid generation for the seawater barrier project. Figure 11.10 is another effective methodology for illustrating likely travel time in a given aquifer and location of nearby extraction wells. This figure displays the results of particle tracking with the groundwater code in a manner that is easily understandable to regulators and project decision-makers. Zones of travel time within specified years, such as 10 to 15 years from the injection barrier, are plotted to show the estimated time required for water to flow away from the injection barrier. This analysis provides an estimate of the time between when water is injected, and when the injected water may reach downgradient extraction wells.

11.10. Conclusion

Interfaced GIS and groundwater model code (e.g. CFEST) provide efficient tools to import available information, display input data maps for review and project uses, generate input simulation data, and efficiently display the analysis results with background information. One of the key capabilities of GIS is efficient display of available information and results for routine and key project decisions. Integrated GIS and CFEST simulations have been performed for several large complex groundwater systems.

The GIS-CFEST interface developed for these projects consists of a nearly seamless link between the data base and the model code, providing for ease of model development and calibration. Rather than providing schematic diagrams and descriptive procedures, this article presents illustrations from field applications to provide a general feel of procedures and the potential outcome of interfacing GIS with a groundwater model code.

References

DCDSTF (Digital Cartographic Data Standards Task Force) (1988) The proposed standard for digital cartographic data. The American Cartographer 15(1):9-140.

Gupta, S. K., C. R. Cole, C. T. Kincaid, and A. M. Monti. Coupled Fluid, Energy, and Solute Transport Model. Office of Nuclear Waste Isolation, Battelle Memorial Institute. Columus, Ohio. October 1987.

CHAPTER 12

Nonpoint Source Pollution Modeling (with GIS)

C.T. Haan and D.E. Storm

12.1. Introduction

Pollution modeling in this chapter refers primarily to the modeling of nonpoint pollution in surface runoff. This type of modeling is commonly denoted as hydrologic and water quality (HWQ) modeling. A good HWQ model must have a good hydrologic model as its foundation. HWQ models incorporate hydrology and water quality parameters. They are used to describe the occurrence and movement of water, nutrients, pesticides and other materials through a hydrologic system. Most HWQ models require large amounts of spatially referenced input data. A geographic information system (GIS) is an excellent way of managing these data. Further a GIS can be used to display modeling results in a manner that is understandable by the lay person. Results can be prepared in a wide array of graphical formats enabling one to better visualize model predictions. This chapter details efforts with nonpoint source pollution modeling and GIS in the United States. There are, however, significant advances being made outside the U.S.

Water quality is generally more difficult to model than water quantity because the quality of water emanating from a watershed is dependent on the pathways the water takes as it moves through the various flow systems to the watershed outlet. Water quality will reflect the characteristics of the material that the water flows over and/or through. Thus, a characterization of the quality of the water emanating from a watershed requires knowledge of the flow paths of the water and the chemical characteristics of the materials the water encounters as it makes its way to the watershed outlet.

The characterization of flow paths and the properties of the various materials the water encounters obviously requires a great deal of data. These data must be spatially referenced since the ability of the flow to transport pollutants is depend-

V. P. Singh and M. Fiorentino (eds.), Geographical Information Systems in Hydrology, 323–338.
© 1996 *Kluwer Academic Publishers. Printed in the Netherlands.*

ent on the current pollutant load of the flow, the availability of pollutants, and the transport capacity of flow. For example, flow that is heavily laden with sediment may not contribute to erosion as the flow traverses an erodible area if the flow capacity for sediment transport is already satisfied. However, clear water flow over the same area at the same rate may cause considerable erosion. In modeling nutrient loading to streams from agricultural areas it is important to spatially reference nutrient availability with respect to the stream of interest. Thus, it is important to have spatial referencing of the flow.

For watersheds that have considerable heterogeneity in their physical and chemical characteristics, a GIS is an excellent tool to manage the input data required for a HWQ model so as to maintain the spatial referencing of these data. The interactions among soil hydrologic properties, erodibility, land use, spatial and temporal distribution of rainfall, nutrient loading, and spatial referencing with respect to channelized flow can be reflected using distributed models and GIS in a way that is not possible with lumped parameter models. Thus, a GIS enables one to better describe the system being modeled and improve site specific parameter estimates for the model. A GIS can also be used interactively to display the results of nonpoint pollution modeling efforts and to show the location of the source and magnitude of the contribution of various pollutants to the overall quality of the runoff.

A GIS used in HWQ modeling may be either vector or raster based. To date the majority of work with HWQ problems has been done using the raster based approach. A GIS is also a natural to use along with a digital terrain model (Bevins and Moore, 1991). The digital terrain model can develop information on slopes, lengths, and areas that can go directly into a GIS for use in developing parameter estimates for a HWQ model. Expert systems are being developed that will assist a model user in determining the proper parameter values to use in a HWQ model. These expert systems will make recommendations based on GIS analysis of the available data.

One thing that must be kept firmly in mind is that any model system, whether it is a fully integrated GIS, expert system, HWQ model or a collection of individual models, should be considered as an analytic tool or decision aid. They should not be used to replace competent judgement. A great danger exists that as models become easier to use, those with little or no understanding of the system being modeled or the models being used will generate neat, professional looking HWQ reports that have not been subjected to competent professional judgement.

12.2. Model and Data Resolution

As with other types of hydrologic modeling, two major concerns are the spatial and temporal resolution required for a specific application and the data available to address the problem. Two broad categories of temporal resolution are event based and continuous simulation modeling. Similarly, two broad categories of spatial resolution are distributed and lumped parameter models.

12.2.1. TEMPORAL RESOLUTION

Event based models simulate the HWQ characteristics of a watershed for a particular hydrologic event such as a single rainstorm. Models of this type do not keep track of the hydrology of the watershed between events. Antecedent conditions existing at the time of the storm of interest must be supplied to the model or must be an implicit assumption. Event based models have found great use as engineering design models especially for flood protection.

Continuous simulation models simulate the response of a watershed to a continuous time series of climatic and other inputs. The outputs from a model of this type are also typically a continuous time series of flow and water quality parameters. Models of this type are especially valuable in estimating total, annual or seasonal pollutant loads and the temporal pattern of the concentrations of these constituents.

12.2.2. SPATIAL RESOLUTION

Models that require single estimates of parameters averaged over the watershed are known as lumped parameter models. Lumped parameters are weighted to reflect a measure of overall watershed response and are not necessarily representative of any particular location within the watershed. While models of this type may give reasonable estimates of total watershed response, especially if model parameters are calibrated to the particular watershed, they can not be used to identify particular areas within a watershed that contribute various levels of pollutants to this total response. Thus, it is difficult to use lumped parameter models to evaluate the effect of treatment measures on a part of a watershed on the total system response.

If a watershed is divided into cells or small areas, parameter inputs are developed for each of these areas, and if the spatial referencing of these parameters is maintained, the model is referred to as a distributed parameter model. The effects of the various parameters are geographically distributed throughout the watershed so that the contribution to the overall pollutant loading that originates within each cell can be identified. In these models the HWQ of each unit is determined as well as interactions between the units. The total watershed response is then an integration and routing of the individual unit responses. Thus, areas within the total watershed that are contributing to the pollutant load and the magnitude of this contribution can be identified.

12.3. Selection of Climatic Inputs

Most event based models require only rainfall as a climatic input since during a particular event lasting from a few minutes to 24 hours other processes such as evaporation which depend on temperature, solar radiation, and wind movement would be of little concern. For larger areas the spatial and temporal distribution of the event rainfall may be of concern whereas for smaller areas this spatial variability may be minimal. Many event based models are used as a basis for engineering design and as such require rainfall input appropriate to the design under consideration. This rainfall is generally specified in terms of a duration and a frequency of occurrence or return period. Basing input on rainfall of a specified duration and frequency of occurrence requires a statistical analysis to insure that the long-term characteristics of the rainfall are incorporated. If the temporal distribution of the rainfall within the duration is of importance, several methods are available for distributing the amount through time (Haan et al., 1994a).

Continuous simulation models may require a time series of rainfall, temperature, wind speed and direction and other variables for long periods using time increments of minutes to one day. These data may be observed data based on historic climatic records or may be stochastically generated using models specifically designed for this purpose (Richardson and Wright, 1984; Pickering et al., 1988).

For continuous simulation models it is especially important that the weather inputs reflect the long term expected behavior of the hydrologic system. Haan et al. (1994b) have shown how variable chemical transport in a natural hydrologic system can result from the normal variation in weather sequences. Generally very long records or many short records must be used to adequately characterize the transport processes. Using only one or two years of data can result in misleading conclusions since the years used may be "unusually" wet or dry. There are no generally accepted criteria for determining a "typical" year since not only annual totals but the temporal distribution of weather variables throughout the year is an important determinant of runoff and pollutant transport.

12.4. Parameter Estimation

The estimation of the appropriate parameters to use in a HWQ model is a very difficult process. For catchments with an adequate data base, observed flow and water quality data can be used as a guide in parameter estimation. Classical parameter estimation techniques can be used in the case of lumped models so that a parameter set that optimizes some objective function may be found.

Criteria typically used are a minimization of the prediction error sum of squares or sum of absolute deviations for a particular model output. For models that generate several outputs - i.e. stream flow, sediment load, nutrient load, etc. - a multivariate objective function involving all of the model outputs and their interactions may be better than a univariate optimization. There is no reason to believe, for example, that a set of parameters based on a minimization of a sum of squares on predicted stream flow would also result in the best estimates of sediment loads. An objective function that includes both stream flow and sediment load simultaneously would be preferred. Haan (1989) discuss multi-objective parameter estimation.

In most cases it is advisable to do a sensitivity analysis on the parameters and eliminate from the optimization those parameters that have little impact on the objective function. Otherwise convergence may be slow or not occur due to nonuniqueness in the parameter estimates. If a particular objective is defined by a subset of the parameters, the values for these parameters may be determined independently from the rest of the model.

In the case of a distributed model, observed data are useful in terms of adding user confidence in the models ability to simulate flow and water quality parameters, but generally are of only limited value in parameter estimation. A distributed model having n cells and m parameters per cell requires mn parameters. The value of m typically exceeds 10 requiring 10n or more parameters to be estimated for each cell. If the watershed is divided into 100 cells, more than 1000 parameters may need to be estimated. There may be many sets of parameters that yield essentially the same predictions. Thus, one can not use an optimization procedure on an objective function to get a unique estimate for the required parameter set. A GIS can be a valuable tool in estimating values for this large number of parameters.

In the absence of observed data with lumped models and generally with distributed models, parameter estimates must be based on the physical setting, on the experience of the modeler, and the accumulated experience of other modelers in the form of tables, charts, and algorithms. Often these algorithms can be programmed so that the data required for parameter estimation can be extracted from a GIS data base and provided directly to the model.

12.5. GIS/Model Integration

Figure 12.1 depicts several levels of integration between models and a GIS. These levels vary from essentially considering the GIS and the model as separate systems to fully integrating the model and the GIS. Currently the structure of HWQ models and GIS systems are quite different preventing a complete integration of the two modeling approaches. The lowest level of integration consists essentially of using the GIS as an aid in developing the input data file for the model. A user then takes the preliminary data file and modifies it to produce a complete input file in the format required by the model. This approach enables one to use an existing GIS and an existing model without modification to either but requires the most user effort.

The next level of integration is to use an interfacing program specifically written to communicate between the GIS and the model. The interface program may serve as a control program issuing commands to the GIS and the model. Output from the GIS is converted into the proper input format for the model and then read into the model. Output from the model may likewise be converted to a GIS format and then displayed by the GIS. All of these operations are carried on under the control of the interface program. This level of integration is largely the current state of the art. It enables one to use existing, well established models such as the HEC suite of models, SIMPLE, AGNPS, various EPA sponsored models, and usually the other models.

A third level of integration occurs when the interface program is incorporated

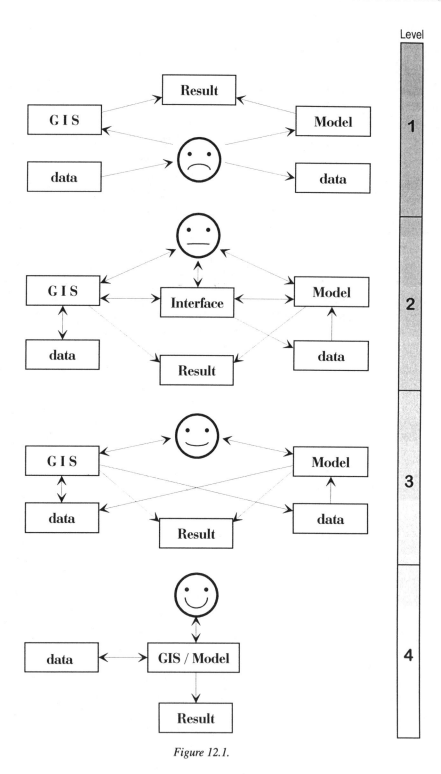

Figure 12.1.

into the model. This requires modification to the input/output routines of existing models or developing special input/output routines for new models. Some programming may also be done within the GIS to alter its input/output structure to make it more compatible with that of the model. If one is making extensive changes to a model or developing a new model, this level of integration would be appropriate.

The highest level of integration occurs when the model and the GIS are essentially a single, integrated unit. One way of achieving this is by programming the model using the programming language appropriate to the GIS being employed. Over time the programming language accompanying some GIS packages has increased in capability permitting this higher level of integration.

A current limitation on using GIS in pollution modelling is that it can not efficiently handle temporal variability in model parameters. In addition, GIS software is generally not powerful enough to solve some of the numerical models currently in use. Over time these limitations will undoubtedly be relaxed as GIS software undergoes further development.

12.6. Model Selection

There are many HWQ models. The various models each have strengths and weaknesses that must be carefully evaluated in terms of a particular application. It is important to use an appropriate model for the problem under consideration. Some models are better at predicting sediment loads, same are better at simulating nutrient loading, some are designed for special circumstances, and some combine surface and subsurface flow. The proper model to use depends on the problem at hand. If one wants to evaluate the impact of some particular practice or combination of practices on a water quality parameter, the model that is used must have parameters and algorithms that are sensitive to the practices being imposed. If one wants to estimate the nutrient loading to a lake, a continuous simulation model rather than an event model may be required. On the other hand the design of the required hydraulic capacity for a grassed filter strip may best be done with an event model. The following steps to be followed in model selection are based on Haan et al. (1993):

1. Identify the quantities that are important to model and the required accuracy.
2. Identify the HWQ processes that govern the behavior of the quantities to be modeled.
3. Identify models that adequately represent the identified hydrologic processes.
4. Identify available data.
5. Narrow model selection to those that have data requirements compatible with the available data or data that can be obtained.
6. Perform sensitivity analysis on the model to determine the most important parameters.
7. Assess the expected variability or uncertainty in the estimated parameters.

8. Assess the expected variability or uncertainty in model outputs resulting from uncertainty in input parameters.
9. Narrow model selection to those models capable of meeting accuracy requirements determined in step 1 in view of the uncertainty determined in step 8.
10. Test the models against any available data. Select those models that produce acceptable results.
11. Make final selection based on ease of use, documentation, output formats (i.e. tables, graphics, etc.), simplicity and familiarity of the user with the model.

For lumped parameter models, these steps can be applied in a straight forward manner. For distributed parameter models, steps 6-8 must be applied to single cells within the overall modeling scheme. Considering multiple cells in steps 6-8 would result in a very intense computing effort and would yield results that would be difficult to interpret.

Step one in the above process seems obvious but is often not given adequate attention. Problems to be solved are generally stated in general terms that must be translated into specific modeling objectives. For example, the problem under investigation might be the apparent eutrophication of a small lake. A panel assembled to address the problem may identify nutrient loading to the lake as a possible contributor. A model is sought to estimate nutrient loading. It would be a formidable task indeed to come up with a complete estimate of all nutrients entering the lake. The problem needs to be narrowed somewhat. Further discussion may indicate that the probable culprits are either nitrogen or phosphorus (or both) loading. Thus, a model is required to estimate the loading of these two nutrients to the lake. The watershed and stream above the lake contain agricultural and urban land uses with upper income residential areas and a couple of industries. All industrial effluent passes through a small waste water treatment facility operated by the local municipality before entering the lake. The agricultural and rural areas have a relatively dense animal population consisting of stables with pleasure horses, two dairies, and three poultry operations. Animal wastes are generally land applied within the watershed. Agricultural land use is limited to pastures and wooded areas, several gardens and a few truck cropping operations.

Anyone familiar with HWQ modeling recognizes that we have gone from a simple statement of desiring to model nutrient loading to a description of a rather complex watershed as far as water quality is concerned. We have determined that we need to estimate the loading of nitrogen and phosphorus. What remains is to determine the temporal scale required for the loading estimates and the accuracy that is required. If we require a continuous record of nutrient concentrations, it will require a more complex model than if we are satisfied with estimates on a daily basis. An annual time scale may be too course in that applications of fertilizers and animal wastes occur on specific dates and the proximity of these dates to rainfall events and the existing antecedent conditions have a major impact on the loss of nutrients in runoff and erosion.

Further, we need to determine why we want these estimates. If some measures are going to be evaluated to reduce the loading, we will need to know the exact sources of the nutrients and we will have to be able to determine the effectiveness of any proposed treatments. We will need to know where these treatments are being imposed, when they are being implemented and there proximity to potential nutrient sources.

A final aspect of step one is to develop an accuracy statement for the modeling. Do we need to be within 10% or 20% of the actual loading or do we just need a rough estimate getting us within 100% of the true loading. Presumably, since a problem exists, we will need a rather reliable estimate and will set our goal at estimating the loading to within 20% of the actual loading. As this discussion develops, the impossibility of unreasonably small tolerance limits will become apparent. Step two consists of identifying important HWQ processes governing the problem. Nitrogen and phosphorus may enter the lake in solution in the runoff water or through adsorption to sediments and organic matter carried into the lake by runoff water. Some nitrogen could presumably reach the lake by moving through the soil profile and then via ground water discharge to streams and the lake itself. If this is an important pathway, it must be included in the modeling effort. The quantities, and temporal and geographic location of nitrogen and phosphorus inputs to the system in the form of animal wastes and fertilizers needs to be identified. Other inputs of these substances from point sources and possibly from septic systems may also need to be identified. Transformations of the nutrients as they interact with the soils, water and organic materials they come in contact must be characterized. To adequately describe these interactions, the flow pathways must be identified so that reaction times and materials contacted can be accurately identified.

Step three requires examining existing models to ascertain which will adequately model the quantities and processes identified in steps one and two with an acceptable accuracy. Often compromises have to be made at this point as a model with all of the desired traits may not exist. One might attempt to develop a model specifically for the application at hand. In general, this is an expensive and time consuming option and often results in an unverifiable model. Generally a model that is well tested over a range of conditions is preferred unless special circumstances exist within the watershed under consideration. In the latter case, a special purpose model may be required. Such a model may have to be developed possibly by modifying an existing model. It is generally preferable to select the simplest model congruent with the modeling objectives. This will ease the modeling burden and reduce the parameter estimation problem. Of course it makes little sense to select a model that requires data that are not available and can not be obtained within existing time and budget constraints. Generally the more complete and complex models have more demanding data requirements than simpler models with more limited capabilities.

Once a model is selected, a sensitivity analysis should be conducted to ascertain the sensitivity of the model to various input parameters in the region of the parameter values corresponding to the watershed under consideration. Sensitivity coefficients can be computed from:

$$S = \frac{\partial O}{\partial I} \qquad S_r = \frac{\partial O}{\partial I}\frac{I}{O} \qquad (12.1)$$

where S and S_r are the sensitivity and relative sensitivity coefficients, respectively, O is the model output and I an input model parameter. For models that produce several outputs, sensitivity coefficients for each relevant output and parameter must be computed. S gives the absolute change in the output for a unit change in the input and has units corresponding to the units of the output divided by the input. S_r is dimensionless making it possible to compare relative sensitivity coefficients directly to ascertain which parameters must be determined with more accuracy. S_r gives the percent change in the output for a one percent change in the input. Parameters that produce low values for S_r can be estimated with less precision than parameters that have high values for S_r without adversely affecting prediction uncertainty. Thus, more effort should be expended in estimating parameters with large relative sensitivities.

Step seven requires an assessment of the expected variability in the input parameters. In general the values for input parameters are not known with absolute certainty. One has an estimate for the input parameters but can not be sure these estimates are 100% correct. One might feel that a particular parameter is known to within 15% of it true value. By combining this knowledge with the relative sensitivity of the model outputs to this particular parameter, one can get an estimate of the uncertainty that is present in the output due to the uncertainty in the input. For example, if S_r is 0.8, then a 15% error in the input parameter would translate to an approximate 12% error in the estimated output. If this error is too great, added effort needs to be made to reduce the uncertainty in the input parameter. It should be noted that the procedure described here addresses uncertainty in the estimates only. Accuracy of the estimates, or the ability of the model to predict what is actually happening, can only be fully assessed if data from the watershed of concern are available. Reasoning such as this should be extended to all of the input parameter and model output combinations.

Those models that indicate an acceptable predictive ability according to the criteria specified in step one, given the expected uncertainty in the input parameters, continue to be good candidates for the problem at hand. The next step is to compare model predictions with available data (if any) on the quantities being predicted. Only by comparing model results with measured results can one be sure of the predictive ability of the model on the watershed of interest. Otherwise one has to use data from other watersheds or published comparisons and assume that these results are transferrable to the watershed of interest.

Completion of all of these steps provides information that should enable one to make a rational decision as to which model should be used for the problem at hand. Often the decision is a default one in that only one model (hopefully at least one) survives all of these scrutinies.

12.7. Available U.S. Spatial Digital Data

The three basic data layers used in nonpoint source pollution models are land use, topography and soils. There are numerous data bases available for use with nonpoint source pollution models. Land use data used in nonpoint source pollution modeling are obtained from satellite imagery, aerial photography, or ground surveys. Satellite imagery typically used to develop land use data include Thematic Mapper (TM), Multispectral Scanner (MSS), Advanced Very High Resolution Radiometer (AVHRR), and *Système Pour l'Observation de la Terre* (SPOT). A few selective sources of satellite imagery and aerial photography are given in Table 12.1.

At a national level, a few examples of classified land use data in digital form include 4-ha data from the U.S. Geologic Survey (USGS) GIRAS data base (Land Use Land Cover or LULC) (Mitchell et al., 1977), and 1-km^2 AVHRR (Loveland and Scholz, 1993; Eidenshink, 1992). Another source of digital land use data is the Topologically Integrated Geographic Encoding and Referencing System, which is commonly referred to as the TIGER files. The TIGER files are a database developed by the U.S. Census Bureau. The database contains data on rivers, roads, railroads, political boundaries, and other related information. GIS World (1992) provides an extensive list of available sources of TIGER data.

The U.S. Geologic Survey provide a variety of digital topographic data that is used with HWQ models. These data are in the form of digital elevation models (DEMs) and digital line graphs (DLGs). DEMs are a grid of point elevations, and are available or will soon be available for 1:24,000 and 1:100,000 scale maps at sampling intervals of 30 m and 60 m, respectively. The entire U.S. database, however, is not available at the time this chapter was printed, but is currently under development. The U.S. Defense Mapping Agency (DMA) also has DEM data available in 3-arc second grids based on 1:250,000 scale maps. DLGs for the Public Land Survey System (PLSS), boundaries, transportation, hydrography, and hypsography are available from USGS for 7.5′ and 15′ quadrangles, and 1:100,000 and 1:250,000 scale maps. At the time of this publication, these data are only available for portions of the U.S.

Digital soils data for nonpoint source pollution modeling in the U.S. are available in two databases, the Soil Survey Geographic Database (SSURGO) and the State Soil Geographic Database (STATSGO) (Lytle, 1993). SSURGO is a detailed soils database with maps based on the USGS 7.5 minute quadrangles (1:24,000 scale). There is only a small portion of the SSURGO database completed at the time of this publication. However, the STATSGO soils database is available for the entire U.S. STATSGO is a more generalized database that was prepared using the USGS 1:250,000 scale quadrangles. STATSGO was created for more regional scale natural resource planning activities, while SSURGO was created for more site specific applications.

There are numerous additional sources of digital data. Other Federal agencies or State agencies are also many times a valuable source of digital and hard copy

TABLE 12.1. Selected sources of satellite imagery and aerial photography used to develop land use data for nonpoint source pollution modeling (Lillesand and Kiefer, 1994).

Source	Product
Aerial Photography Field Office U.S.D.A.-A.S.C.S. P.O. Box 30010 Salt Lake City, UT 84130	U.S.D.A. Agricultural Stabilization & Conservation Service Photographs U.S. Forest Service Photographs National Aerial Photography Program National High Altitude Photography Program
Bureau of Land Management Denver Service Center Division of Technical Services P.O. Box 25047 Denver, CO 80225	Bureau of Land Management Photographs
EOSAT Corporation 4300 Forbes Boulevard Lanham, MD 20706	Landsat Data
EROS Data Center Sioux Falls, SD 57198	U.S. Geological Survey Photographs National Aeronautics & Space Administration Photographs Landsat Data National Aerial Photography Program National High Altitude Photography Program
Satellite Data Service Branch NOAA National Environmental Satellite, Data, & Information Service World Weather Building Room 100 Washington, DC 20233	AVHRR Data
SPOT Image Corporation 1897 Preston White Drive Reston, VA 22091	SPOT Data

data. GIS World (1992) provide a detailed listing of other spatial databases. Many times, however, adequate data are not available for a specific application. When data must be developed, this is typically a very resource and time intensive endeavor. It should be emphasized that the most valuable and important component on any modeling effort is the data. Without adequate data, results obtained from nonpoint pollution models are questionable at best.

12.8. Nonpoint Source Pollution Potential Screening

Nonpoint pollution potential indexes based on soils, geology, topography, land use and other factors have been developed to quickly locate areas that have high and low potential for causing pollution of surface and ground water. These index methods are typically called models but are not models as commonly understood in hydrology. No modeling actual takes place. No algorithms or relationships are used to estimate an output based on an input. Rather various factors are assigned quantitative scores and these scores are combined to form an overall pollution potential rating. An example of this is DRASTIC (Allen et al., 1985) developed by the U.S. Environmental Protection Agency. The factors constituting DRASTIC are:

D	Depth to water
R	(net) Recharge
A	Aquifer media
S	Soil media
T	Topography
I	Impact of the vadose zone
C	Conductivity (hydraulic) of the aquifer.

A numerical ranking system to assess the potential for ground water pollution was developed for each of these factors. The system is based on weights, ranges and ratings. Weights range form 1 to 5 with 5 being the most significant or representing the highest pollution potential. Weights are used to relate the various factors to each other. Ratings range from 1 to 10 with 10 being the most significant. Ranges are dependent on a quantitative measure of the factor being considered. The pollution potential, P, is calculated from:

$$P = \sum_{i=1}^{7} R_i W_i \tag{12.2}$$

where R_i is the rating and W_i the weight for the ith DRASTIC factor.

DRASTIC can be implemented using a GIS by having a data layer corresponding to each of the seven factors. An attribute table for each factor provides the corresponding rating. The GIS is used to prepare a DRASTIC map of the area under investigation so that the most sensitive areas, as far as ground water pollution potential is concerned, can be identified. Evans and Meyers (1990) used this approach to characterize the ground water pollution potential for an area in south-eastern Delaware, U.S.A., and Rundquist et al. (1991) have applied this technique to Nebraska, U.S.A.

12.9. Targeting Critical Source Areas of Nonpoint Source Pollution

The Universal Soil Loss Equation (USLE) has been used in the U.S. since the early 1950's to predict average annual soil loss from agricultural fields (Wischmeier and Smith, 1978). To date, a common GIS application for nonpoint source pollution assessment has been to use the USLE with a delivery ratio to estimate sediment loading to surface waters (Hession et al., 1992; Hession and Shanholtz, 1988). This application, like DRASTIC, do not contain a hydrologic model, but provide an efficient method to calculate reasonable potential loading estimates for targeting critical source areas of nonpoint source pollution. Others have used the USLE along with export coefficients or unit-area loadings to estimate nitrogen and phosphorus loadings (Heidtke and Auer, 1993; Robinson and Ragan, 1994).

12.10. GIS Integrated Hydrologic and Water Quality Models

The more advanced applications of GIS with nonpoint source pollution modeling have been the use and/or partial integration of a GIS with a HWQ model. Most of these applications to date have used the raster based GIS GRASS (U.S. Army, 1991). Other applications of GIS with HWQ models include the use of VirGIS (Hession and Shanholtz, 1988) or ARC/INFO (ESRI, 1994). The following watershed- and basin- scale HWQ models have integrated or used GIS for parameter estimation and/or graphical display: 1) Spatially Integrated Model for Phosphorus Loading and Erosion (SIMPLE) (Chen et al., 1994), 2) AGricultural NonPoint Source (AGNPS) (He et al., 1994; Kang et al., 1992; Srinivasan and Engle, 1991a, 1991b; Vieux and Kang, 1990; Hession et al., 1989), 3) Soil and Water Assessment Tool (SWAT) (Srinivasan and Arnold, 1994), 4) Areal Nonpoint Source Watershed Environmental Response Simulation (ANSWERS) (Rewerts and Engle, 1991), 5) a field scale model for Chemicals, Runoff, and Erosion from Agricultural Management Systems - Water Table (CREAMS-WT) and CREAMS (Kiker et al., 1992; Bekdash et al., 1991), and 6) Finite Element Storm Hydrograph Model (FESHM) (Wolfe and Neale, 1988; Hession et al., 1987)

References

Allen, L., T. Bennet, J. H. Lehr, R. J. Petty. 1985. DRASTIC: A Standardized System for Evaluating Ground Water Pollution Potential Using Hydrologic Settings. EPA/600/2-85/018. Robert S. Kerr Environmental Laboratory, USEPA, Ada, OK, 74830.

Bekdash, F.A., A. Shirmohammadi, W.L. Magette, T.H. Ifft. 1991. Best management practices (BMP) evaluation using GIS-CREAMS linkage. ASAE Paper No. 91-7516. ASAE, St. Joseph, MI.

Bevins, K. J. and I. A. Moore. 1991. Terrain Analysis and Distributed Modeling in Hydrology. John Wiley and Sons Ltd., West Sussex, England.

Chen, Z., D.E. Storm, M.D. Smolen, C.T. Haan, M.S. Gregory, G.J. Sabbagh. 1994. Prioritizing nonpoint source loads for phosphorus with a GRASS-modeling system. Water Resources Bulletin, 30(4):589-594.

Eidenshink, J.E. 1992. The conterminous United States AVHRR data set. Photogrammetric Engineering and Remote Sensing 58(6):809-813.

Evans, B. M. and W. L. Myers. 1990. A GIS-based approach to evaluating regional groundwater pollution potential with DRASTIC. Journal of Soil and Water Conservation 45(2):242-245.

ESRI. 1994. Environmental Systems Research Institute, Redlands, CA.

GIS World. 1992. 1992-1993 International GIS SOURCEBOOK. GIS World, Inc., Fort Collins, CO.

Haan, C. T. 1989. Parametric uncertainty in hydrologic modeling. Trans. Am. Soc. Agr. Engrs. 32(1):137-146.

Haan, C. T., B. Allred, D. E. Storm, G. Sabbagh and S. Prabhu. 1993. Evaluation of Hydrologic/Water Quality Models - A Statistical Procedure. Presented at the 1993 International Winter Meeting, Paper 932505 of the American Society of Agricultural Engineers, St. Joseph, MI

Haan, C. T., B. J. Barfield and J. C. Hayes. 1994a. Design Hydrology and Sedimentology for Small Catchments. Academic Press, San Diego, CA.

Haan, C. T., D. L. Nofziger and F. K. Ahmed. 1994b. Characterizing Chemical Transport Variability Due to Natural Weather Sequences. Journal of Environmental Quality 23(2):349- 354.

He, C., J.F. Riggs, and Y.T. Kang. 1994. Integration of geographic information systems and a computer model to evaluate impacts of agricultural runoff on water quality. Water Resources Bulletin 29(6):891-900.

Heidtke, T.M., and M.T. Auer. 1993. Application of a GIS-based nonpoint source nutrient loading model for assessment of land development scenarios and water quality in Owasco Lake, New York. Water Science Technology 28(3-5):595-604.

Hession, W.C., V.O. Shanholtz, S. Mostaghimi, and T.A. Dillaha. 1987. Extensive evaluation of the Finite Element Storm Hydrograph Model. ASAE Paper No. 87-2570. ASAE, St. Joseph, MI.

Hession, W.C., and V.O. Shanholtz. 1988. A geographic information system for targeting nonpoint source agricultural pollution. Journal of Soil and Water Conservation 43(3):264-266.

Hession, W.C., K. L. Huber, S. Mostaghimi, V.O. Shanholtz, P.W. McClellan. 1989. BMP effectiveness evaluation using AGNPS and a GIS. ASAE Paper No. 89-2566. ASAE, St.

Hession, W.C., J.M. Flagg, S.D. Wilson, R.W. Biddix, V.O. Shanholtz. 1992. Targeting Virginia's nonpoint source programs. ASAE Paper No. 92-2092. ASAE, St. Joseph, MI.

Kang, Y.T. V. Siems, J. Bartholic, C. He, and B.E. Vieux. 1992. Using GRASS WATERWORKS in the Sycamore Creek Watershed: A case study of interfacing a GIS with a water quality model. Poster paper presented at the 47th Annual Meeting of the Soil and Water Conservation Society, August 9-12, 1992, Baltimore, Maryland.

Kiker, G.A., K.L. Campbell, J. Zhang. 1992. CREAMS-WT linked with GIS to simulate phosphorus loading. ASAE Paper No. 92-2016. ASAE, St. Joseph, MI.

Lillesand, T.M. and R.W. Kiefer. 1994. Remote Sensing and Image Interpretation. John Wiley & Sons, New York, NY.

Loveland, T.R., and D.K. Scholz. 1993. Global data set development and data distribution activities at the U.S. Geological Survey's EROS Data Center: American Congress on Surveying and Mapping. American Society for Photogrammetry and Remote Sensing Annual Convention and Exposition. New Orleans, Louisiana, February 1993, Proceedings, Bethesda, Maryland, 2:204-211.

Lytle, D.J. 1993. Digital soils databases for the United States. Environmental Modeling with GIS. Oxford University Press, New York, NY.

Mitchell, W.B., S.C. Guptill, K.E. Anderson, R.G. Fegeas, and C.A. Hallam. 1977. Geographic information retrieval and analysis system for handling land use and cover data. U.S. Geological Survey, Professional Paper No. 1059. 16 p.

Pickering, N. B., J. R. Stedinger and D. A. Haith. 1988. Weather Input for Nonpoint source Pollution Models. Journal of Irrigation and Drainage Engineering, American Society of Civil Engineers, JIDEH 11(4):674-690.

Rewerts, C.C. and B.A. Engle. 1991. ANSWERS on GRASS: Integrating a watershed simulation with a GIS. ASAE Paper No. 91-2621. ASAE, St. Joseph, MI.

Richardson, C. W. and D. A. Wright. 1984. WGEN: a model for generating daily weather variables. U. S. Department of Agriculture, Agricultural Research Service, ARS-8, 83p.

Robinson, K.J., and R.M. Ragan. 1994. Geographic information system based nonpoint source pollution modeling. Water Resources Bulletin 29(6):1003-1008.

Rundquist, D.C., D.A. Rodekohr, A.J. Peters, R.L. Ehrman, L. Di, and G. Murray. 1991. Statewide groundwater-vulnerability assessment in Nebraska using the DRASTIC/GIS model. Geocarto International 2:51-58.

Srinivasan, R., and B.A. Engle. 1991a. A knowledge based approach to extract input data from GIS. ASAE paper No. 91-7045. ASAE, St. Joseph, MI.

Srinivasan, R., and B.A. Engle. 1991b. GIS: A tool for visualization and analyziation. ASAE paper No. 91-7574. ASAE, St. Joseph, MI.

Srinivasan, R. and J.G. Arnold. 1994. Integration of a basin-scale water quality model with GIS. Water Resources Bulletin 30(3):453-462.

U.S. Army Corps of Engineers. 1991. GRASS Reference Manual Version 4.0. USA CERL, Champaign, Illinois.

Vieux, B.E. and Y.T. Kang. 1990. GRASS WATERWORKS: A GIS toolbox for watershed hydrologic modeling. Conference on application of geographic information systems, simulation models, and knowledge-based systems for land use management. November 12-14, 1990, Blacksburg, Virginia.

Wischmeier, W.H., and D.D. Smith. 1978. Predicting rainfall erosion loss - a guide to conservation planning. Agricultural Handbook No. 537. U.S. Department of Agriculture, Washington, D.C.

Wolfe, M.L. and C.M.U. Neale. 1988. Input data development for a distributed parameter hydrologic model (FESHM). In: Modeling Agricultural, Forest, and Rangeland Hydrology. Proceedings of the 1988 International Symposium of the ASAE. December 12-13, Chicago, IL.

CHAPTER 13

Soil Erosion Assessment Using G.I.S.

A.P.J. De Roo

Abstract.   The use of GIS in soil erosion modelling is demonstrated. Several currently used soil erosion models are discussed. Taking the recently developed LISEM soil erosion model as an example, the capabilities of models integrated in a GIS is shown. The main reasons for using a GIS is that runoff and soil erosion processes vary spatially, so that cell sizes should be used that allow spatial variation to be taken into account. Also, the data for the large number of cells required is enormous and cannot easily be entered by hand, but can be obtained by using a GIS. The possibilities of rapidly producing modified input-maps with different land use patterns or conservation measures to simulate alternative scenarios, the ability to use very large catchments with many pixels, so the catchment can be simulated with more detail, and the facility to display the results as maps are further advantages of using a GIS.

13.1. Introduction

Recently, awareness of soil erosion problems all over the world has increased (Lal, 1988). This is indicated by several reports of long-term loss of fertile arable soils and the damaging effects of surface runoff and soil erosion downstream. However, quantitative information on soil erosion rates and the effects of conservation strategies is often not available. Policy makers, farmers organizations and environmental groups need a quantitative evaluation of the extent and the magnitude of the soil erosion pro- blems and the possible management strategies on a regional basis. Therefore, field measurements are necessary. Also, quantitative simulation models of surface runoff and soil erosion, which can be used to evaluate alternative strategies for improved land management, not only in monitored areas, but also in ungauged catchments, are useful.

If these models can be linked to or integrated in a Geographical Information System (GIS), a powerful research tool is available. Unlike a large number of mod-

V. P. Singh and M. Fiorentino (eds.), Geographical Information Systems in Hydrology, 339–356.
© 1996 Kluwer Academic Publishers. Printed in the Netherlands.

els, which need to be fed with input files typed in manually, the power of soil erosion models using a raster or vector GIS database, used to generate the input files for the model and to display the results, is demonstrated. With the GIS it is shown that a detailed description of the catchment morphology by Digital Elevation Models (DEM's) greatly benefits modelling soil erosion, in spite of other variables that cannot be described in such detail.

13.2. Soil Erosion Processes

Soil erosion can be defined as 'removal of soil by forces of nature more rapidly than various soil-forming processes can replace it, particularly as a result of man's ill-judged activities' (Monkhouse & Small, 1970). The terms 'rapid erosion' and 'accelerated erosion' are used to distinguish this form of erosion from 'natural' or 'geological' erosion.

Several factors influence soil erosion and flooding. Numerous research workers have stressed the sensitivity of the some soil types, e.g. loess and in general silt loams, to soil erosion and soil sealing. Therefore, the main cause of soil erosion is the presence of susceptible soils on undulating topography with a poor cover by vegetation or crops. Important factors in the soil erosion process are the climate, slope gradient, land use, urbanisation, ploughing direction, soil stability, soil sealing, infiltration capacity and the recent intensification of agriculture (De Roo, 1993b).

In order to understand how to reduce the magnitude of the world wide soil erosion problem it is necessary to make and test quantitative models of runoff and erosion which can be used to evaluate alternative strategies for improved land management.

13.3. Soil Erosion Models

Models are formulations of processes and logic as represented by the modeller. The formulations may be representations of simple processes or combinations of processes as the modeller deems necessary to solve a specific problem. Therefore, sometimes variables may be included that are not physically measurable. These variables must be derived by a procedure of calibration for a specific case. This does not mean that a model has a low quality, but it is just a compromise between the physical basis of a model and available time and money for model development, data acquisition and computing time.

Model users should clearly define what they expect from their modelling study. Soil erosion models can be used for farm planning, site-specific assessments, project evalu- ation and planning, policy decisions or as research tools to study processes and the behaviour of hydrologic and erosion systems (DeCoursey, 1985; Foster, 1990). Decisions have to be made about which specific objectives are most important in evaluating the problem, and the need for absolute values or relative comparisons between alternative scenarios. After defining their expectations, model users should carefully consider attributes of different models relative to the specific problem to determine which can provide the desired results. Some consid-erations include model purpose, representation, data requirements and availability, ease of parameter es- timation, and both ease and cost of simulation (Leonard & Knisel, 1986).

Because no prediction method is perfect, one of the most difficult issues is the validity of soil erosion models. Foster (1988) states: 'The best measure of validity is this: Does the method serve its intended purpose?'. If the purpose of a model-ling study is to provide global estimates of soil erosion such as 'no risk of erosion', 'slight erosion' and 'severe soil erosion', a different model is needed than if a re-searcher wants to simulate exactly what is going on on a hillslope. For example, if we consider a relatively simple empirical model such as the Universal Soil Loss Equation (USLE) (Wischmeier & Smith, 1978), a regression equation used to cal-culate long-term average annual soil loss from small areas (see below), the model will do a very good job in global estimation, and thus may be considered a 'good' model. However, the model cannot describe the physical processes on a hillslope. May the USLE be considered as a 'bad' model? Summarizing, the purpose for which the model has been developed has to match the needs of a model user. One cannot just classify the USLE as a bad model because it was never developed to describe the physical processes on a hillslope.

If a user requires global estimates/risks of soil erosion in large areas a relative simple model with few variables is probably the best choice. If a more complex model is used, the data base required will be tremendous. If a complex model is run with too few data and a large part of the area is simulated with data 'assumed' to be constant over a certain part of the area, the results may be very unreliable. Therefore, one criterion to select a model may be the presence of readily available data (soil maps, topographical maps, remotely sensed data on land use). The ideal model may be available, but if the user does not have the resources to collect the required data, the model is of little use (DeCoursey, 1985). To overcome the lack of data problem, more recent models such as EPIC and WEPP (see below) include subroutines that generate some of the input data needed such as daily rainfall, air temperature, wind and solar radiation, provided the user identifies the study site.

Also, the most important time scale needed has to be defined. Some models were developed to simulate response for a single, design-type storm event. Other models give weekly or monthly values, or longterm annual values. Single event models may be adequate when considering only surface runoff and sediment yield, but when for example the effects of soil erosion on crop yields need to be con-sidered, continuous (or daily) simulation over a relatively long climatic record is recommended.

Another criterion important for model selection is the need for calibration. If a user needs to predict runoff and sediment yield in one catchment only, and field measurem- ents are available for calibration, a calibrated lumped model may do a very good job. However, if models that need calibration are applied to differ-ent sites or management practices without recalibration, erroneous results can be obtained. Therefore, the use of models which need calibration in ungauged catchments is not desirable.

Finally, the choice between a lumped and a distributed model must be taken carefully. Four major areas which offer the greatest potential for the application of distributed models can be identified (Beven, 1985). These are: forecasting the effects of land-use change; forecasting the effects of spatially variable inputs and outputs; forecasting the movement of pollutants and sediments; and forecasting the hydrological response of ungauged catchments where no data are available for calibration of a lumped model.

Currently used mathematical deterministic soil erosion models

Reviews of existing hydrologic and soil erosion models can be found in Curwick and Jennings (1982), Foster (1982), DeCoursey (1985), Perez-Trejo (1986), Novotny (1986), Bork (1988), Smith and Ferreira (1988), Marsalek (1989) and De Roo (1993b). Several groups of models can be distinguished: single equations, lumped models, empirically-based distributed models and physically-based distributed models. A major division can also be made in hillslope models and catchment models.

USLE, the Universal Soil Loss Equation (Wischmeier et al., 1958; Wischmeier & Smith, 1978), is the best known and most used soil erosion model. The USLE has been developed to estimate interrill soil losses over extended time periods. The simple, empirical equation has been developed from regression analyses of 10.000 plot years of data from natural runoff plots ('Wischmeier' plots) and plots under artificial rainfall simulators in the USA, east of the Rocky Mountains. Therefore, the validity of the USLE outside this part of the USA can be called into question. Further limitations are that the equation does not estimate deposition, sediment yield, channel erosion, or gully erosion. The model is not accurate for a single storm event (Foster, 1982).

The AGNPS   (AGricultural NonPoint Source) (Young et al., 1987; Young et al., 1989; Mitchell et al., 1993) model is a single event total storm model. Runoff is predicted using the Soil Conservation Service (SCS) runoff curve number method. Sediment yields are predicted using a modified version of the USLE. Nutrient movement com- ponents have been adapted from the CREAMS model. AGNPS has been linked to the GRASS GIS (Mitchell et al., 1993). Catchments up to 20,000 ha can be simulated using AGNPS.

The WEPP   (Water Erosion Prediction Project) Hillslope Profile version erosion model (Nearing et al., 1989) is a 'new generation' soil erosion model that can be run as a continuous simulation model and on a single-storm basis. WEPP is based on fundamentals of infiltration, surface runoff, plant growth, residue decomposition, hydraulics, tillage, management, soil consolidation, and erosion mechanics.

WEPP uses equations for sediment continuity, detachment, deposition, shear stress in rills, and transport capacity. Also temporal variations of erodibility variables as a function of above and below ground residue, plant canopy, and soil consolidation are incorporated. WEPP uses the Green/Ampt infiltration equation to calculate overland flow. The model requires a large number of data on management practices, which may be difficult to obtain. Also, the model uses pedotransfer functions using the grain size distribution and the organic carbon content for estimating a large number of variables. However, the validity of these functions is not discussed. The WEPP Catchment version is available since 1995.

ANSWERS, Areal Nonpoint Source Watershed Environment Response Simulation (Beasley et al., 1980), is a distributed erosion model designed to simulate the hydrological behaviour of catchments having agriculture as their primary land use, during and immediately following a rainfall event. Its primary application is in planning and evaluating various strategies for controlling surface runoff and sediment transport from intensively cropped areas. The hydrologic part of the model originates from the USDAHL model (Holtan et al.,1967; Holtan & Creitz, 1969; Holtan et al., 1975). The model uses the Holtan infiltration model, slightly changed by Huggins and Monke. Kinematic wave equations are used to route overland flow and channel flow. The model uses separate equations for detachment and transport on overland flow areas and transport in channels. Limitations of the model are that the model allows no erosion in channels, gully erosion is not included, and groundwater flow is simulated empirically. A strong and unique characteristic of the ANSWERS model is that it considers runoff transmission losses (Curwick & Jennings, 1982), and that the model is easy to link with raster-based Geographical Information Systems and remotely sensed data.

KINEROS, the KINematic EROsion Simulation model (Smith, 1981; Woolhiser et al., 1990) is an event-oriented, physically based model describing the processes of interception, infiltration, surface runoff, and erosion from small agricultural and urban catchments. It uses the Smith/Parlange infiltration model and the kinematic wave approximation to route overland flow. The catchment is represented by a cascade of planes and channels. The strength of this model is the physically-based infiltration equation in contrast to the empirically-based approach in ANSWERS and FESHM, using the Holtan equation (Curwick & Jennings, 1982).

SEM/SHE, the Soil Erosion Model (Storm et al., 1987; Styczen & Nielsen, 1989), is a physically based model which simulates the spatial and temporal variations of the soil erosion and sediment yield in a catchment. It has been developed based on a physical description of the involved processes, to improve models such as ANSWERS and MODANSW. SEM has been incorporated as a separate component in the physically- based, distributed hydrological modelling system SHE.

The sheet erosion equations are to a large extent based on the ROSE model. Unique in erosion modelling are SEM's attempts to simulate rill erosion. Incorporated are sub-processes such as rill initiation, headcut erosion, addition of interrill material to the rill, rill tail erosion, and erosion on bottom and sides, wall collapse and erosion of this material.

MEDALUS (MEditerranean Desertification And Land USe) is a physically based simulation model for vegetation growth, hydrology and soil degradation (Kirkby et al., 1992). The MEDALUS I model simulates a hillslope catena, consisting of successive slope segments. In each individual segment plant growth, evapotranspiration, surface water and sediment transport, and soil water movement are simulated. The model is focused on the Mediterranean, with special attention given to the influence of raindrop impact induced sealing on soil water infiltration and representative types of plants. Two and three dimension (catchment) versions of the model (MEDRUSH) are in preparation.

EUROSEM, the European Soil Erosion Model (Morgan et al., 1991/1992), is a single event process-based model for predicting soil erosion by water from fields and small catchments. It can also be used to assess the risk of erosion and to evaluate the effects of soil protection measures. The equations used in the model to describe the processes of erosion are from many sources and are claimed to be representative of the current state of research in Europe. The equations are linked to the KINEROS model which provides the basis for generating runoff as infiltration excess and for routing runoff and sediment over the land. EUROSEM aims to simulate three situations: (1) A single plane or element for predicting erosion from small fields; (2) Multiple planes or elements for predicting erosion along a slope; (3) Multiple planes and channels for predicting erosion from a small catchment. In the future, the model will be linked to a revised version of the SHE model.

LISEM, the LImburg Soil Erosion model (De Roo et al., 1993b) is a distributed physically-based catchment model of soil erosion developed by the Utrecht University and the Winand Staring Centre Wageningen. The model, integrated in a raster GIS, simulates single rainfall events in small catchments. Both present land use and soil conservation scenarios can be evaluated using the model.

13.4. Soil Erosion Models Using GIS

The advantages of linking a model with a GIS have been described also in De Roo et al. (1989) and De Roo (1993b). Further examples of soil erosion models linked to or integrated in a GIS can be found in Mallants & Feyen (1989), Mitchell et al. (1993), Dryer & Frhlich (1994) and De Jong (1994). The main reasons for using a GIS is that runoff and soil erosion processes vary spatially, so that cell sizes should be used that allow spatial variation to be taken into account. Also, the data for the large number of cells required is enormous and cannot easily be entered by hand, but can be obtained by using a GIS. Not all data are equally available, however. At present most detailed information is available about the terrain form (Kirkby, 1985). To make the best possible use of distributed topographic data a catchment should be modelled by a digital elevation model (DEM). In many cases this DEM will have several thousands of elements, so that entering data by hand is no longer feasible. A DEM can be con- structed by digitizing contour maps. The GIS can compute maps of altitude, slope and aspect, which are all input for the LISEM model. Because detailed field sampling of input variables is not feasible, a limited number of point observations of the soil etc., collected during field exper- iments, are often available. Geostatistical interpolation techniques, incorporated in the GIS, can be used to produce maps from these point observations. Using a method such as block kriging, point observations are interpolated to blocks of the same size as the elements used for simulation. When there are no sufficient field measurements available, the distribution of a desired input variable can be derived from digitized soil or land use maps. A raster-based GIS is the ideal tool to serve needs and fulfil requirements associated with the DEM and the geostatistical in- terpolation techniques.

Further advantages of using a GIS are 1) the possibilities of rapidly produ- cing modified input-maps with different land use patterns or conservation meas- ures to simulate alternative scenarios, 2) the ability to use very large catchments with many pixels, so the catchment can be simulated with more detail, and 3) the facility to display the results as maps. A series of maps can be produced show- ing the variation with time of spatial patterns of soil erosion, sedimentation and runoff over the catchment. These maps can be compared by subtraction to yield maps indicating how erosion or sedimen- tation might be affected by certain con- trol measures within the catchment or they can be viewed successively to create a video of the modelled process. Runoff can also be displayed as an overlay on the landform surface (digital elevation model) (figure 13.1).

The main advantage of incorporating models in GIS is that the 'source code' of the model then resides on the comprehensible abstraction level of one or two lines of source code, a GIS command, per process (e.g. interception, infiltration and sediment routing). Such a high level of abstraction simplifies model modifica- tion, maintenance and reusability of parts of the model in other models. The current implementation of LISEM is less than 200 lines (exclusive comments). The GIS must contain a set of tools to build such models (Van Deursen & Kwadijk, 1990). Therefore PCRaster contains tools to query and report time series, routing tools such as drain, accumulate and kinematic wave (Chow et al., 1988) and equation solvers such as Newton's method.

Besides the availability of the right set of tools to create a model, the GIS must have advanced scripting facilities. While most temporary GI systems are capable of scripting series of commands that could make up a model, often their scripts suf- fer from an inefficient data transfer mechanism and a lack of abstraction level de-

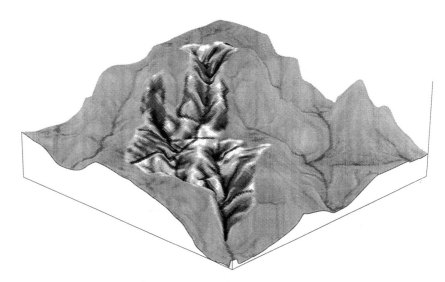

Figure 13.1. Simulated overland flow at 2:30 AM, 27 May 1988, in the Etzenrade catchment (South-Limburg, The Netherlands) draped over the digital elevation model. The background is draped with the 'wetness' index, as used by Beven & Kirkby (1979)

sired for modelling complicated processes. Especially the data transfer between the various commands makes it practically impossible to run models, since each command output is written to the database or file system and then read by some other command despite the fact that most information is only used local in the model. In other words intermediate output that is written is only used in some succeeding step and then discarded. For example, modelling an event of 9 hours with a time step of 15 seconds by a 200 line model would result in a data transfer of 432,000 raster maps.

13.5. LISEM: a Physically-Based Model Integrated in a GIS

13.5.1. INTRODUCTION

The LISEM (LImburg Soil Erosion Model) model, a physically-based hydrological and soil erosion model, is one of the first examples of a physically based model that is comple- tely incorporated in a raster Geographical Information System. Incorporation means that there are no conversion routines necessary, the model is completely expressed in terms of the GIS command structure. This approach was introduced by Van Deursen and Kwadijk (1990) and the principles were demonstrated in RHINEFLOW, a water balance model for the river Rhine (Van Deursen & Kwadijk, 1993). PCRaster (Van Deursen & Wesseling, 1992) is used as the GIS to prototype these ideas. Furthermore, the incorporation facilitates easy application in larger catchments, improves the user friendliness, and allows remotely sensed data from airplanes or satellites to be used. If required, the model can be linked easily with other GIS's.

The development and structure of the LISEM model is based on the experiences with the ANSWERS model (Beasley et al., 1980; De Roo et al., 1989; De Roo, 1993a; De Roo, 1993b) and SWATRE (Belmans et al., 1983), but process descriptions are changed totally. Processes incorporated in the model are rainfall, interception, surface storage in micro-depressions, infiltration, vertical movement of water in the soil, overland flow, channel flow, detachment by rainfall and throughfall, detachment by overland flow, and transport capacity of the flow. Also, the influence of tractor whee- lings, small paved roads (smaller than the pixel size) and surface sealing on the hydrolo- gical and soil erosion processes is taken into account. For a detailed description of the processes incorporated in the model is referred to De Roo et al. (1994b).

After rainfall begins, some is intercepted by the vegetation canopy until such time as the maximum interception storage capacity is met. Besides interception, direct throughfall and leaf drainage occur, which, together with overland flow from upslope areas, contribute to the amount of water available for infiltration. The amount of water remaining after infiltration begins to accumulate on the surface in micro-depressions. When a predefined amount of depressions are filled, overland flow begins. Overland flow rates are calculated using Manning's n and slope gradient, with a direction according to the aspect of the slope. When rainfall ceases, infiltration continues until depression storage water is no longer available. Soil detachment and transport can both be caused by either raindrop impact or overland flow. Whether or not a detached soil particle moves, depends upon the sediment load in the flow and its capacity for sediment transport. When water and sediment reach an element with a channel, they are transported to the catchment outlet. Sedimentation within a channel appears when the transport capacity has been exceeded.

LISEM is written in a prototype GIS modelling language currently developed at the University of Utrecht. The language comprises all PCRaster commands as statem- ents with exactly the same syntax as the command form of this statements. When compiled, an efficient run time mechanism eliminates redundant data transfer.

A large part of the source code of LISEM contains sets of equations using the 'CALC' statement, which can be described as an intelligent calculator for maps. If e.g. the maximum amount of rainfall which is needed to fill a micro-depressions on the soil surface (RETRAIN) needs to be calculated, the following equation, derived from Onstad (1984), is used:

calc("RETRAIN=max(0.329RR + 0.073sqr(RR) - 0.018RR(GRAD),0.0)");

where:		
	RETRAIN	maximum amount of rainfall, in cm;
	RR	the Random roughness of the surface in cm;
	GRAD	slope gradient, in %;
	sqr(x)	take the square of equation x;
	max(x,y)	take the maximum of eqs. x and y.

Since RETRAIN cannot be less than 0, the minimum resulting value is 0, by taking the maximum of 0 and the right side of the equation.

13.5.2. LISEM INPUT

LISEM needs a number of input files and maps to run. These inputs are described below.

Rainfall file
Data from multiple raingauges can be entered in an input data file. A map is used as input to define for each pixel which raingauge must be used. For every time increment during the simulation of a storm, the model generates a map with the spatial distribution of the rainfall intensity. Thus, the model allows for spatial and temporal variability of rainfall. In the future, this approach allows for the input of e.g. radar data indicating rainfall intensity patterns changing in space and time: e.g. to simulate a thunder storm which moves over a catchment.

Tables for the soil water model
Incorporated in LISEM is a modified version of the SWATRE soil water model, which simulates the vertical movement of water in the soil (Belmans et al., 1983). Within the catchment, soil profiles are defined. The vertical soil water movement is simulated by subdividing a soil profile in a user defined number of layers: we have used 14 layers for a 120 cm profile. For each characteristic soil horizon, the measured $K - \theta - h$ relations are read from the horizon specific tables. Optionally, the 'van Genuchten parameters' can be entered.

Maps of relevant topographical, soil and land use variables
To run LISEM, a number of maps are needed in the PCRaster format:
  − a group of maps which describe the catchment morphology:

  • an 'area.map', in which the main catchment is defined;

  • an 'id.map', which defines the spatial rainfall pattern;

  • a map with the locations of the main outlet and subcatchment outlets;

  • a map with the 'Local Drain Direction', which refers to aspect;

  • a map with slope gradient;

  • a map with the Manning's n for overland flow;

  • a map with slope gradient of the main channels;

  • a map with the Manning's n for channel flow;

  • two maps which describe the channel morphology;

  • a map with the location and width of roads;

  • a map with the location and width of wheeltracks from tractors;

— a group of maps needed for the soil water sub-model:
  - map with the soil profile types, referring to the conductivity tables;
  - a similar map, but then for profiles under wheeltracks;
  - maps with the initial soil matric suction for each soil layer;
— a group of maps with soil and land use variables:
  - a map of the Leaf Area Index;
  - a map with the soil coverage by vegetation;
  - a map with the crop height;
  - a map with the random roughness of the soil surface;
  - a map with the aggregate stability of the soil;
  - a map with the soil cohesion;
  - a map with the soil cohesion of channels;

Command file

When the model is run, the user is prompted for the selection of the catchment, the rainfall event, a few tuning parameters and the desired output. Alternatively, the user can specify this information in a command file. This interface empowers the user to:

— Select the catchment by specifying the directory of the topographical, soil and land use map database;
— Select the soil water model parameters by specifying the directory of the soil water tables. Separating the map database and the soil water tables permits optional sharing of the soil water tables between different catchments;
— Select the rainfall event by specifying the rainfall file;
— Select the starting and ending time of the simulation;
— Select the overall simulation time step, and the minimum time step for the soil water sub-model;
— Select a precision factor of the soil water sub-model;
— Select a number of parameters and coefficients used in the detachment and transport formulas, such as settling velocity of the soil particles and a splash delivery ratio. If necessary, a few of these parameters could be used for calibrating the sediment part of the model;
— Select names of the output files: e.g. hydrograph files (main outlet and outlets of predefined subcatchments), runoff maps at several times, soil erosion map and the 'results' file with totals.

13.5.3. LISEM OUTPUT

The results of the LISEM model consist of:

- a text-file with totals (total rainfall, total discharge, peak discharge, total soil loss etc.);
- a ASCII data file which can be used to plot hydrographs and sedigraphs.
- PCRaster maps of soil erosion and deposition, as caused by the event;
- PCRaster maps of overland flow at desired time intervals during the event.

Examples of these are given in table 13.1 and figures 13.2, 13.3 and 13.4.

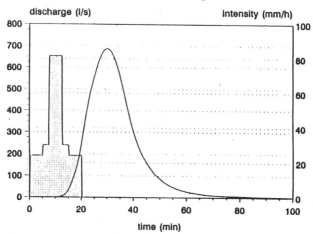

Figure 13.2. Example of a LISEM hydrograph output

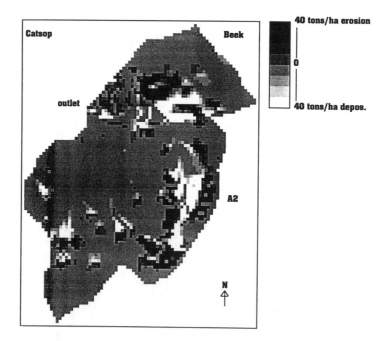

Figure 13.3. Soil erosion and deposition simulated in the Catsop catchment (South-Limburg, The Netherlands) using the LISEM model

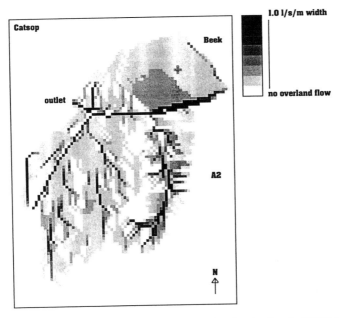

Figure 13.4. Overland flow in the Catsop catchment simulated using the LISEM model

TABLE 13.1. Example of a standard LISEM output text file.

LISEM results catchment area (ha)	: 41.56
total rainfall (mm)	: 13.30
total discharge (mm)	: 1.997
discharge percentage	: 0.150
total discharge (m^3)	: 829.8
peak discharge (l/s)	: 687.6
peak time (min)	: 30.00
total soil loss (ton)	: 0.907
average soil loss (kg/ha)	: 21.83

One of the applications of LISEM is for planning and evaluating various strategies for controlling pollution from intensively cropped areas. Within the LISEM project, several possible scenarios, of which a few control measures are seriously considered to be implemented, are evaluated (figure 13.5). Also, using the model the best possible locations for the measures can be determined.

Maps of soil erosion and sedimentation of the scenarios can be compared by subtraction. These simulations indicated where possible control measures would have the greatest positive and negative consequences. Planners can decide to combine the positive elements from several scenarios and leave out the negative elements and develop new scenarios. The advantage of using LISEM is that this operation can be done very quickly: combining maps in the PCRaster GIS which accompanies LISEM is a standard operation. The map database can easily be updated with new data (such as field and laboratory measurements), followed by new interpolations to produce maps.

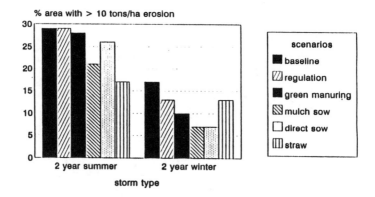

Figure 13.5. The effectiveness of soil erosion scenarios in the Catsop catchment (South-Limburg, The Netherlands)

13.6. Validation, Calibration and Error Propagation Problems in Physically-Based Models

However, before models such as LISEM can be applied successfully, the quality of these models must be known first. To evaluate the quality of simulation models, field research in three small catchments has been carried during several years. Measured and simulated values of discharge and sediment concentration at the catchment outlet and at two sub-catchment outlets. Also, overland flow patterns observed in the field are compared to the simulated patterns. Furthermore, spatial soil erosion patterns are evaluated using [137]Cs (De Roo, 1991; Quine & Walling, 1991). Thus, the distributed model is evaluated in a distributed way. In the three research catchments in Limburg which are monitored during the LISEM project, the following validation data are available and used:

  — measurements of discharge and sediment load at three outlet gauging stations;
  — measurements of discharge and sediment load from six sub-catchments within the main catchments;
  — measurements of the soil pressure head at 12 locations along four slope profiles at seven depths down to 150 cm;
  — 143 spatially distributed ^{137}Cs samples in one of the catchments;
  — data from rainfall simulation studies in the catchments or at the experimental farm Wijnandsrade (validation of splash detachment equations and flow and transport equations);

- runoff and sediment load data from several plots where soil conservation experi- ments have been carried out.

Necessary data for the validation are kept in the PCRaster Geographical Inform- ation System (GIS). The model results are compared with observed data (valid- ation). Statistical criteria determine the 'goodness of fit'. The model user has to decide whether the results are satisfactory. If so, the simulations end and the 'final results' are produced. If the validation is not satisfactory, there are several options:

- Modify the model;
- Re-calibrate the model;
- Change the resolution (pixel-size or simulation time step);
- Collect more data;
- Collect better data (measurement errors);
- Collect different data (other variables);

This procedure is repeated until satisfactory results are obtained. Validation of LISEM is in progress. Detailed validation results will be published in future.

Many sources of error affect the results of distributed hydrological and soil erosion models. The knowledge of error in physically-based models is important to make a correct judgement of a model. Without a perfect knowledge of initial and boundary conditions it is impossible to validate, accept or reject such models.

There can be errors in the theoretical structure of the model, errors in the math- ematical solution methods in the model, calibration errors, and errors in the data. Many scien- tists concentrate on the development of the theoretical structure of the model and try to fill the gaps in our understanding of hydrological and soil erosion processes, which is very important work. However, the error in the data may have a larger effect on the simulation results obtained with the model. Furthermore, ig- noring the errors in the data may lead to the erroneous conclusion that the theor- etical structure of a model is not correct. If obtaining good simulation results at the catchment scale is our main aim, maybe we are approaching the point at which improvement of data collection methods becomes more important than model de- velopment.

Therefore, an important issue in (soil erosion) modelling is the propagation of errors. Complex models contain need large amounts of input data. These data can contain errors, due to measurement errors, interpolation-errors, spatial and tem- poral variations, affecting the results. The error propagation through models can be examined by stochastical modelling combined with physical-deterministic model- ling, to show the magnitude of error in the output of the model arising from uncer- tainties in input variables such as the infiltration capacity of the soil (e.g. De Roo et al., 1992).

13.7. Discussion and Conclusions

Linking or integrating hydrological and soil erosion models to a GIS results in a powerful tool to calculate the effects of land use changes and to explore soil conservation scenarios during single rainfall events on a catchment scale. Models such as LISEM can be driven with hypothetical storms of known probability of return, so that LISEM is a valuable tool for planning cost-effective measures to mitigate the effects of runoff and erosion. With the GIS detailed maps of soil erosion and overland flow can be produced that are useful for planners.

Hydrological and soil erosion models still need to be improved at many points (De Roo, 1993b). However, not only the theoretical structure or the physical/chemical basis of models needs more attention, but also the data problem ('parameter crisis') should receive more attention. It seems to be useless to develop models with numerous variables that vary both in space and time, such that validation or even running the model with reliable data becomes not feasible. Therefore, the propagation of errors in complex models linked to a GIS should receive more attention.

ACKNOWLEDGEMENTS

The Province of Limburg, the Waterboard 'Roer en Overmaas', the Ministry of Agriculture and 14 municipalities of South-Limburg are greatly acknowledged for funding a large part of the research which lead to the development of the LISEM model. Several researchers have contributed to the LISEM project. C.G Wesseling, N.H.D.T. Cremers, R.J.E. Offermans, M.A. Verzandvoort and H. Kolenbrander of the Utrecht University, C.J. Ritsema, K. van Oostindie and J. Stolte of the Winand Staring Centre Wageningen, and F.J.P.M. Kwaad and P.M. Van Dijk of the University of Amsterdam are all thanked for their contributions. Last but not least dr. W.P.A. Van Deursen is thanked for his stimulating ideas which led to the development of the LISEM model.

References

Beasley, D.B., Huggins, L.F. & Monke, E.J. (1980), ANSWERS: A Model for Watershed Planning. Transactions of the ASAE, Vol.23, No.4, 938-944.
Belmans, C., J.G. Wesseling & R.A. Feddes (1983), Simulation model of the water balance of a cropped soil: SWATRE. Journal of hydrology,Vol.63, 271-286.
Beven, K.J. (1985), Distributed models. In: M.G. Anderson & T.P. Burt (eds.) Hydrological forecasting. John Wiley & Sons Ltd., 405-435.
Beven, K.J. & M.J. Kirkby (1979), A physically based, variable contributing area model of basin hydrology. Hydrological Sciences Bulletin, Vol.24, No.1, 43-69.
Bork, H.-R. (1988), Bodenerosion und Umwelt. Verlauf, Ursachen und Folgen der mittelalterlichen und neuzeitlichen Bodenerosion; Bodenerosionsprozesse; Modelle und Simulationen. Landschaftsgenese und Landschaftskologie, Vol.13, 249 p., Braunschweig.
Chow, V.T., Maidment, D.R. & Mays, L.W. (1988), Applied hydrology. McGraw-Hill, pp. 572.
Curwick, P.B. & M.E. Jennings (1982), Evaluation of rainfall-runoff models for the hydrologic characterization of surface-mined lands. In: Singh, V.P. (ed.), Applied modeling in catchment hydrology. Water Resources Publications.
DeCoursey, D.G. (1985), Mathematical models for nonpoint water pollution control. Journal of Soil and Water Conservation, Vol.40, No.5, 408-413.
De Jong, S.M. (1994), Applications of reflective remote sensing for land degradation studies in a Mediterranean environment. Netherlands Geographical Studies, Vol.177, 237 p.

De Roo, A.P.J. (1991), The use of ^{137}Cs as a tracer in an erosion study in South-Limburg (The Netherlands) and the influence of Chernobyl fallout. Hydrological Processes Vol.5, 215-227.

De Roo, A.P.J. (1993a), Validation of the 'ANSWERS' catchment model for runoff and soil erosion simulation in catchments in the Netherlands and the United Kingdom. In: Applications of Geographic Information Systems in Hydrology and Water Resources. IAHS Publication No. 211, 465-474.

De Roo, A.P.J. (1993b), Modelling surface runoff and soil erosion in catchments using Geographical Information Systems ; validity and applicability of the 'ANSWERS' model in two catchments in the loess area of South Limburg (The Netherlands) and one in Devon (UK). Netherlands Geographical Studies, Vol.157, 295 p.

De Roo, A.P.J., L. Hazelhoff & P.A. Burrough (1989), Soil erosion modelling using 'ANSWERS' and Geographical Information Systems. Earth Surface Processes and Landforms, Vol.14, 517-532. John Wiley & Sons, Ltd.

De Roo, A.P.J., L. Hazelhoff & G. Heuvelink (1992), The use of Monte Carlo simulations to estimate the effects of spatial variability of infiltration on the output of a distributed hydrological and erosion model. Hydrological Processes, Vol.6, No.2, 127-143.

De Roo, A.P.J., C.J. Ritsema, P.M. Van Dijk, N.H.D.T. Cremers, R.J.E. Offermans, F.J.P.M. Kwaad, J. Stolte, M.A. Verzandvoort (1994a), Eindrapport Erosienormeringsonderzoek Zuid Limburg. (Final report soil erosion research South-Limburg). Utrecht University / Staring Centre Wageningen.

De Roo, A.P.J., C.G. Wesseling, N.H.D.T. Cremers, R.J.E. Offermans, C.J. Ritsema, K. van Oostindie (1994b), LISEM: A new physically-based hydrological and soil erosion model in a GIS environment: Theory and Implementation. IAHS Publication, No. 224, 439-448.

De Roo, A.P.J., C.G. Wesseling, N.H.D.T. Cremers, R.J.E. Offermans, C.J. Ritsema, K. van Oostindie (1994c), LISEM: A physically-based hydrological and soil erosion incorporated in a GIS. In: Harts, J.J., H.F.L. Ottens, H.J. Scholten & J. Van Arragon (eds.). Proceedings of the Fifth European Conference on Geographical Information Systems. Paris. EGIS Foundation, Utrecht/Amsterdam, 207-216.

Dryer, D. & J. Frhlich (1994), A GIS-based soil erosion model for two investigations areas in the high Rhine valley and the Swiss Jura plateau (NW Switzerland). In: Harts, J.J., H.F.L. Ottens, H.J. Scholten & J. Van Arragon (eds.). Proceedings of the Fifth European Conference on Geographical Information Systems. Paris. EGIS Foundation, Utrecht/Amsterdam, 1032-1041.

Foster, G.R. (1982), Modeling the erosion process. In: Haan et al. (eds.), Hydrologic modeling of small watersheds. ASAE Monograph number 5, 297-380.

Foster, G.R. (1988), Modeling soil erosion and sediment yield. In: Lal, R. (ed.), Soil erosion research methods, 97-117. Soil and Water Conservation Society, Ankeny, Iowa.

Foster, G.R. (1990), Process-based modelling of soil erosion by water on agricultural land. In: Boardman, J., I.D.L. Foster & J.A. Dearing (eds.) Soil erosion on agricultural land. John Wiley & Sons Ltd., 429- 445.

Kirkby, M.J., J. Lockwood, J.B. Thornes, I. Woodward, A. Baird, M. McMahon, P. Mitchell & J. Sheehy (1992), MEDALUS Project Report; Development of a physically based simulation model for vegetation growth, hydrology and soil degradation. Interim report, April 1992, pp. 9.

Lal, R. (1988), Soil erosion by wind and water: problems and prospects. In: Lal, R. (ed.) Soil erosion research methods. Soil and Water Conservation Society/International Society of Soil Science, Wageningen, The Netherlands.

Leenaers, H. (1989), The dispersal of metal mining wastes in the catchment of the River Geul (Belgium - The Netherlands). Dissertation, University of Utrecht, pp.200.

Leonard, R.A. & W.G. Knisel (1986), Selection and application of models for nonpoint source pollution and resource conservation. In: Giorgini, A., F. Zingales, A. Marani & J.W. Delleur (eds.), Agricultural Nonpoint Source Pollution: Model selection and application. Elsevier, Amsterdam. Developments in Environmental Modelling, No.10, 213-229.

Mallants, D. & J. Feyen (1989), Hydrological Systems Modelling of the Rainfall-Runoff Conversion using A Digital Terrain Model. Landschap, 1989, No.4, 277-290.

Marsalek, J. (1989), Modelling agricultural runoff: overview. In: Sediment and the Environment. IAHS Publication no.184, 201-209.

Mitchell, J.K., B.A. Engel, R. Srinivasan and S.S.Y. Wang (1993), Validation of AGNPS for small watersheds using an integrated AGNPS/GIS system. Geographic Information Systems and Water Resources, March 1993, 89-100.

Monkhouse, F.J. & J. Small (1970), A dictionary of the natural environment. Edward Arnold, London.

Morgan, R.P.C., J.N. Quinton & R.J. Rickson (1992), EUROSEM Documentation manual. Version 1: June 1992. Silsoe College, Silsoe.

Nearing, M.A., G.R. Foster, L.J. Lane & S.C. Finkner (1989), A process-based soil erosion model for USDA - Water Erosion Prediction Project Technology. Transactions of the ASAE, Vol.32, No.5, 1587-1593.

Novotny, V. (1986), A review of hydrologic and water quality models used for simulation of agricultural pollution. In: Giorgini A., F. Zingales, A. Marani & J.W. Delleur (eds.), Agricultural Nonpoint Source Pollution: Model selection and application. Developments in Environmental Modelling, 10, Elsevier, Amsterdam, 9-35.

Onstad, C.A. (1984), Depressional storage on tilled soil surfaces. Transactions of the ASAE, Vol.27, 729-732.

Perez-Trejo, F. (1986), Review of existing watershed models for the Delft Hydraulics Laboratory. Colorado State University, Fort Collins, CO 80523, pp.20.

Quine, T.A. & D.E. Walling (1991), Rates of soil erosion on arable fields in Britain: quantitative data from caesium-137 measurements. Soil use and management, Vol.7, No.4, 169-176.

Smith, R.E. (1981), A kinematic model for surface mine sediment yield. Transactions of the ASAE, 1508-1514.

Smith, R.E. & V.A. Ferreira (1988), Comparative evaluation of unsaturated flow methods in selected USDA simulation models. In: H.J. Morel-Seytoux (ed.), Unsaturated flow in hydrologic modeling. Theory and practice, 391-412. Kluwer Academic Publishers.

Storm, B., G.H. Jorgensen & M. Styczen (1987), Simulation of water flow and soil erosion processes with a distributed physically-based modelling system. IAHS Publication No.167, 595-608.

Styczen, M. & S.A. Nielsen (1989), A view of soil erosion theory, process-research and model buiding: possible interactions and future developments. Quaderni di Scienza del Suolo, Vol.II, Firenze, 27-45.

Van Deursen, W.P.A. & J.C.J. Kwadijk (1990), Using the Watershed tools for modelling the Rhine catchment. In: J. Harts et al. (eds), Proceedings EGIS '90, 254-262. EGIS Foundation, Utrecht.

Van Deursen, W.P.A. & J.C.J. Kwadijk (1993), RHINEFLOW: an integrated GIS water balance model for the river Rhine. In: Applications of Geographic Information Systems in Hydrology and Water Resources. IAHS Publication No. 211, 507-518.

Van Deursen, W.P.A. & C.G. Wesseling (1992), The PC-Raster package. Department of Physical Geography, Utrecht University (info: pcraster@cc.ruu.nl).

Wischmeier, W.H. & Smith, D.D. (1958), Rainfall energy and its relationship to soil loss. Trans. Amer. Geophys. Union, Vol.39, 285-291.

Wischmeier, W.H. & Smith, D.D. (1978), Predicting rainfall erosion losses - a guide to conservation planning. U.S. Department of Agriculture, Agricultural Handbook No. 537, Science and Education Administration USDA, Washington D.C., 58 pp.

Woolhiser, D.A., R.E. Smith & D.C. Goodrich (1990), KINEROS: A kinematic runoff and erosion model: documentation and user manual. USDA-ARS, ARS-77.

Young, R.A., C.A. Onstad, D.D. Bosch & W.P. Anderson (1987), AGNPS: Agricultural-Non-Point-Source pollution model: a large watershed analysis tool. USDA-ARS, Conservation Research Report 35, Washington DC, pp 77.

Young, R.A., C.A. Onstad, D.D. Bosch & W.P. Anderson (1989), AGNPS: A nonpoint source pollution model for evaluating agricultural watersheds. Journal of Soil and Water Conservation, Vol.44, No.2, 168- 173.

CHAPTER 14

A Study of Landslides Based on GIS Technology

T.P. Gostelow

Abstract.   Published techniques and examples of landslide hazard mapping are reviewed with emphasis on possible applications using GIS. This is followed by a description of a study in Basilicata, Italy. A GIS methodolgy based on 3-D hydro-geotechnical models is discussed, together with the difficulties of defining hazard assessment from rainfall triggering mechanisms. The digital datsets within the GIS, their sources and details of data conversion are then summarised. Suggestions are made for additional work, and finally some examples of GIS map output are provided which were used to explore the relationships between land-use change and landslide distribution.

14.1. Introduction

This chapter describes an example of how GIS can be used for data storage and preliminary spatial hazard assessment of landslides. The region chosen for study was Basilicata, S. Italy, where large numbers of landslide complexes have reactivated after both earthquakes and extreme rainfall events (Cotecchia 1985, Del Prete et al, 1992). The most damaging have been adjacent to, and within expanding urban areas, which for a number of geographical and historical reasons have been built at the top, and on, valley slopes. Four principal practical questions have arisen for planners and developers, i) are the frequencies of these instabilities increasing ? ii) what are the factors responsible for the movements ?, iii) what is the role of anthropogenic land-use change ?, and iv) where are the greatest hazards ? A fifth, to some extent academic question, was whether the movements could be used to monitor long-term climatic changes in Southern Europe ?

In an attempt to answer these questions a methodology was developed for data storage and spatial analysis using GIS, which could be used for further research, planning, and decision modelling.

357

V. P. Singh and M. Fiorentino (eds.), Geographical Information Systems in Hydrology, 357–388.
© 1996 Kluwer Academic Publishers. Printed in the Netherlands.

14.2. Landslide Hazard Mapping

There have been several, previous attempts and reviews which have identified key data coverages for landslide hazard mapping, eg by Leighton (1976), Nilsen et al (1979), Carrara, (1983), Varnes (1984), Brabb, (1984), and Hansen, (1984). Qualitatively graded maps based on 2-D overlays of thematic factors such as slope angles, distributions of landslide deposits, and the presence or absence of underlying solid rocks susceptible to mass movement were used to help define relative slope stability, landslip potential, or an empirical probabalistic measure, (eg %) of slope failure.

More recent examples of landslide hazard mapping have tended to include a greater range of variables. Lewis and Rice, (1990) considered 172 in a study of first-time slope failure in 638 areas affected by forest roads and timber harvesting in NW California. They found by discriminant analysis that slope angle, horizontal slope curvature, bedrock type and soil colour were most significant, although in a separate study in the same area, Neely and Rice, (1990) concluded that slope angle, distance to a spring discharge, and distance to an uphill slope break were the factors most influential in causing instability.

Siddle et al, (1991) used existing landslide distribution and morphology, as a basis for hazard mapping in the industrial area of S. Wales, UK. Twelve factors of relevance to instability (Table 14.1) were compiled into 2-D maps at 1:10000 scale showing zones with a consistent set of conditions. The data were gridded for computer storage, and graded by their degree of association with areas of existing landslides. A second analytical step combined the ratings into a simple algorithm. Various combinations of weighted factors were tried subjectively to give the best fit with actual conditions, and a landslip potential (LP) was defined as:

$$LP = \frac{R1(R2 + R3 + R4)}{300} \tag{14.1}$$

where R1, R2, R3, and R4 were ratings for slope angle, superficial deposit type, superficial deposit thickness, and groundwater potential respectively. South Wales has had a long history of industrial development accompanied by underground coal extraction (resulting in vertical subsidence and tilting), but surprisingly these factors were not included in the index. Their map of LP suggested that slope instability hazards mostly followed natural linear features in the landscape reflecting trends in the underlying solid sedimentary geology and zones of groundwater discharge.

A comparable study carried out by Carrarra et al (1991) in a rural drainage basin in Umbria, Italy, used PC ARC/INFO GIS for data storage. The investigation began with the digitisation of each pre-existing landslide for a 1:10000 scale vectorised map coverage. Slope, or terrain units within the sub-basins (catchments) of the main drainage basin were then derived from a digital terrain, (elevation) model, (DTM, DEM) and associated with 40 morphological, geological and vegetational attributes. Discriminant analysis was applied to these attributes in an attempt to

TABLE 14.1. Slope instability factor types and basis for zonation, (after Siddle et al 1991)

Factor Name	Basis of Zonation
1. Slope aspect	Slopes defined by the four cardinal directions, their intermediate points and areas of no slope.
2. Slope angle	Portrayed in increasing 5° increments.
3. Lithostratigraphic units	Argillaceous, arenaceous, mixed lithologies, and stratigraphic units.
4. Superficial deposit type	Glacial soils, alluvium, peat. Residual soils and colluvium types assessed from primary mapping, location on slope and underlying lithologies.
5. Superficial deposit thickness	Isopachs prepared from geomorphological considerations and many data sources.
6. Dip vs. ground slope	Dip direction (derived map) compared with slope aspect.
7. Faulting	Proximity to fault and size of fault.
8. Groundwater potential	Location of points of groundwater emergence, together with an evaluation of potential groundwater catchments.
9. Erosion potential	Presence or absence and river trained areas.
10. Mining strains	Severity of total strains from a total extraction, (derived) map. direction.
11. Mining tilt	Relationship of tilt direction (from total extraction map) to ground slope direction.
12. Excavation and filling	Type of activity and location.

classify the units into those which were stable or unstable, (based on the presence or absence of existing landslide deposits). Four were found to have the greatest discriminant power where landslides were present, ie

1. Hydrogeological conditions, ie presence of impermeable materials overlain by permeable rocks.
2. The presence of certain lithostratigraphical layers.
3. The relative percentage of forested area.
4. The slope unit form (shape) and the size of the sub-drainage basin.

The resulting maps of landslide hazard were divided into four relative stability classes, but their distribution suggested there were no clear patterns of unstable and stable areas. This may have been because the polygons were defined by arbitrary slope or terrain units, ie the existing landslides were poorly related to the underlying topography, which was not shown by contours, and hence difficult to visualise. Comparatively few active landslides were found in the basin, although large areas of the map were ultimtely classified as unstable. Geotechnical assessments of the four classes were not made, and it might also be difficult, (using the 4 factors above), to justify a hazard zonation of 'areas affected by landsliding and predicted as stable', without further qualification.

Difficulties perhaps arose with this GIS mapping approach, because 2-D slope unit classes were used to define 3-D landslide objects and geotechnical systems. Nevertheless, it is of interest that the hydrogeology and land-use were recognised as important instability factors and that the landslide distributions seemed to be 'clustered', rather than following linear features (as in South Wales, UK). This difference in spatial distribution probably reflects a contrasting regional hydrogeology, and in Umbria could be due, in part, to surface saturation (and low effective shear strengths) from local geological features such as faulting (rather than a lithostratigraphical boundary), or a series of topographic hollows acting as groundwater discharge zones. The suggestion that slope unit 'form' was of greater significance than angle suggests that the latter might have been the case.

An assessment of landslide risk was also made in this study, but suffers from similar drawbacks to the example from S Wales, ie many (if not most) of the landslides were triggered in the past, in the case of the UK dating back to the Late-Glacial period, or to different climatic conditions during the Holocene. The factors influencing landslip potential were almost certainly different than at present (nivation, pluvial conditions, no urbanisation, deforestation, wildfires etc). In addition, in both studies there was a lack of distinction between potential movements associated with first-time and pre-existing slides, and a means of portraying, or taking into account the different time dependent triggering mechanisms which were operating in the unstable areas.

Programmes of identifying and mapping existing landslides have also been carried out in Basilicata, Italy, especially since the 1980 earthquake. Fulton et al, (1986) argued that systematic mapping should be introduced, and put forward some suggestions for the types of maps which are needed. They demonstrated there are clear economic advantages for the introduction of a proper, geographically referenced digital database for recording landslide distribution, and areas susceptible to instability. Such data are required for regional structure plans, and for monitoring project impacts. However, as discussed, there are a wide range of factors which contribute to instability, (Table 14.1) and the most significant seem to vary from area to area. Which of these should be taken into account ?, and how can they be represented on maps, and used in decision models or by engineers ? Before describing the approach developed in this study, which was based on GIS technology, a brief review of the geology, land-use, style and extent of the landsliding in Basilicata is given.

14.3. Geological Setting, Land-use and Landsliding in Basilicata

Basilicata can be divided into three main structural areas; from east to west, i) the Apulian foreland, ii) the Bradanic trough, and iii) the Southern Apennine mountain chain. The foreland consists largely of uplifted Mesozoic limestone blocks and is represented by a small area to the east of the region. The Bradanic fore-trough is an elongated NW-SE trending sedimentary basin consisting of Plio-Pleistocene sediments with thicknesses approaching 3000m. The S. Apennines mountains were formed as a result of Alpine orogenesis and consists of a series of nappes separated by thrust faults. Seven tectonic units have been recognised, and in common with other alpine zones consist of ophiolites, volcanics, limestones, sandstones and a number of argillaceous units, including varicoloured 'scaley clays' and flysch sediments.

The region can be divided into a mountain zone, arbitrarily defined as land over 700 metres, a hill zone of 300 to 700 metres, and plains below 300 metres. They occupy 45%, 45% and 10% of the surface area respectively, which in total is some 1 000 000 ha. The hill zone has been the most intensively cultivated, and because of this, has generally become the most unstable with regard to mass movement. Apart from landsliding, the other chief geomorphological effect of land-use change has been the development of badland landscapes, (calanchi) which consists of unvegetated bare clays, rills and deep gulleys. According to Del Prete (1993) nearly 300 000 ha of land is now uncultivated or marginal as a result of erosional development and poor agricultural practices. After geomorphological surveys in the early 1980's, 26 000 ha were found to be covered by deep-seated landslides and 184 000 ha by erosion/shallow landsliding processes. In addition more than 1000 km of the hydrological network was shown to be causing active vertical and horizontal stream erosion. However, in many cases of slope instability, it has been urbanisation, a gradual enlargement of the calanchi, with high intensity rainfall events associated with a Mediterranean climate, which have increased susceptibilities to mass movements (Gostelow and Gibson, 1993, Del Prete, 1993).

Recent landsliding in Basilicata has been mainly restricted to the outcrop of the argillaceous lithologies which occur in both the Apennines and the Bradanic foretrough regions. Although first-time failures occur, Cotecchia, (1985) has suggested that the most damaging have been reactivations of old colluvial deposits which have been stable for a long time. Of particular interest was his observation that many of them tended to be found adjacent to carbonate aquifers where sudden and large variations in spring discharge were recorded. Earthquakes, and extreme rainfall events have contributed to this association, and landsliding has frequently been combined with flooding. The meterological conditions which caused disasters in Nov 1925, Sep 1929, Feb 1931, Sep 1934, Nov 1944, Nov 1946, Nov 1959, Jan 1961, Jan 1972, and Nov 1976 were especially severe and are summarised by Caloiero and Mercuri, (1982).

14.4. Landslide Complexes at Grassano, Pisticci and Ferrandina

In this GIS study, emphasis was placed on the environental conditions responsible for landslide complexes next to urban areas. Jones (1992) has recently reviewed a number of cases, and shows that out of 131 towns in Basilicata, 115 have been affected by slope instability problems. Three typical examples can be found at Ferrandina, Pisticci and Grassano which lie in the lower parts of the Basento valley in the eastern Bradanic foretrough, part of the region.

The lithological succession beneath, and surrounding, the towns generally consists of a cemented conglomerate (the Irsina formation), loose, fine sands (Monte Marano formation) and stiff, but weakly lithified, grey blue clays of medium plasticity, (20%-30%) with undrained strengths (unweathered), of between 100 kPa-300 kPa, (the Apennine formation). Subsurface investigations at Pisticci, (Guerricchio and Melidoro, 1979) and Grassano (Cotecchia and Del Prete, 1986), have shown that the slopes beneath, (ie the building foundations) and immediately surrounding the towns are flatter, (c. $3^o - 7^o$) and are covered with a layer of ancient cohesionless debris (colluvium), consisting of reworked conglomerates and sands, beween 2 and 15m thick. At Grassano and Ferrandina, the debris occupies a platform, 400m-500m wide, which slopes at between 5° and 7°, steepening to nearly 14° immediately below an in situ conglomeratic slope crest which stands out as an escarpment on both sides of the valley, (Figure 14.1 shows the crest at Grassano). The majority of the buildings at Grassano have been sited on the platform, whilst at Pisticci and Ferrandina they are also found on in situ conglomerates/sands behind the slope crest. In both towns construction has taken place close to the escarpment edge, and at Pisticci buildings are found on the upper parts of one of the slides of 1688, where 400 lives were lost (Cotecchia and Melidoro, 1974).

Within the Bradano foretrough a prominent springline has developed on all the valley sides, at the junction of the M Marano sands and the underlying blue clays. The hydrogeological significance of this boundary is illustrated by the conditions at Grassano, (figure 14.1) where two major springs, possibly fault controlled, discharge from the edge of the debris platform into large gullies which are eroding back across the platform towards the slope crest. The break of slope which occurs at the platform edge possibly reflects a period of renewed downcutting following uplift, and this, together with past land-use practices has resulted in desiccated bare soils, slaking, piping and gulleying. The development of a calanchi (badland) zone on the blue clays is now a serious problem here, and elsewhere in the Basento valley. Below Grassano the gulleys are spaced evenly across the slope with similar orientations, suggesting they are structurally controlled. Increased water pressures associated with springs downslope of the discharge area also appears to have been responsible for major instability at the edge of the platform after a storm on the 21st November 1976, which damaged the cemetery there (Cotecchia and Del Prete, 1986). Other movements downslope of the platform edge have been recorded (Del Prete et al 1992), but there have been no significant slides in historic time, of the conglomeritic escarpment slopes, or areas upslope of these discharge points, or zones of groundwater saturation.

Reconstruction of the slide geometry below the town by Cotecchia and Del Prete, (1986) suggested there was a deep-seated slip surface within the blue clays up to 60m below ground level, which had caused a considerable rotation and lateral movement of bedding. However, this was based on limited borehole evidence, and a later examination of descriptive logs and core could not find any signs of

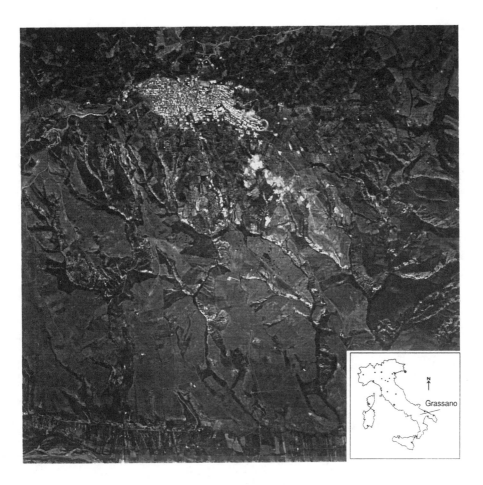

Figure 14.1. Aerial photograph showing the position of Grassano, major gulleys and springs which start at the debris (colluvial) platform surrounding the town. The pale areas are steep, unvegetated clay slopes, (calanchi) eroded through a combination of downcutting and neotectonic uplift. The conglomerate escarpment forms a ridge at the northern margin to the built-up area, and this is also used by the road which enters from the west. The floodplain of the Basento River can be seen at the bottom of the photograph

large scale disruption in the clay. The sequence consisted of reworked colluvial layers of sand and sand blocks, which seemed to overlie an undisturbed stratigraphic sequence between the M Marano sands and the blue clays. The age of the colluvium is unknown, but natural geological exposures around the town suggested there was a major phase of movement followed by a colluvial reworking at upper levels. Deep post-slip weathering/oxidation had occurred, which left a char-

acteristic red/brown colour to depths of 30m. Within this weathered zone, immediately below the ground surface, there was a loose silty surface layer on top of
the platform which may have been a wind-blown loessic deposit. This evidence,
and the presence of at least three river terraces at lower levels suggested the mass
movement was of considerable antiquity, perhaps dating well back into the Qua-
ternary. Figure 14.2 is a view of Grassano from the upper river terrace level which
illustrates how the valley side slopes, (since gulleyed and degraded in response to
a lower river base level) must have originally looked when they were adjacent to
the river at that time.

Figure 14.2. View of the hilltop town of Grassano, Basento valley, Italy showing ancient, degraded
blue clay slopes and calanchi

Pisticci, which has a greater groundwater catchment area, has had a longer history of instability. Guerricchio and Melidoro, (1979) have referred to a large landslide of 1688 which destroyed the village and produced long fissures on the south
side of the site which have since contributed to further instabilities, (Figure 14.3).
In one example, landsliding occurred in 1976 after the same heavy rainfall event
mentioned above and was probably first-time, (Frana Rione Croci, Figure 14.3).
Cross sections through the slides, has suggested that like Grassano the movements
were confined to the sands, with slip surfaces possibly following a weak bedding
plane at the stratigraphic transition with the underlying blue clays.

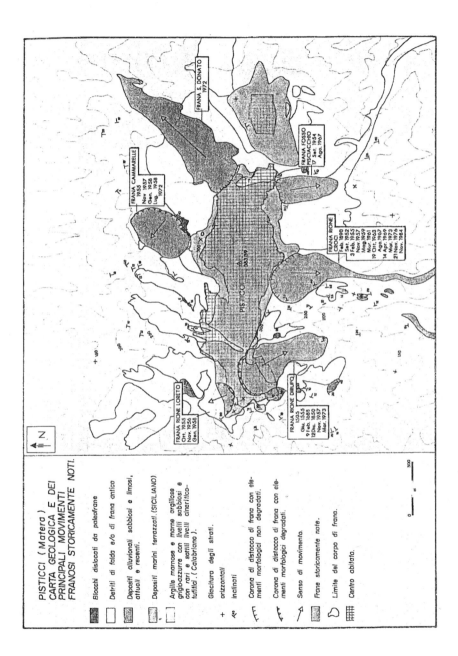

Figure 14.3. Dates of reactivated landslide movements below the town of Pisticci, Basento valley, Italy, (after Del Prete, 1993)

At Ferrandina the geological situation is similar. Cotecchia and Melidoro, (1974) described a slide 2km north of the town which occurred on January 20th 1960 after heavy rain, (Frana Castelluccio, figure 14.4). It was not entirely clear whether it was first-time or reactivated, but the backscarp was close to the escarpment edge and a spring is now discharging from its centre, ie the movement originally took place upslope of a groundwater discharge or saturated area.

All of the landslide reactivations have their landslide backscarps directly adjacent to the built-up areas, and they represent some 15%-30% of the ancient colluvial deposits surrounding the towns, (figures 14.3 and 14.4). Dates of movements were obtained from newspaper and commune archives (shown for Pisticci and Ferrandina only, Figures 14.3 and 14.5), and an attempt was made to relate them to rainfall records. The dates refer mainly to reactivations, although in some cases the descriptions of the events were not clear, and there may have been some first time element involved in the failures.

14.5. Hydro-geotechnical Models of Rainfall Infiltration and recharge on Natural Slopes

14.5.1. GENERAL

The active orogenic and geomorphological setting of the study area has resulted in a set of complex surface, and subsurface geotechnical boundary conditions which can not be represented easily in either conventional paper maps or 2-D GIS digital output. In addition, 'static' 2-D map themes do not provide a user with the 3-D physical insights which are often necessary to carry out 'what if' stochastic and/or deterministic GIS interrogations, either for cost-benefit decision analysis or for project planning and engineering. A new approach was thus adopted which assumed that landslides were mechanical failure responses within broadly defined 3-D hydrogeological systems. For the rainfall triggering mechanism considered here, simplified conceptual 3-D, hydro-geotechnical slope models, capable of linking different climatic/rainfall scenarios to groundwater conditions, weak materials, shear strength loss and slope movements/landslides were developed. Turner, (1989) previously discussed the practical advantages of using 3-D geological models of this kind for studies of contaminant transport, and suggested they were necessary for subsurface characterisation, visualisation and deterministic physical modelling in the GIS environment. Raper (1989) made the useful distinction between sampling limited and definition limited 3-D models, where the former were defined only by boundaries and the latter by a given set of definitions. Within the context of this study, landslides were thus treated as sampling limited geo-objects which existed within spatially referenced definition limited hydro- geotechnical models or classes. Spatial limits of the models, (and their sub-divisions) were represented by conventional 2-D polygons, buffered where boundaries were imprecise, and uncertainty, (if known) by the use of 2-D cross-sections, (Mason et al 1994). However, this latter option was attempted at only one location in Basilicata.

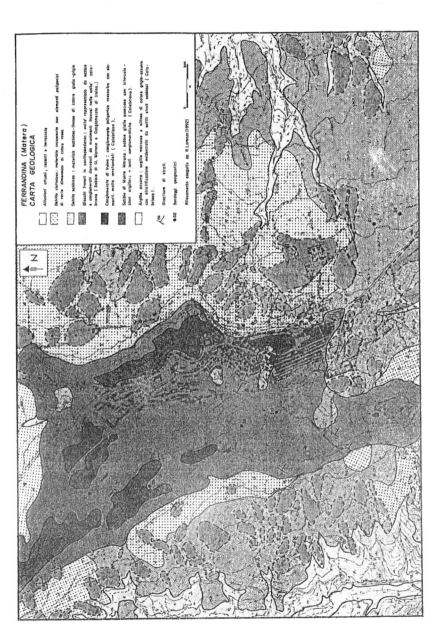

Figure 14.4. Geology map of the area around the hilltop town of Ferrandina, Basento valley, Italy, showing insitu conglomerates, sands and ancient landslide detritus. (after Del Prete, 1993)

Pre-existing landslides, found at Ferrandina, Pisticci and Grassano were treated as 'failed parts' of the modelled areas, and were represented and stored using either conventional 2-D boundaries or attributed points. There are a wide range of published landslide classifications, but individual types are constrained by the model definitions, and hence the only useful additional datasets/subdivisions were estimates of failure depths, and whether soil or rock form part of the slipped mass. In some terrains there was evidence of multiple slope movements and in these circumstances the hydro-geotechnical models also provided a convenient basis for distinguishing intact slopes from areas of past and active deformations.

Most landslides/slope deformations are associated with topographically driven local recharge-discharge, rather than intermediate or regional groundwater flow patterns, (Toth, 1971), and hence in this study, with the emphasis on the geological conditions in Basilicata, only 2 basic hydro-geotechnical 3-D models were used, with the following definitions, ie

i) Where shallow first-time translational slides in soils are found within a weathered soil of relatively high hydraulic conductivity overlying materials of lower conductivity.

ii) Where deep-seated pre-existing mass movement complexes involve an upland aquifer which forms part of the solid geology and is adjacent to weak argillaceous sediments with low frictional strengths and low hydraulic conductivity.

These classes were mapped, by identifying the local recharge-discharge areas using existing geological map boundaries. Hodge and Freeze, (1977) and Rulon et al (1985) have explored the effects of geology on groundwater flow regimes in relation to slope instability, and provided more detailed examples showing the importance of low conductivity units on the distribution of subsurface water pressures. Further refinement of the flow domain characteristics used here will become desirable as geographical databases are developed, (eg the inclusion of confined groudwater conditions, parameter variations, anisotropy, heterogeneity etc), but establishing the distribution of the 2 classes above, has provided a convenient means for making initial, wide-area climatically induced landslide hazard assessments in this geologically complex region.

The actual position of both ancient, (stable) and reactivated mass movements in relation to all hydrogeological flow paterns through natural slopes, has been frequently found to coincide with areas of groundwater discharge, (Toth, 1971), ie with soil/rock profiles which have the greatest water pressures. These areas can be identified and mapped, as a separate 2-D coverage by locating springs, springlines, or seepage faces. Seepage vectors are directed upwards and outwards, and piezometric levels measured vertically, increase with depth relative to the ground surface. In contrast, the catchments, aquifers, or landslide blocks supplying springs have seepage vectors orientated downwards, and piezometric levels measured vertically, decrease with depth relative to the ground surface. Effective shear strengths are thus generally lower in discharge areas and there are a wealth of published examples to show they are associated with pre-existing landslide complexes, (eg Zaruba and Mencl, 1969). In Basilicata, there was also some evidence that relative susceptibility to slope movements may be reflected by the shape of spring hydrographs, but further monitoring is required around the urban areas to explore this possibility, (Gostelow and Gibson, 1993).

14.5.2. HYDRO-GEOTECHNICAL MODELS FOR TRANSLATIONAL SLIDES

There is a widespread literature describing the movement of water from the ground surface through an unsaturated zone to an unconfined saturated zone, or perched water table. For example, using a one-dimensional model, Freeze (1969) has shown there are two separate mechanisms of effective stress change/strength loss which could cause first-time failure within the modelled areas,

i) A loss of suction in the unsaturated zone
ii) A rise in the water table

Several, interelated physical parameters and boundary conditions have been shown to control the extent and timing of the changes, ie the attainment of ponding conditions at the ground surface, soil aggregation, macropores, the antecedent moisture content, the rate of recharge, the depth to the saturated zone, rainfall intensity, and the soil matrix hydraulic conductivity.

A large number of publications have also considered two dimensional hydrological models of hillslopes consisting of a homogeneous weathered soil (generally cohesionless) overlying an unspecified bedrock with a much lower hydraulic conductivity. They have been used in both hydrological and slope instability modelling using the infinite slope limit equilibrium equation, (Montgomery and Dietrich, 1994). Recent examples have also included alternative, deterministic approaches to the problems of timber harvesting on first-time landslide susceptibility, (see above) which have been put forward by Sidle, (1992), Sidle and Terry, (1992). However, whilst these models give some insight into the physical processes and timing of unstable events, it is often found that the actual position of groundwater discharge along a valley side is controlled by a combination of both geology, and/or convergent slope elements, (Tsukamoto et al, 1982).

14.5.3. DEEP-SEATED MASS MOVEMENT COMPLEXES WITH A SURFACE AQUIFER

The first common feature of most deep-seated, geologically controlled landslides are caprocks which act as aquifers. Active and ancient slides tend to be located where the aquifers are in an upland position, and are, or have been subjected to orographic rainfall. The second common feature is a high hydraulic gradient from the aquifers to points of discharge, (steep slopes) and slip surfaces which are below the permanent water-table. The third common feature is a seepage face, sometimes represented by springs, (usually structurally controlled) and the fourth is the presence of underlying weak materials. The fifth feature is a frequent link between an effective rainfall trigger, the aquifer/landslipped mass, transient water levels, groundwater discharge, and mass movement history.

Geological structure may be an important control on reactivations, particularly in earthquake prone areas like Basilicata. For example, during the 1980 normal faulting earthquake ground rupture took place along a 40 km length of a fault (the Irpinia fault), and further aftershocks occurred up to 15 km to the east, associated with an antithetic fault plane (Pantosti et al, 1993). Muir Wood (1994) has shown that such movements have been associated with a recognisable hydrological signature, and that normal faults predominate in the shallow expulsion of meteoric water. He suggested that accompanying fractures, and their apertures could exert a considerable influence on subsequent readjustments to groundwater flow which might even oppose steady-state gravitational fluxes. Extensive alteration to geomechanical and hydrogeological properties must have occurred in this large area of Basilicata, which may now have an increased susceptibility to mass movement from rainfall events.

14.6. Hydro-geotechnical models and their relationship to topography

Within the models, hydraulic gradients and shear stresses, and hence effective strengths and potential instability are controlled by relative relief and slope angle. In this study, digital elevation data were collected for the Basilicata administrative region, rather than for a single groundwater catchment, and the emphasis has been on pre-existing rather than first-time slides. The current DEM in the GIS has a resolution based on a grid of 200m, and is hence generally unsuitable for providing slope angle data for detailed 3-D deterministic modelling, as discussed above. Nevertheless, at this regional scale, simple slope classes, derived from the DEM, have been used to classify broad areas in relation to the two models. Future developments should include a classification of slope thresholds within the modelled classes, improved resolutions and quantification through concepts of DEM geometric signatures, (Pike, 1988).

14.7. Triggering Mechanisms, Decision Analysis, Hazard and Risk

The basic data coverages of hydro-geotechnical 'class', landslide 'object', and gridded slope angle can be used within a GIS environment to investigate and model potential hazard and risk from all the time dependent triggering mechanisms. In Basilcata they fall into four groups, ie earthquakes, weathering/erosion, rainfall and anthropogenic factors, such as land clearance, ground reshaping etc. However, the potential for 'what if' interrogation of the datasets through these mechanisms is considerably improved for planning when they are also set within a socioeconomic context, (Fell, 1993).

Freeze et al (1990) have discussed the applications of decision analysis to potential engineering projects in which hydrogeology has an input. They developed four separate models:

i) A decision model based on an economic risk-cost benefit function.
ii) An engineering reliability model to to represent the performance of the engineering solution.
iii) A simulation model for groundwater flow.
iv) An uncertainty model that takes into account both geological (boundary) uncertainty and hydrogeological parameter uncertainty.

The decision making process is concerned with optimising project benefits, in the case of Basilicata for development within the urban areas. These must be considered in relation to expected costs arising from the probability, (P) of failure from slope deformation over a given time period; each instability mechanism represents a potential economic loss in cost/benefit planning. Post-project/investment performance, (model ii, above) can be assessed by some measure of ground deformation/ remedial costs. In other words many of the towns might be considered as poor performers with regard to recently completed construction and agricultural projects.

The GIS structure has thus been organised to store the datasets for i) to iv) above, so that preliminary subsurface boundary conditions, deterministic 3-D hydrogeological modelling, and geotechnical simulations can be developed and investigated for optimal engineering design. An ideal approach linking the models to planning would thus commonly involve four steps:

1. Identification of the triggering events or sequences of events which may lead to slope failure.
2. Identification of specific features of a landslide, or susceptible area which might initiate failure.
3. The liklihood of combinations of 1 and 2
4. The economic, social and environmental costs of each combination in 3.

In this study an attempt was also made to capture the datasets necessary for a user to follow through these steps with a probabalistic approach to rainfall triggering mechanisms. However, in common with other studies of this kind there were two difficulties.

The first was with step one, ie the identification of the events or sequence of events which trigger mass movement. For example pre-existing landslides of different geometries and geology, will tend to respond differently to an extreme storm rainfall. In each modelled polygon, a preliminary level of hazard may be defined by estimating the probability (P) of encountering an estimated triggering rainfall and its antecedent conditions within a specified period of time (n) for different return periods (T) using the formula,

$$P = 1 - (1 - \frac{1}{T})^n \tag{14.2}$$

Those rainfall conditions can then be mapped and compared with conditions at comparable polygons elsewhere. It is known that rainfall extremes have certainly been responsible for first time and reactivated mass movement in Basilicata, but even with historical data, the magnitudes and significant durations of those extremes remain unclear, ie 1 day, 16 day, 128 day (a winter period) etc. The mean annual maxima for a range of durations were thus mapped spatially as an index by kriging, and gridded so that the return period of any particular event over a landslide or modelled polygon could be obtained from an appropriate rainfall growth curve. For example, in the often used Gumbel extreme value method a rainfall triggering event R_T of return period T years is given by

$$R_T = R_{av} + \sigma(0.78y - 0.45) \tag{14.3}$$

where R_{av} is the mean of all the extreme values, σ is the standard deviation of the series and y is the reduced variate and a function of T.

In general, those areas with the highest average, annual extremes of any duration might be expected to be most susceptible to climatically controlled mass movement, although other factors within the GIS, (see below) must also be taken into account.

Although an empirical rainfall triggering concept is convenient, and has been widely used, there is a fundamental difficulty with applying it to pre-existing landslides which must be mentioned as a cautionary note. This concerns the possible non-linearity between a trigger and landslide deformation, ie what is being observed may not be predictable, (with an acceptable degree of accuracy). One reason for this could be connected to the modelled system, which never returns to precisely the same state, so it never repeats past patterns of deformation, following the same rainfall event. An infinite variety of states are possible, which may only approximate that initial condition. For example, the equation,

$$x_{t+1} = Cx_t(1 - x_t) \tag{14.4}$$

represents a number of physical processes where a state at time $t + 1$ depends in a non-linear way on a state at an earlier time t, and a constant C. A sequence of movements may thus cause a kinematic effect, ie through a change in morphology or lateral stress transfer, which results in new emergent shear surfaces which are more brittle, and activated by different climatic events, perhaps with a lower rainfall threshold. Alternatively, a sequence of movements within a colluvial deposit, or landslide complex, following a certain rainfall time series, may slowly raise the factor of safety, increasing the threshold. Another possibility might be a corresponding reduction in the factor of safety in the adjoining in situ material, resulting in first-time failure. The 1976 slide disaster in Pisticci, described earlier, probably involved a first-time element, and may have originated in this way.

Most of the towns in Basilicata, are surrounded by multiple, pre-existing landslides. Historical evidence suggests that only one or two of these have been activated by an extreme rainfall event at any time, and they are, apparently, selected at random from both a landslide group, and towns. Once movement has been initiated, it tends to re-occur at that site, but this may not always be the case, and a major difficulty for planners is to predict where and when a new sequence of react-ivations will be initiated. The historical data collected in Basilicata suggests that the feedback processes reflected by C and previous states, may be significant, and a high degree of uncertainty still prevails. Anthropogenic influences are clearly im-portant, (see below) and further consideration should be given to this factor in in-dividual landslide cases through the analysis of the hydro-geotechnical polygons which surround urban areas.

The second difficulty has concerned step two, ie the aquisition of all the digital data necessary to complete the system. In this investigation, a pilot area was thus chosen to illustrate some of the principles involved, (using pre-existing areas of mass movement), but additional historical and hydrogeological data are also re-quired so that more rigorous analyses can be attempted. In addition, the approach needs to be extended more widely to include other examples of pre-existing land-slides, so that the geological boundaries and parameters which are necessary to undertake stochastic/deterministic hazard and risk assessments can be stored and accessed. Some of these datasets have been partly captured already, but further col-laborative work (ideally, with the Basilicata Regione) is still required in these areas to bring the GIS up to a standard necessary for real decision modelling. Environ-mental data for remaining hazardous landslide complexes can be added gradually as, and when necessary.

14.8. Digital Datasets and ARC/INFO GIS

ARC/INFO GIS software was used for the project and allowed integration of envir-onmental data in several forms. Digital vector mapping in ARC has been merged with tabular data in the relational database INFO. In addition, raster maps and im-ages, attributed raster maps, and tabular data from other databases such as OR-ACLE were included within the same system. The majority of the maps are 2-dimensional, but TIN (Triangulated Irregular Networks) and Grids have enabled 3-dimensional surfaces to be represented. Topological structuring of the spatial data allowed query and analysis based upon spatial relationships as well as sim-pler relational queries on tabular data. Original data included digitised paper maps, raster-scanned paper maps and diagrams, aerial photographs, tabular data from re-ports, text notes, digital terrain models, and pre-existing digital maps imported from other systems.

The ARC/INFO software was run on a network of SUN workstations, with the software provided from a central SUN 630 server machine. The project data was integrated on a SUN IPX with a 40-Mhz SPARC CPU, with 24Mb dynamic memory and a 1.3Gb SCSI external storage module. Data exchange facilities were provided with network access to 150Mb QIC and 1Gb DAT tape drives, a CD-ROM drive, DEC VAX tape facilities, and access to the UK JANET (Joint Aca-demic Network) and the Internet.

The majority of the data exchange made using the ARC/INFO EXPORT format for data, and ORACLE EXP format for tabular data. A diverse variety of other digital transfer formats were used. For example, much of the vector digitised data was imported from INTERGRAPH's DGN format files, and the raster images for the maps were imported from INTERGRAPH's RLE format files.

14.9. Sources of GIS data

Data from a variety of sources were integrated into the GIS. Paper maps covered geology, topography, climate, seismic hazards and known landslide occurrences, but digital maps were used wherever possible because costs were lower than with the other data types. Digital data included :

1. Landuse and soil mapping provided by the EC CORINE project.
2. Comune socioeconomic data provided by the Italian Seismic Survey.
3. Rainfall data provided by Agrobios (Metaponto, Italy) and the Institute of Hydrology (UK).
4. Digitised contour and height data for Grassano, provided commercially.
5. A digital terrain model provided by the Italian Geological Survey.
6. Seismic/earhquake information from 2. above

Borehole log images and tabular data were sourced from reports issued by the Universities of Basilicata and Bari, aerial photographs were provided for the town of Grassano, and geological data compiled from a variety of paper reports obtained from Italy.

14.10. Data Conversion

14.10.1. GENERAL

Data were converted by vector digitising, raster scanning, map generation from lists of coordinates, import from digital transfer files, and manual keying-in of tabular and coordinate data. A general problem was accurately registering the data to a common coordinate system, particularly with smaller-scale (1:100,000 to 1:250,000) maps where map linework was drawn to an rectangular grid expressed in degrees for latitude and longitude. In addition, data such as the spheroid to be used during projection conversion was not defined on some maps.

The Universal Transverse Mercator (UTM) projection, zone 33, was chosen as the base projection to which all maps were converted. The UTM projection is conformal (i.e. at any point on the map the scale is the same in all directions) and distances, directions and shapes on the map within a restricted East-West band (such as the area of Basilicata) remain quite accurate. Projection conversion parameters were provided by ESRI-Italia for conversion from latitude/longitude figures into UTM, and formed the basis of most projection conversions of the data. In some cases, a later localised warping of parts of the maps was required for accurate registration of one map to another. However, the crucial environmental data coverages were believed to be sufficiently accurately registered for the purposes of data analysis.

14.10.2. DATA CONVERSION DETAILS

The majority of paper maps were converted to ARC vector coverages. This conversion was made by manual stream digitising of paper maps on a A0 digitising table attached to a UNIX workstation running INTERGRAPH MicroStation software. The INTERGRAPH DGN linework or point data files were then converted into ARC coverages using the ARC IGDSARC translator. Attribution of the ARC coverages was created using customised 'forms' menus in ARC/INFO, followed by projection of the coverages to UTM. The paper maps which were to act as locational backdrops for other maps were converted to raster images and scanned at 512 dpi on an Intergraph scanner to create RLE format files. These were then converted to TIFF format, generalised to 300 dpi, compressed and registered to map coordinates in UTM.

Raster images were also created for some borehole log diagrams. They were scanned at 400dpi on a flatbed scanner in TIFF format and compressed before saving to display within the GIS. Aerial photographs were scanned as 8-bit greyscale images on a flatbed scanner in TIFF format and registered as backdrops for urban study areas.

Tabular data such as details of landslip occurrences, borehole descriptions, and geotechnical properties were manually keyed in. For larger tables held in ORACLE, customised forms in SQLFORMS were used. For smaller tables held in INFO, the INFO 'add' facility was used to enter data. Relational definitions were then set up to link the tabular database to GIS graphics coverages.

14.11. Regional Hazard Mapping in Basilicata using GIS

14.11.1. METHODOLOGY

An underlying methodology has been established which is based on 3-D hydrogeotechnical boundary models as mapping units. Published examples of landslides, and mapping in Basilicata has suggested they can be simply based on i) the recognition of areas containing ancient deep-seated pre-existing slide complexes involving an aquifer and weak argillaceous material, (subdivided, if required by their geological settings) and ii) the identification of areas susceptible to first time translational failure, (also subdivided by their geological settings). With both models the level of landslide susceptibility increases with increasing slope angle and water saturation. The former can be quantified using a DTM and classified/grouped as required. The latter is more difficult, but current areas of surface groundwater discharge, (springs) have been shown to be closely associated with climatic instability near urban areas. Further spatial refinements of relative hazard are beyond the present capabilities of these GIS coverages and as discussed, will require additional empirical studies, mapping, monitoring and deterministic development of transient groundwater models.

The second part of the analysis considers the spatial distribution of the wide variety of temporal triggering mechanisms which can lead to instability. The method can thus be summarised in six steps, ie

1. Identify local groundwater discharge areas and boundaries of hydro-geotechnical models based on geology and physical properties.

2. Differentiate between i) intact areas, ii) landslides with no historic movements and iii) slides and slide areas which have a history of mass movement. Relevant map (digital and paper) and environmental attribute data describing the landslides can be stored separately.

3. Identify the time dependent processes, and their spatial distribution, which lead to a primary, potential shear strength loss, under constant total stress conditions, for example through weathering (usually long-term), land- use, (vegetational) change, loss of suction due to man-made pavements, groundwater discharge through aquifers (rainfall, snowmelt), etc.

4. Identify the time dependent processes, and their spatial distribution, which lead to a primary, potential shear stress increase, as a result of a total stress change, ie through erosion (river, marine), land-use change (reshaping), earthquakes (neotectonic uplift/tilting).

5. Identify and map those areas which are most susceptible to a shear stress increase, following a change in total stress, ie those with high existing shear stresses, with steep slopes, and/or with high relative relief.

6. The sixth step, is to compare those areas distinguished in steps 1 and 2, with the spatial distribution of triggering mechanisms, (plus their magnitude and freqency) associated with a shear strength loss and model accordingly.

Steps 3 and 4 are coupled, and in this study the drawbacks and difficulties of only using a probabilistic analysis of rainfall extremes to explore instability were demonstrated and discussed above.

This mix of 'static' and 'coupled dynamic' datasets is not easily summarised for hazard assessment using conventional paper map production. For example, there is a requirement for socio-economic and environmental GISs in sensitive/vulnerable areas such as Basilicata, which can be constantly updated and monitored, preferably in real time for decision modelling. The impacts from natural events or anthropogenic change can then benefit from up to date 'static information', and/or reanalysis of mapped hydro-geotechnical boundary models using appropriate dynamic inputs. The important factors behind the methodology are summarised by a flow chart, (Figure 14.6).

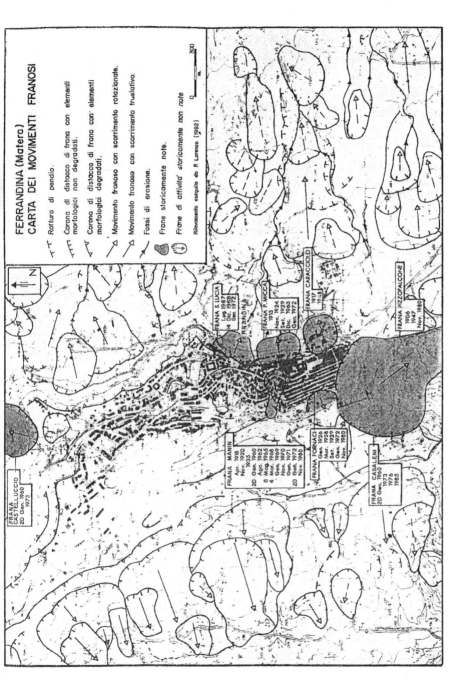

Figure 14.5. Ancient landslide boundaries surrounding the town of Ferrandina showing reactivated movements and their dates, (after Del Prete, 1993)

A typical sequence of maps derived through interrogation can be obtained from the GIS as follows. These examples, printed in black and white, (Gostelow and Gibson, 1994) were used in a preliminary exploration of the relationships between the coupled relationship between land-use change, (using CORINE data) the hydro- geotechnical models, rainfall and a landslide database compiled at the University of Bari for the Regione Basilicata.

14.11.2. CORINE LAND-COVER DATA

The land-use data was provided by the EC DGXI CORINE project group in Brussels. This data was collected and mapped at 1:100 000 scale for large areas of the EC to a standard classification, and forms part of the first multinational, European environmental database (Mounsey 1991). The data was originally sourced from LANDSAT TM and MSS satellite images with appropriate ground control. Fortyfour land-use classes were used, and placed in 3 hierarchies or levels, with the smallest land unit defined as 25 hectares. The land-cover nomenclature is divided on the basis of the three levels, and five basic land-use groups. Additional details and descriptions of the vegetation are provided in the CORINE biotope user's handbook (EC 1991).

At the mapping scales used in this survey, the relative effects of a shear stress increase following ground reshaping, or a shear strength decrease from vegetation removal cannot be accurately quantified for individual sites. However, the broad land-use classes can be re-ordered to provide a comparative spatial guide to increased mass movement susceptibility from these two effects. The approach adopted, considered that land-use associated with the urban areas represented most anthropogenic change, and natural woodland/ vegetation/wetlands the least. The 22 classes which emerged from a review of the CORINE data are as follows:

1. Mine dumps and construction sites
2. Discontinuous urban fabric

---

3. Industrial, commercial and transport units
4. Continuous urban fabric

---

5. Artificial, non-agricultural vegetated areas
6. Burnt areas

---

7. Sparsely vegetated areas
8. Agro-forestry
9. Arable land
10. Pastures

---

11. Annual crops associated with permanent crops
12. Permanent crops

| 13. | Complex cultivation patterns |
| 14. | Land principally occupied by agriculture, with significant areas of natural ve |

15.	Forests
	i) Broad-leaved forest
	ii) Coniferous forest
	iii) Mixed forest
16.	Scrub and/or herbaceous vegetation associations
17.	Open spaces with little or no vegetation
18.	i) Beaches, dunes, sands
	ii) Glaciers and perpetual snow
19.	Inland wetlands
20.	Maritime wetlands
21.	Inland waters
22.	Marine waters

These classes were divided arbitrarily into 6 groups, (separated by a line in the list above). Whilst there might be some argument about the ordering of individual classes, they represent evidence of an increasing degree of slope interference/land use change, which when combined with geological, climatic and topographic data can be used in a GIS to support assessments of regional hazard susceptibility.

14.11.3. EXAMPLES OF GIS OUTPUT

Figure 14.7 illustrates the re-classified CORINE datasets from a selected area within the region. Figure 14.8 illustrates general topography, and with Figure 14.7 shows how most land-use change has occurred in the 300 to 600m hill zone. Figure 14.9 illustrates the general surface geology (lithologies only), and Figure 14.10 is a raster slope map based on a digital elevation model.

Figure 14.11 illustrates a buffer around the lithologies which form upland aquifers on the map extract. The un- classified polygons within the buffered zone are the original CORINE land-use classes, whilst those with shaded infill are the re-classified groups. When compared with figures 14.7 and 14.8, and figures 14.9 and 14.10 this overlay can be used to identify those areas which may be most susceptible to mass movement and erosion from a hydrological triggering mechanism. These will be particularly hazardous where steep slopes and a spring are present (not shown at this scale). When these areas are overlain with the land-use categories and slope classes, they show i) where future change may be need to be monitored carefully, ii) where past change may have accelerated landsliding and slope erosion, and iii) where remedial measures/extra expenditure may be required.

Also shown on this map, (figure 14.11) are recorded landslide positions, (stars) which have been compiled for the Potenza province, (Regione Basilicata) by Palmentola et al (1981). Their database has been included within the GIS and includes

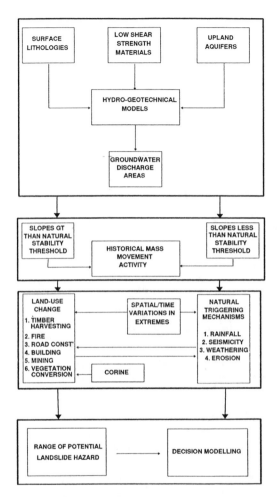

Figure 14.6. Methodology for decision modelling

attributes of slide type, area, and lithology. Although dates for these slope movements are not available, most are known to have taken place in the last 50-60 years. The map is a good illustration of the widespread slope degradation in Basilicata, and shows the strong relationship in the eastern parts, between the edge of the aquifer buffer in the hill zone, the areas affected by land-use change, (levels 2 and 3) and the recorded landslides. However, with the exception of Castronuovo, (not outlined in the CORINE data) there are no obvious clusters of mass movement events around the towns in this part of the region. As might be expected, some of the slides do not relate to the land use or geological boundaries, which suggests that other factors are also important. For example there are several slides outside the buffered zone in the south-east of figure 14.10 near Castelsaraceno which are probably shallow, and topographically controlled.

This database would now benefit from a more accurate raster topographic over-

Landuse type

Discontinuous urban fabric
Non-irrigated arable land
Annual & permanent crops
Complex cultivn patterns
Agro-forestry areas
Broad-leaved forest
Natural grasslands
Moors and heathlands
Transitional woodland-scrub
Bare rocks

Basilicata
CORINE landuse data

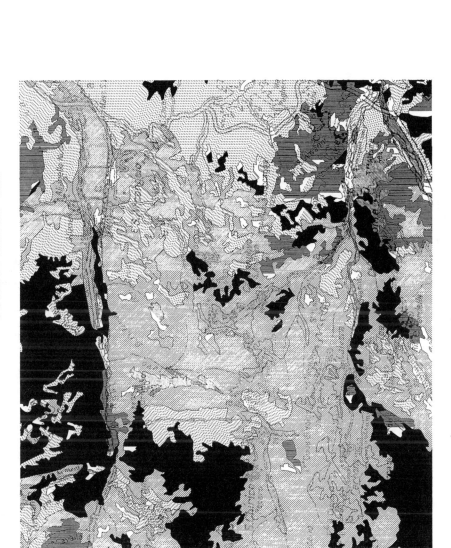

Figure 14.7. Re-classified CORINE data into areas of relative change

Elevation

0 to 300m
300 to 600m
600 to 900m
900 to 1200m
1200 to 1500m
1500 to 1800m
1800 to 2100m
2100 to 2400m

Basilicata
Topographic map

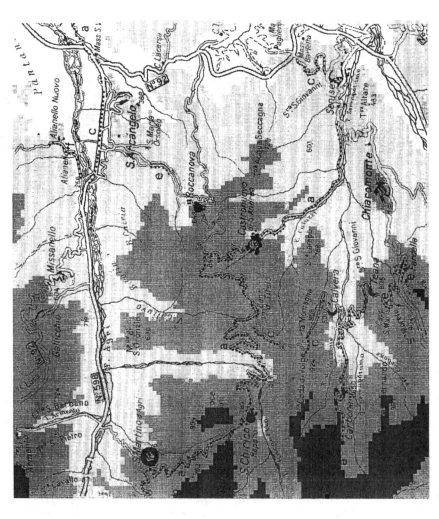

Rock type

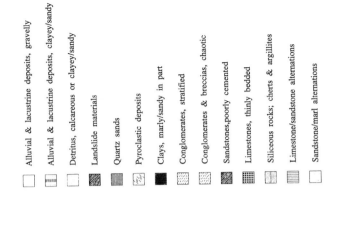

Alluvial & lacustrine deposits, gravelly

Alluvial & lacustrine deposits, clayey/sandy

Detrius, calcareous or clayey/sandy

Landslide materials

Quartz sands

Pyroclastic deposits

Clays, marly/sandy in part

Conglomerates, stratified

Conglomerates & breccias, chaotic

Sandstones,poorly cemented

Limestones, thinly bedded

Siliceous rocks; cherts & argillites

Limestone/sandstone alternations

Sandstone/marl alternations

Figure 10. Lithological map

Basilicata
Lithological map

Figure 14.9. Lithological map

Slope angle classes

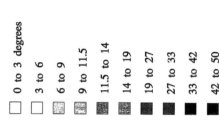

Basilicata
Slope angles map

Basilicata
CORINE landuse data

Landuse hazard classes

Level 1 - low
Level 2
Level 3 - medium
Level 4
Level 5
Level 6 - High

Figure 14.11. Buffered upland aquifers in relation to reclassified land-use classes and landslide positions, (shown as stars)

lay; updating by a programme of continued monitoring of land-use change; the inclusion of mass movement events from 1981-1995; polygons showing erosion, (calanchi); comparable datasets from the Matera Province; and further interrogation.

14.12. Conclusions

Previous attempts at landslide hazard mapping have tended to use a 2-D thematic approach to what is essentially a 3-D dynamic problem. A broad methodology for landslide hazard and risk assessment based on 3-D hydro- geotechnical slope models, topography and existing landslide areas has thus been developed for urban areas in Basilicata using GIS technology. The potential use of the data sets in decision analysis has been discussed, and the importance of historical records for establishing links between ground deformation and triggering mechanisms, such as extreme rainfalls is emphasised. However non-linearity between cause and effect currently presents difficulties with mapping hazard and risk, and it is suggested that further work on collecting ground deformation records throughout the region is urgently needed. The use of mapped, pre-defined 3-D hydrogeological models as classified units, simplifies geological conditions, limits the style of mass movement which might be expected, assists with attribution, and provides the preliminary boundary conditions which are required for decision modelling and engineering simulations outside the GIS environment.

Acknowledgements

The study in Basilicata was carried out for the EC under EPOCH contract PL 0029, and the author would like to acknowledge contributions from PS Naden, A Calver, DW Reed, NS Reynard, (Institute of Hydrology, UK), M Del Prete, M Bentivenga, E Favia, E Giaccari, AL Grignetti, P Lorenzo, D Saccente, V Simeone, V Summa, (University of Basilicata, Italy), JR Gibson, JT Dodds, SMM Fenwick, (British Geological Survey, UK). Published with permission of the Director, British Geological Survey, (NERC).

References

Brabb E.E. (1984) Innovative approaches to landslide hazard mapping. Proc. 4th. Symp. Landslides, Toronto, 1, 307-324

Caloiero D., Mercuri T. (1982) Le alluvioni in Basilicata dal 1921 al 1980. Cons. Naz. Ric. per la Prot. Idrogeol. Cosenza. No 16, 67 pp

Carrarra A. (1983) Multivariate models for landslide hazard evaluation. Math. Geol. 15, 403-426

Carrara A., Cardinalli M., Detti R., Guzzetti F., Pasqui V., Reichenbach P. (1991) GIS techniques and statistical models in evaluating landslide hazard. Earth. Surf. Proc. 16, 5, 427-447

Cotecchia V., Melidoro G. (1974) Some principal geological aspects of the landslides of southern Italy. Bull. Int. Ass. Eng. Geol. 9, 23-32

Cotecchia V., Del Prete M. (1986) Some observations on stability of old landslides in the historic centre of Grassano after the earthquake of 23rd November, 1980 Geolo. Applic. e Idrogeol. 21, 4, 155-167

Cotecchia V. (1985) Progetto Finalizzato del CNR, "Conservazione del Suolo", Sottoprogetto, "Fenomeni Franosi. Geol. Applic. e Idrogeol. 20, 13-23

Del Prete M., Gostelow T.P., Pininska J. (1992) The importance of historical observations in the study of climatically controlled mass movement on natural slopes, with examples from Italy, Poland and the UK. Proc. 6th. Int. Symp. Landslides. Christchurch New Zealand. 3, 1559-1567

Del Prete M. (1993) Rainfall Induced Landslides: General historical review of landslides/climate in Basilicata. Report to CEC EPOCH PL-890112, Universita degli studi della Basilicata, Potenza 237 pp

EC (1991) CORINE Biotopes Manual. CEC Report EUR 12587/3EN, Luxembourg

Fell R. (1994) Landslide risk assessment and acceptable risk. Can. Geotech. Jnl. 31, 261-272

Freeze R.A. (1969) The mechanism of natural groundwater recharge and discharge, 1. One dimensional, vertical, unsteady, unsaturated flow above a recharging or discharging groundwater flow system. Wat. Res. Res. 5, 1, 153-171

Freeze R.A., Massman J., Smith L., Sperling T., James B. (1990) Hydrogeological decision analysis. 1. A framework. Groundwater, 28, 738-766

Fulton A.R.G., Jones D.K.C., Lazzari S. (1986) The role of geomorphology in post disaster reconstruction: The case of Basilicata, southern Italy. in 'International Geomorphology' John Wiley and Sons ltd, London.

Gostelow T.P., Gibson J.R. (1993) Rainfall induced landslides in selected Mediterranean mountainous zones of Italy, Spain and Greece: The application of Geographic Information systems to hazard mapping. British Geological Survey Technical Report WN/93/36, 108pp

Gostelow T.P., Gibson J.R. (1994) CORINE land-cover data: Its application to regional landslide susceptibility mapping in Basilicata, Italy. Proc. Int. Conf. Veg and Slopes, ICE, Oxford, T Telford, London, 222-237

Guerricchio A., Melidoro G. (1979) Fenomeni franosi e neotettonici nelle argille grigio-azzure Calabriane di Pisticci (Lucania) con saggio di cartografia. Geol. Applic. e Idrol. 14,1 105-138

Hansen A. (1984) Landslide hazard analysis, in 'Slope Instability', J Wiley and Sons, London, 523-602

Hodge R.A.L., Freeze R.A. (1977) Groundwater systems and slope stability. Can. Geotech. Jnl. 14, 466-476

Jones D.K.C. (1992) Landslide hazard assessment in the context of development. in 'Geohazards, Natural and Man-made', ed by GJH McCall et al. Chapman Hall, London 117-141

Leighton F.B. (1976) Urban Landslides: Targets for land-use planning in California. Geol. Soc. Am. Spec. Pap. 174, 37-60

Lewis J., Rice R.M. (1990) Estimating erosion risk on forest lands using improved methods of discriminant analysis. Wat. Res. Res. 26, 1721-1733

Mason D.C., O'Conaill M.A., Bell S.B.M. (1994) Handling four-dimensional geo-referenced data in environmental GIS. Int. J. Geog. Inf. Sys. 8, 2, 191-215

Montgomery D.R., Dietrich W.E. (1994) A physically based model for the topographic control on shallow landsliding. Wat. Res. Res. 30, 1153-1171

Mounsey H.M. (1991) Multisource, multinational environmental GIS: Lessons learnt from CORINE. in 'Geographical Information Systems, Principles and Applications', ed by DJ Maguire, 2, 185-200, Longmans Press, London

Muir-Wood R. (1994) Earthquakes, strain-cycling and mobilisation of fluids. Geol. Soc. Lond. Spec Pub. No 78. Geol. Soc., London, 85-98

Neely M.K., Rice R.M. (1990) Estimating risk of debris slides after timber harvest in northwestern California. Bull. Ass. Eng. Geol. 27, 281-289

Nilsen T.H., Wright R.H., Vlasic T.C., Spangle W.E. (1979) Relative slope stability and land-use planning in the San Francisco Bay region, California. USGS Prof Pap 944, 96 pp

Palmentola G., Sigillito V., Vignola N. (1981) Catalogo dei fenomeni franosi della parte meridionale della Provincia di Potenza in Basilicata. Rep. Regione Basilicata, Universita degli studi di Bari, 54pp

Pantosti D., Schwartz D.P., Valenise G. (1993) Paleoseismology along the 1980 surface rupture of the Irpinia fault: Implications for earthquake recurrence in the Southern Apennines, Italy. Jnl. Geophys. Res. 98, B4, 6561-6577

Pike R.J. (1988) The geometric signature: Quantifying landslide-terrain types from digital elevation models. Math. geol. 20, 491-511

Raper J.F. (1992) Key 3-D modelling concepts for geoscientific analysis. in '3-D modelling with Geoscientific Information Systems'. ed AK Turner, Kluwer Academic Pub, Dordrecht, 215-232

Rulon J.J., Rodway R., Freeze R.A. (1985) The development of multiple seepage faces on layered slopes. Wat. Res. Res. 21, 11, 1625-1636

Siddle H.J., Jones D.B., Payne H.R. (1991) Development of a methodology for landslip potential mapping in the Rhondda valley. in 'Slope Stability Engineering' Thomas Telford London, 443 pp

Sidle R.C. (1992) A theoretical model of the effects of timber harvsesting on slope stability. Wat. Res. Res. 28, 7, 1897-1910

Sidle R.C., Pearce A.J., O'Loughlin C. (1985) Hillslope stability and land use. Am Geophys. Un. Wat. Res. Mono. 11 140pp

Sidle R.C., Terry P.K.K. (1992) Shallow landslide analysis in terrain with managed vegetation. "Erosion, debris flows and environment in mountain regions", IAHS Publ. 209, 289-298

Toth J. (1971) Groundwater discharge: A common generator of diverse geologic and morphologic phenomena. Bull. Int. Sci. Hyd. 16, 1, 7-24

Turner A.K. (1992) Applications of 3-D geoscientific mapping and modelling systems to hydro-geological studies. in '3-D modelling with Geoscientific Information Systems' ed AK Turner, Kluwer Academic Pub, Dordrecht, 327-364

Tsukamoto Y., Ohta T., Noguchi H. (1982) Hydrological and geomorphological studies of debris slides on forested hillslopes in Japan. IAHS Publ. 137, 89-98

Varnes D.J. (1984) Landslide hazard zonation: A review of principles and practice, UNESCO press, Paris, 63pp

Zaruba Q., Mencl V. (1969) Landslides and their control. Elsevier, Amsterdam, 205pp

CHAPTER 15

Land-Use Hydrology

C.A. Quiroga, V.P. Singh and N. Lam

15.1. Introduction

The land surface is subject to continuous change due to natural and man-made causes. As the landscape in a watershed is altered in both space and time, the factors that influence the hydrologic response of the watershed also change. Evaluation of the relationship between landscape changes and such factors is one of the goals of the study of land-use hydrology.

Considerable work has been done in this area over the last 3 decades. Unfortunately, two constraints have characterized most efforts: (1) the inability to keep track of changes as they actually occur, especially in regions with rapidly changing landscapes; and (2) a lack of quantitative understanding of the effects caused by such changes. As a result, most modeling efforts tend to be rather descriptive and/or rely too much on lumped parameters, making an accurate prediction of hydrologic response difficult (Wigmosta and Burges, 1990). Worse, even if the models used are conceptually sound, there is usually not enough data to calibrate them, much less to validate them, which in turn means that the conceptual soundness of such models cannot be established accurately. Finally, there are often the constraints due to the lack of readily available spatial analytical tools to study the various types of data that are frequently encountered in hydrologic modeling.

As seen throughout this book, remote sensing (RS) and GIS have become powerful tools for managing and analyzing geographic data to levels of coverage and accuracy not possible before, especially land-use and land-cover data. RS techniques, for example, can allow the automation of much of the data capturing and manipulating process. Multispectral data can be utilized for land use and land cover classification, using supervised and/or unsupervised classification algorithms. Several techniques for multitemporal analysis are also available (Campbell, 1987). Vegetation indices, such as the normalized difference vegetation index (NDVI) derived from NOAA's AVHRR images, can be a powerful tool to estimate

389

V. P. Singh and M. Fiorentino (eds.), Geographical Information Systems in Hydrology, 389–414.
© 1996 *Kluwer Academic Publishers. Printed in the Netherlands.*

biomass or vegetative cover changes on a daily basis. This, in turn, can facilitate the modeling of some hydrologic processes, including evapotranspiration and infiltration (Schultz, 1993). On the other hand, GIS techniques can be used to perform spatial operations such as overlaying, neighborhood, and connectivity operations, which have the potential to alleviate and systematize the data manipulation process required in most hydrologic modeling efforts. With this constraint lifted, it has now become possible to concentrate on improving existing modeling capabilities.

This chapter presents a summary of efforts that have been undertaken in this area. A review of basic concepts on land-use systems from a hydrologic perspective is provided in Section 15.2. Approaches for quantifying hydrologic effects of land-use changes are presented in Section 15.3. A more detailed description of specific effects is included in Sections 15.4 with special reference as to how GIS has been used in the modeling process, or, in some cases, how GIS can be potentially used. Section 15.5 describes specific management issues that may arise when a GIS is used as a tool for land-use hydrology.

15.2. Conceptualization of the Land System

A land system can be conceptualized in many different ways. For hydrologic modeling, it is of interest to determine the distribution and movement of moisture. From this perspective, then, it may be reasonable to model the land system either by use or by structure. An example of a moisture movement model in which the land component of the hydrologic cycle is conceptualized by use is shown in Figure 15.1. According to this model, water use is distributed among the various types of land use present, and a basic assumption is made in the sense that water does not change use. Such an assumption is frequently made in continuous hydrologic modeling, for which volumes are more important than rates and timing.

In some cases, especially for ungaged, remote basins or highly homogeneous basins, the land system structure within each land-use type can be assumed to be uniform. In many other cases, however, especially where there is a significant variability in land use types, it may be required to make the land system structure explicit. Consider the case of Figure 15.2, in which the land system has been divided into three structural subsystems: the vegetation subsystem, the permanent feature subsystem, and the soil subsystem. The permanent feature subsystem refers, in general, to man-made structures, but it can also be extended to include natural geologic controls such as rock formations.

Each subsystem is assumed to play a specific, independent, and static role in the movement of moisture throughout the watershed. In real life, however, all these subsystems are interrelated, making the separation of their roles extremely difficult. Further, if there are modifications in the characteristics of any one of the subsystems as a result of land-use changes in the watershed, the characteristics of at least one of the other subsystems are also likely to change. This, in turn, may result in changes in the distribution and movement of moisture in all land-use types,

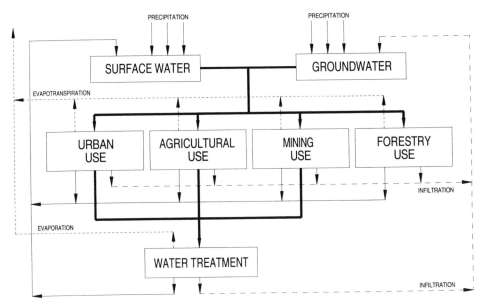

Figure 15.1. Schematic Distribution Of Moisture According To Land-Use

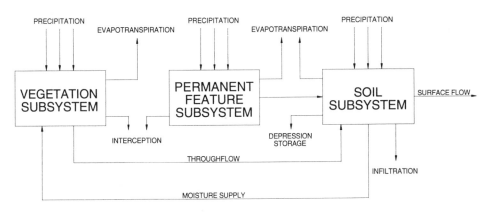

Figure 15.2. Movement Of Moisture In The Land System (adapted from Singh, 1989)

which means that the entire cycle is altered. For example, deforestation may mean not only the removal of most of the vegetation subsystem, but also the complete transformation of the soil subsystem, and the creation of a permanent feature subsystem. However, the fact that part of the forest is transformed into, say, agricultural areas also implies the creation of water requirements to satisfy agricultural needs which, in turn, may imply changes in the entire hydrologic cycle.

Depending on the specific relationship between land structure and land use, the hydrologic response of a watershed may vary. A survey of hydrologic effects due to watershed changes for various land use types is shown in Table 15.1. For example, open-pit mining usually implies the complete removal of vegetation over large areas for extended periods of time. As long as the mining operation is in place, it is possible that the vegetation subsystem may not exist. Under these conditions, the hydrologic response of the watershed is defined mainly by the effect of the drainage configuration on peak flow, lag time, sediment yield, and surface water quality. When landscape restoration is performed, either as a result of termination of the mining operation or as a result of environmental regulations, additional hydrologic components, such as direct runoff volume and groundwater supply, are affected.

Obviously, adequate land-use hydrologic simulation depends on how accurately the land use change component itself is modeled. In a GIS environment, this implies that the GIS package have spatiotemporal capabilities. Unfortunately, as mentioned in Chapter 4, current systems do not support temporal linkages between spatial objects and, consequently, their potential as tools for hydrologic simulation is still somewhat limited. However, it is expected that as more experience is gained by hydrologists in the development, population and use of geographic databases, and as GIS packages begin supporting temporal variables, such a limitation will be lifted.

15.3. Hydrologic Modeling Strategies

Two types of models are available according to the period of time hydrologic processes are assumed to take place for: event and continuous simulation models. Regardless of the type of model, however, parameters are normally estimated at specific points in time, and the hydrologic model is run to simulate the corresponding hydrologic responses. Most models assume unchanging land use and land cover characteristics, which means that the kind of results they produce actually corresponds to a watershed in equilibrium. If the watershed is changing slowly, the use of such an assumption may still be valid. However, if the watershed is changing rapidly, or if, as a result of the change in land use, additional undetected hydrologic processes are triggered or altered, the use of such an assumption may translate into significant levels of uncertainty and errors in simulation results.

Event-based models. Event-based streamflow simulation models normally deal with the relationship between rainfall and runoff. Other hydrologic processes are either ignored, lumped or very roughly estimated. Event-based models normally compute direct runoff hydrographs (DRH's) either using a unit hydrograph (UH) or an instantaneous unit hydrograph (IUH), or a physically based approach based on the shallow water-wave theory. In the traditional approach, most parameters remain spatially lumped. If sufficiently long records of watershed data are available, regression analysis techniques can be used to correlate changes in parameters with spatial indicators of land use change. In this approach, temporal effects are neg-

TABLE 15.1. Some Watershed Changes and their Possible Hydrologic Consequences (adapted from Singh, 1989)

Watershed Change	Direct Runoff Volume	Peak Flow	Lag Time	Flood Frequency	Low Flow	Sediment Yield	Surface Water Quality	Ground water Supply	Ground water Quality
Agricultural									
Contour farming	X	X	X	X		X	X		
Terrace farming	X	X	X	X		X	X		
Furrow farming	X	X	X	X		X	X		
Tillage operation	X	X	X	X	X	X	X		
Drainage	X	X	X	X		X	X	X	X
Urban									
Imperviousness	X	X	X	X		X	X		
Drainage	X	X	X	X		X	X		
Population							X	X	
Tree planting	X	X	X	X	X	X	X		
Industrialization							X	X	X
Forest									
Afforestation	X	X	X	X	X	X	X	X	X
Fires	X		X				X		
Drainage	X	X				X	X		
Application of chemicals							X		
Highways									
Drainage		X	X	X		X	X		
Bridges		X	X	X		X			
Embankments		X	X	X		X			
Clearing of snow and ice							X		
Mining									
Restoration of landscape	X	X	X	X		X	X	X	X
Drainage		X	X			X	X		
Mined material disposal							X	X	X
Point changes									
Dams		X	X	X		X			
Channel improvement	X	X	X	X		X			

lected, and spatial variability is lumped. An example of this type of application is found in the Colorado Unit Hydrograph (CUH) method, developed by the Denver Urban Drainage and Flood Control District (Singh, 1989). The CUH method is a variation of the Snyder method for which the C_t coefficient is computed as

$$C_t = a\frac{7.81}{I^{0.78}}S^{-0.2} \tag{15.1}$$

where I is percentage of impervious area, S is slope (ft/ft), and a is a correction factor for high or low slopes. Additional corrections must be included for high or low density of sewered areas. In general, C_t, which is a parameter used to evaluate time to peak and, by extension, lag time and time of concentration, decreases as I increases.

By following a different modeling approach, Richter and Schultz (1988) obtained a similar conclusion regarding the inverse relationship between time of concentration and percentage of imperviousness. They considered a spatially distributed model with a 'time-rain-concentration' diagram to account for translation effects, coupled with a linear response function to account for storage effects. They selected three basins in Germany with similar geomorphologic, geologic and climatic conditions, and assumed that changes in hydrologic response could only be attributed to changes in percentage of the impervious area. They used LANDSAT TM data to determine types of land use in each of the basins selected.

In some other cases, land-use data is, at best, incomplete, but historical records of rainfall and runoff are available. Under these circumstances, UH or IUH parameters can be estimated using rainfall-runoff data at specific points in time, which, in turn, can allow for the definition of trends in the values of such parameters. In this approach, temporal effects are taken into consideration, but spatial effects are totally neglected. Consequently, relations obtained are necessarily local in nature. Many applications in Hydrology have followed this approach. A typical example is that of Bhaskar (1988). He used the Clark instantaneous unit hydrograph method to make projection of changes in hydrologic response in the Beargrass Creek watershed, Louisville, Kentucky. He derived the Clark IUH parameters using maximum annual historical flood data prior to 1973, and projected trends of these parameters, using regression analysis, to estimate the shape of the unit hydrograph in 1990. He validated the results with observed rainfall-runoff data between 1973 and 1983, year in which the streamflow gages were decommissioned, and obtained less than 4% discrepancy between observed and projected peak flow rates. The relationship between year and the sum of T_c (time of concentration, expressed in hours) and K (storage attenuation constant, also expressed in hours) turned out to be of the form

$$T_c + K = e^{50.92449 - 0.02470279\ year} \tag{15.2}$$

$$T_c + K = e^{65.00405 - 0.03199750\ year} \tag{15.3}$$

for the south and middle forks of Beargrass Creek, respectively. These trends correspond to a very specific urbanization process (from 17.5% urbanized in 1944 to 60% urbanized in 1973) and must be considered, therefore, local in nature.

If data regarding the spatial variability of land-use changes is available, the hydrologic modeling of the watershed can be significantly improved by using spatially distributed models. Regression analysis could still be used, but their potential is limited as land-use patterns rarely follow a nice, smooth mathematical function. A better approach is to conceptualize the relationship between land-use change and the corresponding hydrologic response at the smallest spatial unit. Clearly, for this approach to be used efficiently, large amounts of geographic data must be handled, and GIS can be used to assist in this effort.

Continuous simulation models. For continuous streamflow simulation, the emphasis is more on the integration of all components in the land phase of the hydrologic cycle. As most models do account for individual hydrologic processes explicitly, the number of parameters defined is usually much larger than for event-based models. Evapotranspiration, infiltration, interception, depression storage, subsurface flow and baseflow are normally included. In the traditional approach, the watershed is divided into subbasins, following a similar approach as in event-based models. Each subbasin is simulated separately, under the assumption that water does not move across subbasin boundaries. While this assumption may be valid for the surface-flow component of most natural watersheds, it clearly ignores the changes in drainage patterns and water movement caused by human activity, which usually goes beyond natural water divides. Obviously, in this approach, land-use and land-cover patterns are separated along subbasin boundaries.

An alternative approach that is gaining acceptance with the advent of RS and GIS is to divide the watershed into areas that produce homogeneous hydrologic responses, even if they are not located within the same subbasin. These land units are called hydrologic-response units (HRU's). Depending on the particular application, the number of parameters used for the definition of these HRU's may vary. For example, in the case of the IRMB model, developed by the Royal Meteorological Institute of Belgium (Bultot, Dupriez and Gellens, 1990), each HRU is defined in terms of area, interception capacity, maximum available water capacity, and albedo. In the case of the U.S. Geological Survey's Precipitation-Runoff Modeling System, HRU's are characterized according to altitude, land slope and aspect, land cover, soil types, geology, and even climatic patterns (Jeton and Smith, 1993). In the case of the application of Cruise and Miller (1993), HRUs are characterized by land slopes, soil types, and land use. A more detailed description of this concept is provided in Section 15.4.3.

15.4. Hydrologic Modeling Applications

15.4.1. URBAN/BUILT-UP AREAS

Urbanization usually results in increases in peak and volume of direct runoff, decreases in lag times, and changes in water quality due to domestic, commercial and industrial activities. It also causes a reduction in low flows. These phenomena have been widely reported in the existing literature especially for the past four decades (Richter and Schultz, 1988; UNESCO, 1974; Soil Conservation Service, 1972), and will likely remain so as long as population pressures and economic activities continue provoking increases and changes in urbanized areas.

Most applications of GIS to urban hydrologic modeling are related to storm-water management.

In general, the GIS is used to handle topographic, soil, and climatic data; road and drainage networks; subdivision boundaries; and building, houses and drive-way outlines, as shown in Figure 15.3. The GIS can also be used for postprocess

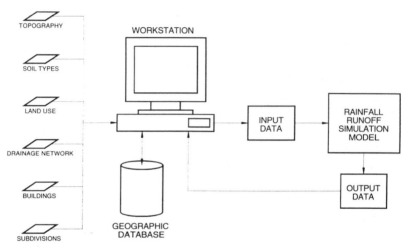

Figure 15.3. GIS-Urban Runoff Model Linkage

spatial display and analysis. An example of this type of application is that of Meyer et.al. (1993). They used a raster GIS, IDRISI, in conjunction with a spatially distributed urban watershed model, RUNOFF, to evaluate the effect, particularly as it pertains to flood risk, of various storm-management strategies in small urban watersheds. RUNOFF is similar to the RUNOFF block of the U.S. Environmental Protection Agencys Storm Water Management Model (SWMM), and is used to simulate runoff from overland flow and other areas drained by minor structures. They used data from the Greenbriar subdivision in Fort Collins, Colorado, as a case study. Using the GIS, they subdivided the entire area into subwatersheds, and derived information needed by the RUNOFF model, including percentage of impervious areas for each subwatershed, soil distribution, length of the drainage network, and rainfall depth distribution for 10-year, 50-year, 100-year, and 500-year storm events. Since they used a raster GIS, they had to develop a procedure to deal with uncertainties associated with building and road edges, as well as transitions. Their solution consisted of several encoding formats, including (a) single value format; (b) edge probability format, for situations such as the definition of imperviousness of house edges; and (c) central line format, for linear features such as roads. Then, they used the RUNOFF model to compute runoff hydrographs and water surface elevations at subwatershed outlets. With these results, they estimated water surface elevations at other points, and used again the GIS to create maps that showed the spatial distribution of flood depths. Figure 15.4 shows an example

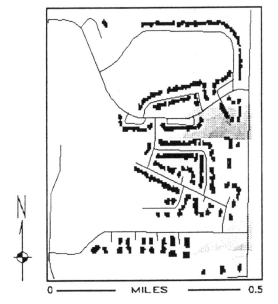
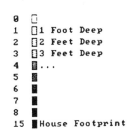

Figure 15.4. 500-year-flood zone-risk map for Greenbriar subdivision in Fort Collins, Colorado (after Meyer et.al., 1993)

of such spatial distribution of flood depths for the 500-year storm event. Because of lack of observed data, they could not validate their results and, consequently, questions may remain with respect to the accuracy of the flood distribution maps. While this is a fairly common problem, as data records are usually short in length, if they exist at all, the methodology followed clearly illustrated the potential for the use of GIS as a practical management tool.

Models such as the RUNOFF model are suitable for areas that are already urbanized. In many cases, however, it is required to measure the impact of changing land use from rural or forest to urban. As this process usually takes place over long periods of time, both event and continuous simulation must be considered and, as a result, the use of hydrologic-response units (HRU's) may be required. This was the approach followed by Wigmosta and Burges (1990) in developing criteria to minimize hydrologic impacts of urbanization in small gauged and ungauged watersheds. In their application, the HRU's are called flow production zones, which are essentially subbasins assumed to be homogeneous, based on slope, soil and litter layer characteristics, and main runoff mechanism. They utilized a GIS to do the overlaying process needed to derive such flow production zones, and to compute parameters needed for modeling.

Wigmosta and Burges's model is both event-based and continuous, and allows for land-use changes to be represented by making successive runs, assuming in each case a different land- use condition. For each homogeneous subbasin or production zone, seven hydrologic components are simulated, as shown in Figure 15.5. Obviously, the relationship among these seven hydrologic components var-

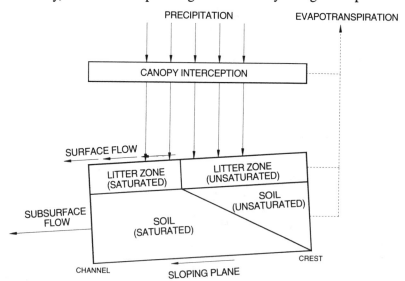

Figure 15.5. Land System, Including Components for Canopy Interception, Evapotranspiration, and Surface, Litter, Unsaturated and Saturated Flow (adapted from Wigmosta and Burges, 1990)

ies according to subbasin and according to the particular land-use condition being modeled. Each subbasin is simulated separately, and the results from all subbasins are routed using a simple, physically-based scheme.

Canopy interception is simulated by making rainfall enter an interception storage until this storage is filled to capacity, which is defined according to the type of canopy. Interception storage, as well as surface detention, are depleted by evapotranspiration. Rainfall that is not intercepted (or throughfall) reaches the ground where it makes contact either with litter or soil. Throughfall that lands on the litter zone is temporarily detained, increasing the average soil moisture, which, in turn, determines the amount of percolation. Percolation is also assumed to be a function of the type of litter, and the infiltration rate from the litter zone to the soil. Infiltration is simulated using Philip's model. Throughfall that lands on the saturated litter zone creates saturated overland flow. In the soil unsaturated zone, soil moisture is assumed to be constant, and infiltrated water is routed at a rate equal to the soil hydraulic conductivity. In the soil saturated zone, a simple storage routing scheme that assumes the energy gradient to be the same as the slope of the underlying impermeable layer is used. Throughfall that lands on bare saturated soil generates Horton overland flow. Overland flow is routed using a nonlinear storage model that allows rainfall intensity and the saturated zone to be time-variant.

Wigmosta and Burges applied their model to a second-order, 21-ha basin located in the Novelty watershed near Seattle, Washington. As in the case of Meyer et.al. (1993), discussed earlier, they did not validate their model with observed data. In their analysis of the model behavior, however, they perceived tendencies consistent with flow-production mechanisms, which is an indication of the soundness of their modeling approach. For the case studied, they compared two situations: from practically zero urbanization in 1978, to light-density urbanization in 1982. For the predevelopment stage, they divided the second-order forested basin into 2 first-order subbasins. Each first-order subbasin was further subdivided into 5 homogeneous zones, according to slope, litter and soil characteristics, and predominant runoff mechanism (subsurface and/or saturated overland flow), for a total of 10 flow production zones. For the postdevelopment stage, the drainage network changed by the addition of roadside ditches, and land use now included forest, lawn, pasture, and roads. The total catchment area increased from 21 to 30 ha, and the number of flow production zones increased to 17. The total simulation time was 11 years, from 1972 to 1982. As an illustration of the model results, Figure 15.6 shows the monthly means for rainfall, evapotranspiration, runoff volume and subsurface flow volume for the predevelopment stage, as well as the simulated variations for the postdevelopment stage. The results obtained, mainly increments in runoff volume and decrements in evapotranspiration and subsurface flow volumes, are typical consequences of urbanization processes.

In the two applications described above, the GIS and the hydrologic model were operated separately. The GIS served mainly as a tool to handle geographic data and to link this data to attribute tables. Input data files needed by the hydrologic models are normally produced using a third, specifically designed program that queries the database and that may use some of the functions already available within the GIS. The hydrologic model is then run and, in the end, it produces output data files which are transferred to the database so that the GIS can display the results. At each point in time, it is implicitly assumed that water boundaries and flow topology are fixed. For management and optimization purposes, however, it may be of interest to determine hydrologic responses at any point within the watershed, which means that it becomes necessary to redefine the drainage network and to measure new areas and parameters each time. Some special-purpose applications have started to look into ways to automate this process. Greene (1993), for example, developed a set of procedures to automatically define and measure hydrologic response areas, the inlet to which they drain and the storm drain located immediately downstream, just by selecting specific locations in the screen using the GIS. Once the system defines the subwatershed boundaries and the corresponding drainage network, it retrieves values from the database for all associated attributes, and computes parameters needed for hydrologic modeling.

15.4.2. AGRICULTURAL AREAS

Agricultural activities usually result in profound changes in land cover and hydrologic response characteristics. Changes in vegetation cover and surface slope usu-

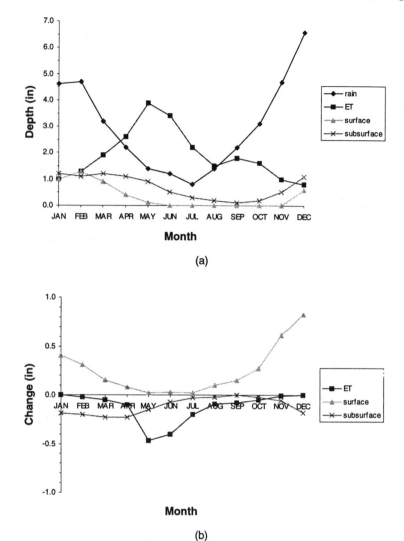

(a)

(b)

Figure 15.6. Eleven-Year Mean Annual Balance Results for (a) Predevelopment and (b) Postdevelopment Minus Predevelopment (after Wigmosta and Burges, 1990)

ally have an impact on evapotranspiration, overland flow, channel flow, and infiltration. Very frequently also, changes in vegetation cover and surface slope result in higher erosion levels. Not surprisingly, there is a continued interest in improving farming practices so that smaller runoff rates, higher absorption, less soil erosion, and greater economic returns can be achieved.

Comparisons among various management practices have been continuously reported in the literature. Most studies normally deal with effects on hydrologic response, particularly runoff volume, low flows, times to peak, erosion and nonpoint-source pollution, on a seasonal or yealy basis, due to specific soil conservation practices. Before computers, much of the work was conducted on experimental stations, with very little modeling. With this procedure, soil conservation practices such as plow-planting, for example, were shown to reduce total runoff, peak flows, and erosion more effectively than tillage (both straight-row and contour), without affecting yields significantly (Harrold et.al., 1967a and 1967b; McGuinness et.al., 1960). Terraced farming was shown to reduce surface runoff, but it was also detected that poor management practices could actually reverse the effect (Richardson, 1972). The advent of computers accelerated the modeling process allowing the analysis of additional management alternatives but, in general, results have tended to confirm that nonstructural, i.e. agronomic, soil conservation practices are more effective in controlling erosion and nonpoint-source pollution than structurally oriented ones (see, for example, Razavian, 1991).

In general, studies like the ones described above have been conducted on a local basis, i.e, without considering the effect of surrounding areas on the local hydrology. Some other studies have included the effect of changes on land-use or on climate on a larger scale but, in general, they have had a tendency to overlook minute management practice details. Continuous simulation models such as the U.S. Department of Agriculture Hydrograph Laboratory Model (USDAHL) (Crow et.al., 1983; Langford and McGuinness, 1976) have been used for the modeling process. In some other cases, statistical analyses have been undertaken to detect historical trends. Potter (1991), for example, used daily streamflows, maximum annual peak flows, climatic data, and annual estimates of harvested acreages to establish the cause of changes in hydrologic behavior on the East Branch of the Pecatonica River in Wisconsin. Between 1940 and 1986, this watershed had experienced decreases in floods, increases in base flow, increases in recharge due to snowmelt events, and decreases in the standard deviation of daily flows. By ruling out climatic variations, he was able to associate these changes with the adoption of various soil conservation practices in the region. This kind of a conclusion clearly supports the statement given at the beginning of the chapter, in the sense that changes in land use usually result in the hydrologic cycle being altered. In most cases, however, reliable detection of such changes requires the manipulation of huge amounts of data for which GIS can be a powerful tool.

So far, most applications of GIS to hydrologic modeling in agricultural areas have been related to erosion and nonpoint-source pollution simulation. This subject is already covered in Chapters 13 and 14. Nonetheless, some mention is appropiate here from a land-use perspective. He et.al. (1993), for example, used GIS in conjunction with the Agricultural Nonpoint Source Pollution model (AGNPS) to evaluate the impact of various management practices on water quality in the Cass River watershed in Michigan. The AGNPS model is an event-based, spatially-distributed model that simulates runoff, sediment and nutrient yields from essentially agricultural watersheds. The model is very flexible in that it has the capability of evaluating hydrologic responses both at the watershed level and on a cell-by-cell basis, making it suitable for comparing the effects of various agricultural practices within the same watershed.

Data requirements include land use and land cover, water features, topography, and soil. To build input data files compatible with the AGNPS model, they wrote a special module called GRASS WATERWORKS. Taking into consideration various data sources, they included in the module three methods of handling data: user input, database extraction, and GIS spatial functions. In a similar fashion as other applications described earlier, He et.al. used the GIS both for data processing and postprocess display, which allowed them to detect critical areas within the watershed, particularly with respect to soil erosion and total nitrogen in runoff. They detected, for example, exceptionally high sediment and nitrogen levels in high-slope areas and near tributaries. Figure 15.7 shows, as an example, the spatial distribu-

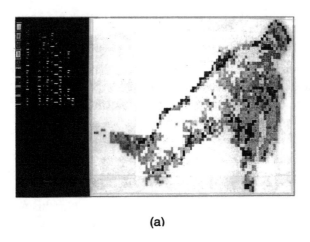

(a)

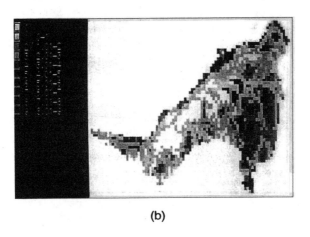

(b)

Figure 15.7. Spatial Distribution of (a) Soil Erosion and (b) Total Nitrogen in Runoff in the Cass River Watershed, Michigan, due to a Hypothetical 25-year Storm Event of 3.7 mm in a 24-h Period (after He et.al, 1993)

tion of simulated soil erosion and total nitrogen in runoff due to a hypothetical 25-year storm event of 3.7 in. in a 24-h period, assuming poor management/moldboard plow tillage conditions. In all scenarios, good management practices showed a decrease in nitrogen and sediment yield. The simulated effect of tillage, however, was not so uniform. Because of the way the AGNPS model handles fertilization availability, tillage practices are assumed to turn over the soil completely, forcing less nitrogen to be available in the soil surface to be washed away. As a result, no-till practices end up producing more nutrient runoff than tillage practices. This may explain why no-till practices resulted in less erosion, but in more nutrient runoff, while moldboard plow tillage resulted in more erosion, but in less nutrient runoff.

De Roo et.al. (1989) used an extended version of the spatially distributed, event-based Areal Nonpoint Source Watershed Environment Response Simulation model (ANSWERS) with a GIS to compare conservation scenarios in intensively cropped areas in the Catsop watershed, Limburg, The Netherlands. The GIS was used to construct maps of altitude, slope, and aspect based on DEM data, as well as of specific information based on field measurements. A special-purpose program converted the maps of input variables into an ANSWERS input file. Results from the model were stored in the GIS, and maps could be produced to indicate changes over time of soil erosion, sedimentation and runoff.

In the extended version of the ANSWERS model, the restriction of 20 soil and land-use types, which was the maximum permissible in the original version, was lifted. With the new limit, now possible with the help of GIS, the maximum number of soil and land-use types was set equal to the number of grid cells (4275). The implications in model response were evident. De Roo et.al. reported increases of 46% in runoff, 42% in peak discharge, and 36% in soil loss. These results are an indication of the sensitivity of the model to spatial lumping and, by extension, to grid-cell size. Other studies have reached similar conclusions. Vieux and Needham (1993), for example, used a GIS to derive input parameters for the AGNPS model, and observed that, when grid-cell sizes were allowed to increase, stream length became shorter and, as a result, simulated sediment yield increased by as much as 32%. Chairat and Delleur (1993) observed decreases in water peak discharge up to 33% when grid-cell sizes were increased from 30 m to 90 m. This problem of defining optimum grid-cell sizes remains largely unsolved and limits, therefore, current capabilities to assess the effects of land-use change.

GIS has also been used as an integrator tool to compute spatially lumped parameters. For example, Hill, Singh and Aminian (1987) used an in-house raster GIS to obtain U.S. Soil Conservation Service curve number (SCS-CN) values for the Amite River basin in Louisiana. The GIS in this case was used to perform image processing of Landsat TM data to obtain land use patterns, which, in turn, allowed the computation of representative SCS-CN values for the entire watershed. These CN values were used to evaluate the performance of the Watershed Hydrology Simulation model (WAHS) (Singh, 1989). In a similar application, Luker, Samson and Schroeder (1993) used a GIS to compute SCS-CN numbers needed by the

TR-20 model to determine runoff volumes for the upper Fish River subwatershed in Alabama.

15.4.3. FORESTED AREAS

Perhaps more than with any other type of land use, changes in forest characteristics have various, potentially enormous, effects on watershed hydrologic response. Chapters 13 and 14 deal with this problem in detail. Here, mention is made of the hydrologic effects of forest-related activities from a land-use standpoint.

Hydrologic effects due to forest land uses have been widely documented. Deforestation results in declines in evapotranspiration and elimination of interception storage, forcing rainfall to reach the ground without any attenuation. Often, the forest litter is removed, which results in a reduction in infiltration capacity. As a result, direct runoff increases, and since surface roughness is usually reduced, higher peak flows and reduced times to peak tend to occur. In Georgia and West Virginia, for example, a linear relationship between reductions in forest cover and first-year water yield have been found (Jones and Holmes, 1985). Deforestation has also been associated with intensification of daily ranges in soil temperature, air temperature, and relative humidity. It has even been associated with the decline in local rainfalls in tropical areas (Baker, 1993).

Clearing methods have an impact on runoff and sediment load. Mechanically-cleared forests tend to produce much more runoff and sediment load than manually-cleared areas because removal of the vegetal cover is more thorough. Increases up to 500% in runoff and up to 5000% in soil eroded have been reported in the literature (Lal, 1981). Poor construction of the roads to drag cut timber is particularly responsible for environmental damage. Often, troughlike ditches, which may run up and down steep hills, are created resulting in further erosion and risk of landslides. Sediment loss due to log skidding may be significant, although its amount depends on the type of soil and soil moisture. Granite-derived soils, for example, have been found to be more susceptible to erosion than basaltic soils (Jones and Holmes, 1985). At the same time, however, there is ample evidence to support the notion that sediment-control techniques can be effectively implemented and that their cost can be manageable, especially when it is taken into consideration that forest is the best cover for minimizing overland flow and erosion (Hewlett, 1982).

Once an area has been deforested, it may take years before it can achieve a new hydrologic equilibrium. Ruprecht and Schofield (1989), for example, observed a continuous increase in streamflow as a percentage of rainfall, even 8 years after clearing (Figure 15.8). They attributed the phenomenon to increased groundwater-saturated source areas caused by the removal of native deep-rooted vegetation, which originally extracted significant amounts of water from the groundwater system. Changes in the balance between water flow and sediment production may imply drastic changes in the drainage network and stream morphology, which fre-

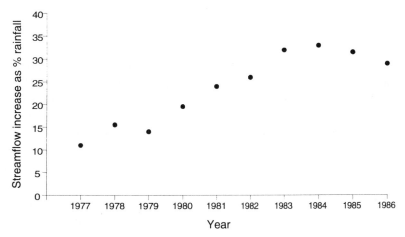

Figure 15.8. Increase of Streamflow in a watershed in Wight, Australia, after clearing (after Ruprecht and Schofield, 1989)

quently take years to develop fully. Cases of massive landscape changes resulting from successive channel head formation by tunnel erosion due to forest clearing done several decades ago have been documented (Dietrich and Dunne, 1993). In general, the relationship between changes in water discharge, sediment load and channel morphology can be explained using Schumms equations (Schumm, 1969)

$$Q_w \propto \frac{w, \lambda, d}{S} \tag{15.4}$$

$$Q_s \propto \frac{w, \lambda, S}{d, P} \tag{15.5}$$

where Q_w is a water discharge, Q_s is bed material load, w is channel width, λ is channel wavelength, d is channel depth, S is channel gradient, and P is sinuosity. Evidently, as Q_w and Q_s change, the remaining variables, which describe channel geometric characteristics, are also expected to change.

Long-term effects such as these require the use of continuous simulation models. As with other types of land use, GIS can be utilized to manipulate maps and to produce input data needed by the models. Cruise and Miller (1993), for example, utilized a raster GIS in conjunction with remotely sensed data and the GLEAMS model to simulate the effects of changing land use on water quality in the Rosario River basin in Puerto Rico. The GLEAMS model is an extension of the Chemical, Runoff, Erosion from Agricultural Management Systems model (CREAMS) that allows simulation of runoff, erosion, and pesticide transport from small watersheds. In particular, the GLEAMS model can simulate pesticide flow through the vadose zone.

Cruise and Miller based their study on the use of hydrologically homogeneous regions or HRU's. To that end, they derived land use digital information from spectral data obtained with NASAs Calibrated Airborne Multispectral Scanner (CAMS). They also used DLG topographic data and digitized soil maps. Using the GIS, they overlaid the topographic, soil and land-use maps to determine clusters or regions of similar physical characteristics. Based on uniformity of topography, land use and runoff characteristics, they divided the basin into seven HRUs. For each HRU, they used the GLEAMS model to simulate daily runoff and sediment yield for the period 1986-1989. They also performed separate simulations within each region for each soil type and land use class. This way, they were able to measure the relative impact of each HRU, soil type and land-use class on hydrologic response at the outlet. They computed the total impact on sediment yield by adding the results from each HRU. Based on the gradient and sinuosity of the Rosario River (1.6% and 1.55, respectively), they assumed the channel to be stable, which allowed them to neglect sediment yield due to channel erosion. Such assumption also allowed them to use unchanging values for channel gradient which, in light of the discussion provided earlier, would not have been possible if the simulation time had been much longer than 48 months, or if most of the area had been cleared during the same period.

As opposed to other recent applications involving GIS, including some described earlier, Cruise and Miller used observed data to calibrate their model. Figure 15.9 shows a comparison between observed and simulated results for runoff

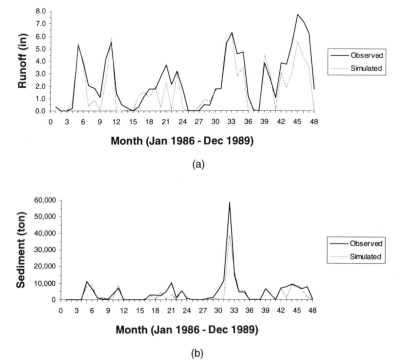

(a)

(b)

Figure 15.9. Observed vs. Simulated Results for (a) Runoff; and (b) Sediment Yield, in the Rosario Watershed, Puerto Rico, Using the GLEAMS Model (after Cruise and Miller, 1993)

and sediment yield. The overall agreement between them supports the notion that sediment yield processes were adequately represented, and that the use of HRUs, defined with the help of a GIS, was sound. Interesting to note is the fact that for the Rosario watershed, 95.4% of the area was covered by forest, 4.3% by agricultural lands, and 0.3% by urban areas. Despite this distribution, after 48 months of simulation, agricultural lands accounted for nearly 17% of the sediment load (50% during base flow periods), which is a clear indication of the impact of forest clearing on the hydrologic response of a watershed.

In a somewhat similar application, Jeton and Smith (1993) used a GIS to assist in the definition of HRU's needed by the U.S. Geological Survey's Precipitation-Runoff Modeling System (PRMS) for hydrologic simulation of two mostly forested watersheds in eastern California and western Nevada. They characterized HRU's according to altitude, slope, aspect, land cover, soil types, geology, and even climatic patterns. To accomplish this, they regrouped source-data variability within each data layer (slope, aspect, land cover, and so on) into new categories according to hydrologic-response characteristics and sensitivity to climatic factors. This allowed the reduction of categories to manageable levels. Categorical lumping was thus created, although it must be pointed out that such lumping was deliberate and selective. For example, 9 different original type classes of land cover became 6; 35 different type classes of soil became 4. This greatly simplified the modeling procedure. Figure 15.10 shows, as an example, the HRU's derived for the East Fork Carson River basin.

Figure 15.11 shows the comparison between observed and simulated annual mean runoff for water years 1969-1990 in East Fork Carson River. Results indicate a good correlation, even though an analysis of individual HRUs would be needed to establish partial effects.

15.4.4. MINING AREAS

Surface mining usually involves major changes in topography and land cover. Since most vegetation is removed, peaks and volumes of direct runoff are usually greater. Also, since drainage works are built, infiltration and base flow decrease. Mining operations are also associated with increases in sediment and pollution of water bodies (Betson, Bales, and Deane, 1981; Gregory, et.al, 1984).

Mining is frequently done in remote, originally undisturbed, areas for which little, if any, hydrologic data is available. Under these circumstances, hydrologic modeling must be done without calibration and validation for local conditions, which means that a significant degree of uncertainty is always associated with simulation results. On the other hand, many parties are usually involved in the operation and regulation of mining activities, which means that models must be as reliable, standardized and simple to use as possible.

An example of this type of application is described by Ross and Tara (1993). The Florida Institute of Phosphate Research (FIPR) developed a mine reclama-

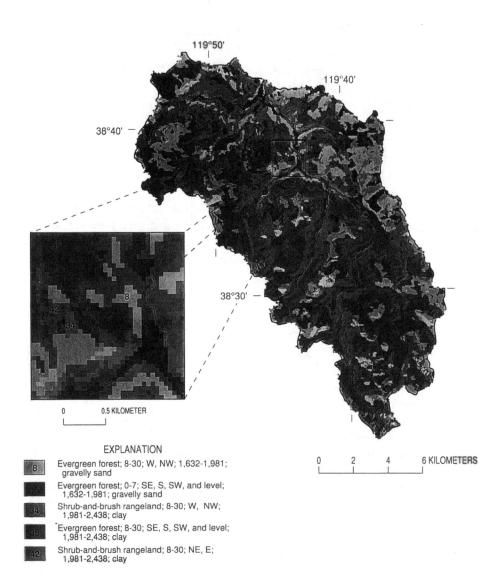

EXPLANATION

Evergreen forest; 8-30; W, NW; 1,632-1,981;
gravelly sand

Evergreen forest; 0-7; SE, S, SW, and level;
1,632-1,981; gravelly sand

Shrub-and-brush rangeland; 8-30; W, NW;
1,981-2,438; clay

Evergreen forest; 8-30; SE, S, SW, and level;
1,981-2,438; clay

Shrub-and-brush rangeland; 8-30; NE, E;
1,981-2,438; clay

Figure 15.10. HRU Spatial Distribution for East Fork Carson River Basin, California (after Jeton and Smith, 1993)

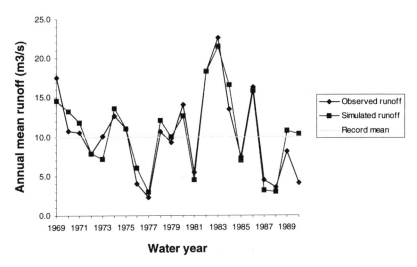

Figure 15.11. Observed and Simulated Annual Mean Runoff for Water Years 1969-1990 in East Fork Carson River, California (after Jeton and Smith, 1993)

tion modeling package to be used in reclamation design and permitting of phosphate mine sites. The package was designed to be run in stand-alone microcomputers, taking into consideration the ample variability among users, which include state and local regulatory agencies, mining companies, and private consultants. The FIPR hydrologic model (FHM) combines a GIS, Spatial Analysis System (SPANS); a surface water model, U.S. Environmental Protection Agency's Hydrological Simulation Program - Fortran (HSPF); a groundwater model, U.S. Geological Survey's Modular Ground-Water Flow Model (MODFLOW); and a specifically designed evapotranspiration simulation module. The model also includes a user-friendly interface designed, in part, to minimize user subjectivity in parameter evaluation. This was accomplished by including default values and checks on the acceptable range of user-defined parameters. The FHM model can be used in two modes: rainfall-runoff event mode, and continuous simulation mode. For the rainfall-runoff option, the FHM uses only the surface water component. For the continuous simulation option, the groundwater and the evapotranspiration components are also used. For management purposes, nine rainfall options can be defined, including historical records and various design storms.

The GIS serves three objectives. First, it provides the conversion of digital landforms to a standardized format (georeferencing). Second, it computes input data needed by the hydrologic models by means of map overlay and spatial analysis. Third, it allows postprocessing graphical display of simulation results. The reclamation site defines the area of interest, which may comprise more than one watershed. Several subbasins can also be considered within each watershed. Land use, soil type and slope are averaged over the watershed and then applied to each individual grid, subbasin or parcel description. This way, the GIS serves as a link between the surface water component, which is based on arbitrarily shaped subbasins drained by an equally arbitrarily shaped stream network, and the groundwater and evapotranspiration components which are based on uniform grid cells.

15.5. Management Issues Affecting Land-Use Hydrologic Modeling

15.5.1. REGIONAL AND CULTURAL EFFECTS

Human activities depend heavily on the availability of land and water. With the use and abuse of natural resources, hastened all over the world in recent decades by increases in population and other pressures, the need for adequate planning and management of such resources has become critical. In many cases, regulatory bodies have been formed to deal with the planning process and the enforcement of policies and rules. These policies and rules, however, are not universal. For them to be effectively used, local environmental and social circumstances must be taken into consideration. Watershed management, of course, is no exception. From the standpoint of land-use hydrology, this means that the conceptualization and modeling process must be adapted to local conditions, rather than the opposite.

Consider, for example, the effects of urbanization on sediment yield and channel erosion. In temperate regions, most cities are paved and lawned, which means that bare soil is a relatively rare occurrence. As a result, sediment yield tends to be low, which, in turn, results in channel erosion downstream of urban settings. By contrast, in many tropical cities, the proportion of bare surface is much higher, and storms provide an erosive power up to 16 times that encountered in temperate regions (Ebisemiju, 1989). Not surprisingly, much more sediment is usually delivered to streams and, as a result, a significant portion of this sediment is deposited in the channel section, despite greater peak flows caused by the removal of vegetation. Greater peak flows tend to cause channel erosion, but in the long run, sediment deposition, which usually takes place in the falling stages of hydrographs, is enough to compensate for the erosion. Under these circumstances, therefore, an urban storm-water management system should include a sediment yield and transport module. A GIS can be used to track the spatial distribution of bare soil and sediment production zones so that proper measures can be devised.

A similar situation occurs in agricultural lands. Because precipitation tends to be much more intense and frequent in the tropics than in temperate regions, the slope threshold for large scale erosion to occur is much less in the tropics: 3.5% in cultivated fields in Rhodesia, for instance, versus 14% in western Europe (Ebisemiju, 1989). If HRU's are used, the corresponding criteria to define slope categories must necessarily change. Deforestation is another example. Strictly speaking, changes in land cover and land use causing deforestation usually result in decreases in interception, evapotranspiration, infiltration and ground water recharge, as well as increases in erosion, runoff response to rainfall, and surface water yield. However, in areas where clearing is done as a result of direct population pressures, apparently unrelated factors such as soil fertility, population distribution, migration patterns, educational levels, and so on, should also be taken into consideration if a land-use management system is to be used to derive data for long-term hydrologic planning. Clearly, the land-use hydrologic simulation can be affected by the operation of such land-use management system.

15.5.2. SPATIAL AND TEMPORAL RESOLUTIONS

Two problems directly related to watershed management are those of defining optimum spatial and temporal resolutions. As described before, studies suggest that spatial resolution plays a significant role in the computation of peak discharges and sediment yield, and, yet, the exact relationship between the smallest spatial unit used and these hydrologic parameters remains to be completely defined. From a management perspective, this lack of confidence frequently precludes the use of modeling tools to decide on issues such as regulating land use in relatively small plots. Consider, for example, a local authority in charge of imposing controls regarding where and to what extent forest can be cleared. If such an authority does not have the appropiate tools to simulate and measure the effects of forest clearing with an accuracy of, say, less than one hectare, how can it grant or deny permissions to clear small plots reliably? How can it effectively penalize infringers? One of the goals of recent research efforts is that appropiate modeling tools may become available so that a systematic use of RS and GIS, coupled with accurate hydrologic models, can assist in management issues such as these.

Temporal resolution actually compresses two factors: (a) the smallest temporal unit used or needed by a hydrologic model for stability and convergence purposes, and (b) the frequency with which land use and other physical watershed characteristics are updated. Of interest here is the second factor. Ideally, land-use changes should be continuously monitored over time. In practice, though, a minimum frequency has to be established in order to guarantee an adequate coverage over time. Unfortunately, the definition of adequate is a relative one, as it depends on the rate of change of land-use characteristics. In most cases, it also depends on financial considerations. Budget restrictions can, and, quite frequently, do restrict frequency of coverage to very low levels. This limitation is particularly severe in areas with rapidly changing landscapes. In this case, as opposed to the spatial resolution problem described above, the technology may already exist, but financial restrictions preclude access to that technology. Clearly, judicious engineering and management practices are particularly critical to define the kind and frequency of data that will render the most useful information possible.

15.6. Summary

In order to make an accurate prediction of hydrologic response due to landscape changes, it is required (1) to be able to track changes as they actually occur; and (2) to quantitatively understand the effects caused by such changes. RS and GIS have become powerful tools for detecting, managing, and analyzing geographic data to levels of coverage and accuracy not possible before, especially land-use and land-cover data. This has allowed to concentrate research efforts on improving existing hydrologic modeling capabilities.

Land use variability can be treated in several ways. For example, regression analysis can be used to detect trends on the values of parameters used for hydrologic modeling. This approach, however, is limited as land-use patterns rarely follow a nice, smooth mathematical function. A better approach is to conceptualize the relationship between land-use and the corresponding hydrologic response using appropriate spatial units. Examples of such spatial units include the so-called homogeneous response units (HRUs), which have become viable as a direct result of the application of GIS techniques to hydrologic modeling.

In most cases, the GIS and the hydrologic model are operated separately, with data transfer being handled manually. The GIS serves mainly as a tool to manage geographic data and to link this data to attribute tables. Input data needed by the hydrologic models are normally produced using a third, specifically designed program that queries the database. The hydrologic model is then run and, in the end, it produces output data files which are transferred to the database so that the GIS can display the results. However, all this process can be automated by defining procedures within the GIS to automatically define and compute hydrologic parameters, as well as to execute the hydrologic model and display the corresponding results. The implications for watershed management can be enormous in that it may become possible to integrate all aspects related to hydrologic modeling into a comprehensive system so that better simulations can be obtained and, therefore, better management can be achieved.

References

Baker, R., 1993. Environmental Management in the Tropics, Lewis Publishers, 217 p.

Betson, R., Bales, J., and Deane, C., 1981. Methodologies for Assessing Surface Mining Impacts: Final TVA Report, Tennessee Valley Authority, Report No. WR 28-10550- 108, Norris, Tennessee, 99 p.

Bhaskar, N., 1988. Projection of Urbanization Effects on Runoff Using Clark Instantaneous Unit Hydrograph Parameters, Water Resources Bulletin, AWRA, February, Vol. 24, No. 1, pp. 113-123.

Bultot, F., Dupriez, G., and Gellens, D., 1990. Simulation of Land Use Changes and Impacts on the Water Balance - A Case Study for Belgium, Journal of Hydrology, Vol. 114, pp. 327-348.

Campbell, J., 1987. Introduction to Remote Sensing, The Guilford Press, 551p.

Chairat, S. and Delleur, J., 1993. Effects of the Topographic Index Distribution on the Predicted Runoff using Grass, Proceedings on the Symposium on Geographic Information Systems and Water Resources, AWRA, March, pp. 285-292.

Crow, F., Ghermazien, T., and Pathak, C., 1983. The Effect of Land Use Parameters on Runoff Simulation by the USDAHL Hydrology Model; Transactions of the American Society of Agricultural Engineers, Vol. 26, No.1, pp. 148-152.

Cruise, J. and Miller, R., 1993. Hydrologic Modeling with Remotely Sensed Databases, Water Resources Bulletin, AWRA, December, Vol. 29, No. 6, pp. 997-1002.

De Roo, A., Hazelhoff, and Burrough, P., 1989. Soil Erosion Modelling using 'ANSWERS' and Geographical Information Systems, Earth Surface Processes and Landforms, Vol. 14, pp. 517-532.

Dietrich, W., and Dunne, T., (1993). The Channel Head, in Channel Network Hydrology, Beven, K., and Kirkby, M. (eds), John Wiley & Sons, pp. 175-219.

Ebisemiju, F., 1989. The Response of Headwater Stream Channels to Urbanization in the Humid Tropics, Hydrological Processes, Vol. 3, pp. 237-253.

Greene, R., 1993. Hydrologic Modeling with a Spatial Database, Dissertation, Louisiana State University, Baton Rouge, LA, 143 p.

Gregory, J., Skaggs, R., Broadhead, R., Culbreath, R., Bailey, J., and Foutz, T., 1984. Hydrologic and Water Quality Impacts of Peat Mining in North Carolina, Water Resources Research Institute, North Carolina State University, Report No. 214, Raleigh, NC, 215 p.

Harrold, L., Triplett, G., and Yonker, R., 1967a. Loess Soil and Water Loss from No-Tillage Corn, Ohio Report, Vol. 52, No. 2, pp. 22-23.

Harrold, L., Triplett, G., and Yonker, R., 1967b. Watershed Tests of No-Tillage Corn, Journal of Soil and Water Conservation, Vol. 22, No. 3., pp. 98-100.

He, C., Riggs, J., and Kang, Y., 1993. Integration of Geographic Information Systems and a Computer Model to Evaluate Impacts of Agricultural Runoff on Water Quality, Water Resources Bulletin, AWRA, Vol. 29, No. 6, pp. 891-900.

Hewlett, J., 1982. Forests and Floods in the Light of Recent Investigation, Proceedings of the Canadian Hydrology Symposium, Fredericton, New Brunswick, Canada, pp. 543- 559.

Hill, J., Singh, V., and Aminian, H., 1987. A Computerized Data Base for Flood Prediction Modeling, Water Resources Bulletin, AWRA, Vol. 23., No. 1, pp. 21-27.

Jeton, A., and Smith, J., 1993. Development of Watershed Models for Two Sierra Nevada Basins Using a Geographic Information System, Water Resources Bulletin, AWRA, Vol. 29, No. 6, pp. 923-932.

Jones, R., and Holmes, B., 1985. Effects of Land Use Practices on Water Resources in Virginia, Bulletin 144, Virginia Water Resources Research Center, Virginia Polytechnic Institute and State University, Blacksburg, Virginia, 116 p.

Lal, R., 1981. Deforestation of Tropical Rainforest and Hydrological Problems, in Tropical Agricultural Hydrology, Lal, R., and Russell, E. (eds), John Wiley & Sons, pp. 131- 140.

Langford, K., and McGuinness, J., 1976. Using a Mathematical Model to Assess The Hydrological Effects of Land-Use Change, U.S. Department of Agriculture ARS-NC-31, June, 38 p.

Luker, S., Samson, S., and Schroeder, W., 1993. Development of a GIS Based Hydrologic Model for Predicting Direct Runoff Volumes, Proceedings of the Symposium on Geographic Information Systems and Water Resources, AWRA, March, pp. 303-311.

McGuiness, J., Harrold, L., and Dreibelbis, F., 1960. Some Effects of Land Use and Treatment on Small Single Crop Watersheds, Soil and Water Conservation, Vol. 15, No. 2, pp. 65-69.

Meyer, S., Salem, T., and Labadie, J., 1993. Geographic Information Systems in Urban Storm-Water Management, Journal of Water Resources Planning and Management, Vol. 119, No. 2, March/April, pp. 206-227.

Potter, K., 1991. Hydrological Impacts of Changing Land Management Practices in a Moderate-Sized Agricultural Catchment, Water Resources Research, Vol. 27, No. 5, May, pp. 845-855.

Razavian, D., 1990. Hydrologic Responses of an Agricultural Watershed to Various Hydrologic and Management Conditions, Water Resources Bulletin, AWRA, Vol. 26, No. 5, pp. 777-784.

Richter, K., and Schultz, G., 1988. Aggravation of Flood Conditions Due to Increased Industrialization and Urbanization, Proceedings of the International Symposium on Hydrological Processes and Water Management in Urban Areas, International Hydrological Programme, UN-ESCO, April 24-29, pp. 495-503.

Richardson, C., 1972. Changes in Water Yield of Small Watersheds by Agricultural Practices, Transactions of the American Society of Agricultural Engineers, Vol. 15, No. 3, pp. 591-593.

Ross, M., and Tara, P., 1993. Integrated Hydrologic Modeling with Geographic Information Systems, Journal of Water Resources Planning and Management, Vol. 119, No. 2, March/April, pp. 129-140.

Ruprecht, J., and Schofield, N., 1989. Analysis of Streamflow Generation Following Deforestation in Southwest Western Australia, Journal of Hydrology, Vol. 105, pp. 1- 17.

Schumm, S., 1969. River Metamorphosis, Journal of the Hydraulics Division, Proceedings of the American Society of Civil Engineers, Vol. 1969, HY1, January, pp. 255-273.

Schultz, G., 1993. Application of GIS and Remote Sensing in Hydrology, HydroGIS 93: Applica-

414

tion of Geographic Information Systems in Hydrology and Water Resources, Proceedings of the Vienna Conference, April 1993, IAHS Publication No. 211, pp. 127- 142.

Singh, V., 1989. Hydrologic Systems, Watershed Modeling, Vol. II, Prentice-Hall, 320 p.

Singh, V., 1992. Elementary Hydrology, Prentice-Hall, 973 p.

Soil Conservation Service, 1972. Hydrology, SCS National Engineering Handbook, Sec. 4, U.S. Department of Agriculture, Washington, D.C.

UNESCO, 1974. Hydrological Effects of Urbanization, Studies and Reports in Hydrology No. 18, Paris, France, 280 p.

Vieux, B., and Needham, S., 1993. Nonpoint-Pollution Model Sensitivity to Grid-Cell Size, Journal of Water Resources Planning and Management, Vol. 119, No. 2, March/April, pp. 141-157.

Wigmosta, M., and Burges, S., 1990. Proposed Model for Evaluating Urban Hydrologic Change, Journal of Water Resources Planning and Management, Vol. 116, No. 6, November/December, pp. 742-763.

CHAPTER 16

Design of GIS for Hydrological Applications

G. Mendicino

Abstract.   After having analysed some of the main criteria for the design of the GIS in the present chapter, a wide range of commercially available packages, which are suitable for hydrological application, are considered.

16.1. Introduction

The acquisition and management of hydrological information at various levels present a whole series of problems linked to both the pace of development in environmental dynamics and to the complexity of interrelations between the different factors which describe the hydrological cycle in its widest terms. Advancing technological developments in the field of automatic data acquisition, in remote sensing, in telecommunications, and in automatic data elaboration, have encouraged the development of sophisticated systems suitable for checking, managing and simulating many of the main environmental phenomena, especially those of a hydrological nature. In fact, the use of apt methodology and instruments for the acquisition and management of territorial and hydrological information in organic and rational form highlights the need for a decision support system for different working strategies.

In every organizational structure, or in every place where choices are made, information systems exist, even if in a not entirely envolved form. However, a greater quantity of available information does not necessarily mean an increase in taking more correct decisions. Often, in fact, data-bases are so complex in their use that recovering suitably co-ordinated information is difficult. A similar consideration can often be extended to model banks which are too complicated to be easily used. Data contained in such systems, indeed, are not connected to geographical references, implying the impossibility of cross-referencing the multi-natured information of any such zone.

V. P. Singh and M. Fiorentino (eds.), Geographical Information Systems in Hydrology, 415–436.
© 1996 Kluwer Academic Publishers. Printed in the Netherlands.

The ever more pressing need to refer to georeferenced information (double-value information: the first understood in spatial definition terms; and the second based on the definition of attributes), associated with systems capable of managing, elaborating, and cross-referencing information, has inclined specialists in the hydrological field to strongly align themselves with those who propose informatics techniques with the aim of favouring the development of the so-called Geographic Information Systems (GIS).

The foregoing should not cause misunderstandings to arise between what an informatics system represents and the structure of an information system. In this regard it should be stressed that the former requires specific infomatics skills (keeping the data concept in its most generic alphanumeric form), whereas the latter requires experience with design and use of GIS in the particular problems encountered by the system itself.

In this case, therefore, the development and management of a GIS should be entrusted to experts in the hydrology sector who, with respect to informatics personnel, guarantee a flexible system, which can be dynamically constructed and reconstructed when necessary, with the aim of extracting the required information and elaborations, according to the predetermined objectives of that specific case. The use of GIS, therefore, cannot be limited to any single application, since its versatility allows it to sustain different analysis and management activities, whilst referring to the same data base of the system (Colosimo and Mendicino, 1989).

In the hydrological applications field, the conviction should anyway remain clear that a GIS, despite its multiple potential, is not a data source, but only an information manipulation instrument. According to this position it can be seen how GIS, favouring new forms of data input and output display, is pressing researchers to modify their way of analysing and describing different hydrological phenomena through ever increasing integration with simulation models and these new calculation instruments.

16.2. Feasibility & Design

Correct GIS design should initially include a detailed feasibility study which is not always easily attainable because of uncertainty in some of the factors which are relevant to the analysis. In particular, a study of GIS feasibility should carefully evaluate the kind of system to be used and which functions the system is capable of developing in order to satisfy the aims and objectives of the research to be carried out and the applications to be effected. The principles and general features which the applied system must guarantee are fixed beforehand. A GIS, in particular, must:

- facilitate the correct operation of the functions and direct error elimination;
- possess integration and modularity characteristics; and
- be designed according to application standards to reduce and simplify updating of the products, even considering the need of personalisation produced by individual application problems.

This leads to the definition of a study of the functional requirements of the system, which summarize the identification of GIS products, the data it requires and the GIS functions which should be available (Mark, et al.,1993). These latter, in particular, should include (Nyerges, 1993):

- spatial and attribute data entry functions;
- spatial and attribute data management functions;
- spatial and attribute data manipulation functions;
- spatial and attribute data analysis functions; and
- spatial and attribute data display/output functions.

Once a study of the functional requirements of the system has been completed, and problems regarding data capture, topology capture, large data volumes, vector-raster database, overlay analysis, front-end language and query language are solved, one can proceed to the research phase of the GIS product and then to its acquisition. This phase can be developed by means of direct interaction with a high number of vendors to whom the nature of the data contained in the data-base of the system, the functions necessary to elaborate the data present in the data- base, and the products which must be provided by the system, should be illustrated. After multiple demos of the different graphic representations and data structures (raster or vector), system functionalities, hardware compatibility, available interfaces to data sources and network and communication protocols, it is necessary, before proceeding to the choice of one of the analysed systems, to compare the different alternatives available on the basis of performance, output quality and price, always referring to the particular application problem.

16.3. GIS Software

With the aim of helping users with the choice of the most suitable products to solve the particular application problem, in this section a listing of some of the major GIS packages available commercially and in the public domain is shown. The information in this listing is directly provided both by the respective GIS publishers or vendors and the GIS users.

• AllyMAP

AllyMAP was South Africa's first locally developed Geographical Information System and has been on the market for approximately 10 years. It is a PC-based package and is a purpose-built geographical information system integrated with a relational database. A customized data structure allows the creation of a digital database with both graphical and non-graphical elements to provide rapid relational access to a great variety of data combinations and permutations. Some of the principal features include:

- Logical user-friendly operation;
- Fast and efficient capture and feature classification;
- Integration of CAD-GIS;

- Standard database compatibility;
- Built-in network support;
- Spatial analysis;
- Thematic mapping;
- Network topology;
- On-line reporting; and
- Built-in programmatic tools.

• Atlas GIS

Atlas-GIS is a PC-based package on DOS and Windows platforms which turns statistical and geographic data into meaningful information for decision-making and presentation purposes. Atlas GIS, in particular, is a information mapping package powerful enough to be called GIS. It offers an advanced database design. Open Lotus 1-2-3, Microsoft Excel or dBASE- compatible files directly, and link them to the map. Query client-server SQL databases using the built-in Q+E query builder or perform relational SQL queries on Atlas or any other database tables. The principal features of this package can be summarized as:

- On-screen presentation (view entire map, view maps and tables, view multiple maps, etc.);
- Map presentation tools (map layers and theme types);
- Database manager;
- Selecting features and data;
- Geographic analysis tools (assign data by location, buffers and bands, data calculations, summary statistics, total data inside or outside polygons); and
- Output and printing tools.

Atlas GIS can be integrated with the Atlas Import/Export program which is a file conversion utility that translates eight major geographic file formats to the Atlas geographic file format. It also converts geographic data and attribute data simultaneously, avoiding to lose information or have to perform manual editing.

• ARC/INFO

ARC/INFO is a popular, full-featured UNIX- and VMS-based GIS package with a large established user base. The core ARC/INFO package includes capabilities for vector data entry, editing, display, analysis, modeling, and management; DATABASE INTEGRATOR, for links to ORACLE, INGRES, INFORMIX, and SYBASE DBMSs image integration tools in the IMAGE INTEGRATOR; ArcTools, an off-the-shelf, customizable graphic user interface; and the ARC Macro Language (AML). Optional extensions further enhance ARC/INFO's capabilities in the following areas:

- COGO, coordinate geometry extensions;
- NETWORK, network analysis and routing extensions;
- TIN, terrain modeling extensions;
- GRID, raster modeling extensions;

— ArcScan, scanning data entry tools;
— ArcStorm, feature-based data management tools; and
— ArcExpress, performance enhancement extensions.

PC ARC/INFO brings sophisticated geographic information management, analysis, and mapping to the personal computer in the DOS environment. PC ARC/INFO software's integrated modules provide map digitizing, data conversion and transfer, relational database management, map overlay, display, query, interactive graphics editing, address geocoding, and network analysis. PC ARC/INFO features ARCSHELL, and easy-to-use menu interface; full compatibility with dBASE, the leading database management system for DOS; TABLES a simple database management system; sophisticated spatial analysis capabilities; and network analysis and geocoding capabilities. A geoprocessing system for 286, 386, or 486 computers running DOS, PC ARC/INFO software may be run independently or networked with a host ARC/INFO system.

• ArcView
ArcView is a highly affordable, easy-to-learn desktop mapping and GIS tool that enables users to quickly select and display different combinations of data for creatively visualizing information. Available for PC-compatibles, Apple Macintosh computers, VMS-based computers, and a variety of UNIX workstations, ArcView works directly with ARC/INFO, ArcCAD, and PC ARC/INFO databases.

• ArcCAD
ArcCAD extends the AutoCAD data model to incorporate ARC/INFO GIS functionality. It extends the AutoCAD data model for creating topological and spatial database relationships, and for communicating with entities of an AutoCAD drawing database. ArcCAD is a complete GIS that creates and works with geographic data directly in native ARC/INFO format. ArcCAD runs inside AutoCAD on DOS and Windows platforms.

How Detailed Information can be Obtained
A substantial amount of user support and help can be obtained through the e-mail user forum ESRI-L. To subscribe to ESRI-L send a one line message (SUBSCRIBE esri-l <your email address>) to:
>   listserv@esri.com

To post a question to the list send a message to:
>   esri-l@esri.com

A usenet newsgroup, however, does not currently exist.

• CARIS
CARIS (Computer Aided Resource Information System) is a state of the art digital mapping and Geographic Information System (GIS) capable of compiling, storing and analyzing spatially oriented information for a wide range of users. The

system supports full topology and a high accuracy three dimensional data structure integrated with raster/vector processing and presentation techniques. Textual attributes are supported via generic database interface supporting a number of RDBMS products including RDB, INGRES and ORACLE. CARIS is designed to utilize emerging technology to meet the current and future needs of information users in disciplines such as topographic mapping, hydrographic charting, marine sciences, environmental sciences, geology, agriculture, forestry, municipal management, transportation and cadastral mapping. CARIS presently supports a number of UNIX and PC platforms including Hewlett Packard, SUN, Silicon Graphics, DEC Alpha, IBM (R6000 series), Data General, SCO/ODT based Pcs as well as 486 PCs under MS Windows 3.1 . Through the use of strict quality control procedures, and adherence to computer industrial standards, CARIS has been developed around a single software library. This insures that the functionality and data available on one platform is the same on all other platforms supporting the same features. CARIS is modular in nature and can be configured to meet a variety of needs and budgets. Basic packages are comprised of:

- CARIS EXPLORER supports single file data compilation and editing;
- CARIS CARTOGRAPHER provides extensive digital cartographic functionality for single or multi-files;
- CARIS TOPOLOGICAL MAPPING extends the Cartographer package by supporting the creation of topology;
- CARIS ACCESS provides extensive analysis functionality and data presentation techniques of a GIS database; and
- CARIS GIS incorporates the features of CARIS Topological Mapping and CARIS Access.

Additional options include CARIS DTM, CARIS Image Processing, CARIS Network Analysis as well as a powerful Application Programming Interface based in C and C++ for tailoring CARIS.

• ERDAS

ERDAS is a raster image processing and GIS with PC, UNIX, and VMS version. This package has capabilities for geometric correction and classification of remotely sensed information such as satellite images and scanned aerial photography as well as analysis of existing raster data. The package has different modules:

- Core module (view programs, file management programs, annotated programs, color modification programs, cursor control programs, demo programs, line printer programs, utility programs);
- Image processing module (enhancement programs, classification programs, geometric correction programs, radiometric correction programs);
- GIS module (analysis programs);
- Tapes module (input programs, output programs, utility programs);
- Polygon digitizing module (digitizing programs, grid programs, measurement programs, projection programs);

- Scanning module;
- Video module;
- Topographic module (terrain analysis programs, surfacing programs);
- 3-Dimensional module; and
- Software toolkit module.

ERDAS can be directly interfaced with ARC-INFO by means of ERDAS/ARC-INFO Link software.

• ER Mapper

ER Mapper is an advanced image processing software package. It uses the latest generation X-Windows based graphical user interface and run under the OPEN-LOOK and MOTIF window managers. ER Mapper will run on DEC Alpha/OSF, DEC MIPS/ULTRIX, HP PARISH/HPUX, Silicon Graphics MIPS/IRIX, Sun SPARC/SunOS 4.1, Sun SPARC/Solaris 2.x and Windows NT, supporting networks and allowing images to be displayed on multiple X displays. ER Mapper, with a complete suite of 200 processing algorithms, provides full integration of vector and raster data, full dynamic mosaicing of datasets with different cell sizes without regridding and ER Mapper's Dynamic Links allow users to directly display data from external systems without manually converting the data. ER Mapper has dynamic links to many GIS systems and DBMS systems supporting more than 80 data formats. This software furthermore shows features very close to ERDAS.

• GENASYS

GENASYS is a topological, vector based GIS developed specifically for the UNIX and X-Windows environment. It includes: client-server RDBMS interface; hardware independent, user-customisable, windows-based interface builder; user orientated single presentation environment. The last release of this system GIS shows some interesting aspects:

- The Genius graphical user interface (GUI) which separates the presentation environment from the processing engines and data;
- Virtual Image Viewing Environment (VIVE) which allows the user to view and manipulate a wide range of image data including aerial photography, satellite imagery, scanned maps, engineering plans and photographs in conjunction with other georeferenced data; and
- Cartographic composition environment which includes in the module GenaMap grids and graticules in any projection and in a variety of layouts, enhanced drawing tools, output of integrated raster and vector maps.

Genasys has developed a series of related products which complement GenaMap's GIS features:

- GenaRave is a GIS-based raster to vector conversion package which offers enhanced productivity tools for generating GIS-ready data;

- GenaDoc is a document image management system which offers the ability to manage the capture, storage, retrieval, printing and display of large numbers of scanned document images;
- GenaCell is a module which provides raster/grid modeling; and
- GenaCivil is a module which provides hydrological and TIN modeling.

• GISPlus

GISPlus is an integrated computer mapping and spatial database management system running on PC-DOS platforms which provides a broad array of functions for the storage, retrieval, managements, analysis, and dysplay of geographically referenced data. In particular, this package has a unique integrated spatial database manager which stores geographic data using efficient topological data structures. GISPlus moreover includes a built-in high performance database engine optimized for GIS applications. It, in fact, provides a full range of data query and selection mechanisms (point-and-click, spatial queries, query by example and SQL interface). This package provides four core functions:

- Computer mapping;
- Geographic database management;
- Data analysis; and
- Application development;

and can be considered as a platform for all types of analysis, by means of the analytical tools that can be directly applied to GIS databases (Contouring/TIN, location/allocation, shortest path, linear models, traveling salesman, dynamic segmentation, choice models, matrix adjustment, polygon overlay, distance calculation, statistical models, gravity models, spatial correlation, kriging, etc.).

• GRASS

GRASS is a public domain package with a large established user base which actually contributes to the code that is incorporated into the new versions. This package was designed and developed to operate under UNIX operating system. It has been ported to numerous hardware environments including: Sun computers, Intergraph Interpro workstation, Silicon Graphics workstations, Apple Macintosh (A/UX), PC 386's and 486's (with Microport or SCO UNIX), DEC workstations, Apollo workstations, HP 9000, IBM PS-2 (AIX), IBM 6000 workstations, Tektronix workstations, AT&T 3B2 workstation. The great number of GRASS's commands can be summarized as follows:

- General data (file) management commands;
- Display (monitor) graphics;
- Paint paper graphics;
- Raster data manipulation and analysis;
- Vector data manipulation and analysis;
- Site data manipulation and analysis;
- Imagery (multi-spectral) data manipulation and analysis; and

— Manipulation of external data.

The latest version (4.1) is available via ftp (includes full sample database containing SPOT, DEM, and other data sets).

How Detailed Information can be Obtained

A GRASS discussion list is available, email a subscription message to:

    grassu-request@moon.cecer.army.mil:

    SUBSCRIBE,GRASSU,<your email address>

For list of distributors see:

    ftp site: moon.cecer.army.mil (129.229.20.254) in /grass directory or

    moon.cecer.army.mil:grass/grassInfoCenter/GICdocs/AcquirSoft.ps.Z

• IDRISI

IDRISI is a grid-based geographic information and image processing system developed, distributed and supported by The IDRISI Project, a non-profit organization within the Graduate School of Geography at Clark University in Worcester, Massachusetts. IDRISI is designed to be affordable and easy to use, yet provide professional-level analytical capability on PC-DOS platforms. IDRISI is a collection of over 100 program modules that are linked by a unified menu system. It provides an extensive suite of tools for image processing, geographic and statistical analysis, spatial decision support, time series analysis, data display, import/export and conversion. In particular, the modules can be summarized in three groups:

— Core modules, providing fundamental utilities for the entry, storage management and display of raster images;

— Analytical ring modules providing major tools groups for the analysis of raster image data (geographic and statistical analysis, image processing, time series/change analysis, decision support); and

— Peripheral modules associated with data conversion utilities between IDRISI and other programs and data storage formats.

By using independent modules linked by a set of simple data structures, the system allows users to develop their own modules with minimal regard for the internal workings of IDRISI modules in the core set (in the IDRISI-L FTP site can be found about twenty modules developed and donated by users).

How Detailed Information can be Obtained

A IDRISI discussion list is available and supported by the Department of Geography and Environmental Planning at Towson State University (Baltimore). To subscribe to IDRISI-L send a one line message (SUBSCRIBE IDRISI-L<your email address>) to either the Internet address:

    MAILSERV@TOE.TOWSON.EDU

or the Bitnet address:

    MAILSERV@TOWSONVX.BITNET

The IDRISI-L FTP site is operating on a MicroVAX that is part of of Towson State University's VAX cluster. The anonymous FTP site is located at:

MIDGET.TOWSON.EDU

• ILWIS

ILWIS (Integrated Land and Watershed management Information System) is a GIS program developed by the International Institute for Aerospace Survey and Earth Science. This PC-DOS package includes both raster and vector capabilities. The data input subsystem allows the capture of data in both spatial and non-spatial (attribute) models. The database subsystem consists of a commercially-available database linked to a graphics database and a set of conventional GIS routines. Digital image processing procedures are also included in another subsystem. Data analysis and modeling include conventional GIS data analysis and manipulation capabilities such as map overlaying, reclassification, proximity analysis, optimum corridor and other cartographic modeling techniques. ILWIS's capabilities can be summarized by means of the following modules:

- Utilities module (general, screen applications, software operation, georeferencing maps and images, colour modification, annotation);
- Information module (digitizing, polygonization, vector to raster conversion, raster operations);
- Internal database and interface module; and
- Remote sensing module (display programs, enhancement programs, classification programs, statistical programs, geometric correction programs and image arithmetic programs).

• INTERGRAPH Modular GIS Environment

MGE package runs on PC (Windows NT), Macintosh (as MicroStation), and UNIX platforms. A full featured package which competes directly with the popular ARC/INFO. MGE is a vector-based package with raster-supporting modules designed to operate in conjuction with Intergraph's Microstation 32 CAD package. MGE can be interfaced with numerous relational data bases, including ORACLE, INFORMIX, INGRES, and DBII. A related module to MGE is Modular GIS Analyst (MGA) that supports spatial analysis functions.

• MacMap

MacMap is a french, vector based GIS running on Macintosh computers. This package is a purpose-built geographical information system integrated with a relational database. MacMap allows animation and data interchange according to the Macintosh's standard (copy and paste). There are a series of related products which complement MacMap's GIS features:

- Pack Dessin is a set of functions for digitizing points and segments, control points, automatic errors detection, network definition, topological consistency;
- Pack Maj allows a real time and automatic link with external databases;
- Pack Berten is a set of functions for statistical analysis;

- XMap is a set of functions for the entry, storage management, and display of new external data;
- vDef is a set of functions for map viewing; and
- MapCalculator provides a full complement of spatial operators and geoprocessing tools.

Different data formats can be converted to MacMap format by means of external modules, but it is not allowed the vice-versa.

• MAPII

MAPII is a grid-based software package designed for map viewing, marking, measuring, transforming and mapmaking. It is a Mac implementation of Dana Tomlin's Map Algebra. The map operations in MAPII contained can be grouped as:

- Local overlay operations (include operations for which the resulting cell values are independent of cell location and can be used for explicit reclassification of map zones, for map arithmetic and map superimposition);
- Focal operations (are those for which the output values are dependent on cell values in a defined neighborhood and can be used to compute terrain characteristics);
- Zonal operations (produce output map in which values depend on regional values); and
- Map geometry operations (these operations are used to geometrically rectify maps by grid sampling, grid alignment, grid rotation, grid resizing, and stretching or rubbersheeting).

In addition to cartographic image processing, MAP II has remote sensing image processing capabilities.

How Detailed Information can be Obtained

All MAPII products (program, reference manual and tutorial manual) are available from John Wiley & Sons, Inc.

• MapBox

Map Box is the first implementation of Dana Tomlin's Map Algebra for cartographic modeling. This software provides a toolkit of integrated Map Algebra functions to solve spatial analysis problems. The device-independent design of MapBox allows it to run on a wide range of PC's with different graphics hardware. MapBox is also available for UNIX workstations (Sun, Silicon Graphics). Versions for Intergraph users are sold by Intergraph as part of the MGA package. MapBox contains a large set of modeling functions. They include operations such as reassignment of map attributes, overlay of maps, measurement of distances and directions, calculation of areas and volumes, computation of slopes and aspects, determination of proximity and travel time, characterization of shapes, delineation of views, simulation of flows, and allocation of paths. MapBox can be used to display and analyze digital maps (raster layer) using standard PC hardware. This software does not include digitalization or plotting functions, but can be created procedures or macros.

• MapGrafix

A Macintosh GIS package with limited raster data handling capabilities, e.g. image as a backdrop. It has recently been integrated with a module GAIA (Geographic Access Information and Analysis), developed by Island Insitute of Maine, a grid-based software designed for map viewing, marking, measuring, trasforming and mapmaking. MapGrafix is an interesting package in that it has no internal DBMS of its own but accesses various commercial databases. In this way, it is possible for this package to utilize existing databases (as long as the attributes are geographically referenced) without the need for conversion. This package is not topologically based and therefore cannot perform such topological functions as optimal path determination, etc. Additional modules include:

- MapLink (a translator for the conversion of 13 different data types such as DLG, TIGER, and DXF);
- MapView (a map projection module);
- MapCon 3D; and
- MapTrans (a EPS translator).

• MapInfo

MapInfo is a desktop mapping software translated into several languages and ported to several platforms (Windows, Macintosh, Sun and HP Workstations). MapInfo opens Lotus 1-2-3, Microsoft Excel or dBASE- compatible files directly, and link them to the map. The vector maps with data associated can be directly created inside the program or imported by means of DXF or ASCII formats. The principal features of this package can be summarized:

- On-screen presentation (view entire map, view maps and tables, view multiple maps, etc.);
- Map presentation tools (map layers and theme types);
- Database manager;
- Selecting features and data;
- Geographic analysis tools (assign data by location, buffers and bands, data calculations, summary statistics, total data inside or outside polygons); and
- Output and printing tools.
  This system can be integrated with MapBasic, a programming language like-BASIC, to develop macros and applications. The last version of MapInfo is able to:
- incorporate raster imagery as backdrop;
- new editing functions; and
- superimposition of graphics and histograms on vector maps.

How Detailed Information can be Obtained

A Map Info discussion list is available To subscribe to MapInfo-L, send e-mail to majordomo@csn.org writing the following message:

SUBSCRIBE MAPINFO-L

Do not put your name on the subscribe line since this is a majordomo server, and not a list server or mailbase server.

• MOSS

Map Overlay and Statistical System (MOSS) is a public domain software written in FORTRAN with a large established user base. MOSS is the data analysis component of a Geographic Information System (GIS) originally developed by the Western Energy and Land Use Team (WELUT). Curently, MOSS is being developed under the direction of the U.S. Bureau of Land Management with cooperation from the U.S. Fish and Wildlife Service, the U.S. Bureau of Indian Affairs, the U.S. Geological Survey, the U.S. Forest Service, the Soil Conservation Service, the Minerals Management Service and the U.S. Army Coprs of Engineers. MOSS (vector data processing) is typically utilized as part of complete GIS which includes other capabilities such as data entry (Automated Digitizing Systems, ADS), raster data processing (Map Analysis Package, MAP), and cartographic output (Cartographic Output Systems, COS). MOSS, and PC-MOSS in particular, represents only one element of a GIS. The elements of map digitizing and map plotting are not provided in the PC-MOSS 1.0 version. However, PC- MOSS is accessed via a front-end called GIS which contains options which are currently not available. These options may be available in the future as demand requires and resources permit. At present only MOSS is available. Although the lack of a direct capability for map digitizing, projection changes, and cartographic output constitutes a limitation to the use of PC- MOSS, the MOSS IMPORT command is available in PC-MOSS and provides one mechanism for entering map data into the system. Data from other PC based digitizing programs could be converted to the MOSS IMPORT format for use in MOSS. Any data transfer capability between a minicomputer running MOSS and a PC running PC-MOSS would permit transfer via the IMPORT format. A tape drive interface for the PC or a serial communications link would provide the necessary capability. Unfortunately, the MOSS EXPORT command is not available in PC-MOSS. This precludes such a simple approach to moving MOSS data into other PC systems (such as CAD systems) for cartographic output.

How Detailed Information can be Obtained

Both PC and UNIX versions available, but only the PC version (exe and source) is available via internet (ftp to csn.org (128.138.213.21) and look in the COGS/MOSS directory for mosssrc.zip and mosssrc2.zip). A copy of the moss programmers manual has been uploaded to dsc.blm.gov in the ftp/pub/gis directory.

• OSU MAP

OSU MAP is a grid cell-based GIS running on PC-DOS platforms. This software is based on the Map Analysis Program originally created by Dana Tomlin. A set of functions is available as:

— Utilities functions;
— Data output functions;
— Data analysis functions (local operations, reclassification, distance analysis, shape analysis, statistical analysis, terrain analysis);
— Data encoding functions; and
— Stand-alone utility programs.

How Detailed Information can be Obtained
e-mail:
  dmarble@magnus.acs.ohio-state.edu

• Oz GIS

OzGIS is a software system for displaying geographically referenced data, such as Census data, as maps and diagrams on screens, printers and plotters.

Digitised map data (e.g. Census boundaries, GIS) and attribute data (e.g. sales, environmental) are accepted as ASCII files and preprocessed (e.g. amalgamation, line thinning, polygon construction, subsetting) before display. Maps of polygons, lines and points can be displayed according to one or two attributes along with various overlays. About 130 menus provide extensive options for interactively designing the layout and appearance of the map, and for attribute handling, classification, interrogation and saving maps. Maps can be output on plotters, printers and various file types. Applications such as territory definition and retail site catchment analysis are supported. Versions are available for both DOS and Windows 3.1 . Oz-GIS is distributed as a "shareware" system. It is very extensive, (it is the product of over ten man-years effort), but also very cheap.

How Detailed Information can be Obtained
OzGIS is available for evaluation via anonymous ftp from:
  Internet: [192.88.110.20] WSMR-SIMTEL20.Army.Mil

• pMAP

pMAP is PC-based package which uses a raster data structure and a non-topological vector structure for geographical " backdrop". This package consists of a map database containing base maps and derived information layers, internally referenced to a common geographic location. pMAP supports more than fifty operations, including:

— Data entry/input devices;
— Raster/vector integration (raster to vector, vector to raster, register vectors over raster);
— Spatial data exchange formats;

- Data management;
- Data analysis (measurement, generate buffers, map analysis functions, local neighborhood operations, surface analysis, network analysis, polygon operations, miscellaneous);
- Data display (multiple maps on single plot, user-defined georeference grids, cartographic elements, 3-D display); and
- Output device support.

Special versions of this package (aMAP and tMAP) are available for academic or tutorial purposes, but are limited by file size restrictions.

• SICAD

In the range of systems which include SICAD as the central GIS solution, SICAD/open can be considered as standalone system for geographic vector data and raster data processing. SIDAC/open is based on the RISC workstation RW family from Siemens Nixdorf Informationssysteme. It will be also available on HP-700 and IBM 6000 paltforms. SICAD/WinCAT is a PC-based object-oriented desktop mapping, presentation and management system for environmental applications. In particular, SICAD/open includes the following modules:

- SICAD-GDB-X stores the graphics and non-graphics data in a relational database management system (Informix, Oracle); and
- SICAD-Base provides the framework on which all application modules are built. This module also includes graphical primitives needed for interactive work such as digital data entry, surveying features, updating of cadastral maps and landregistry plans. Application modules for various needs are overlayed into SICAD-Base, such as:

- SICAD-Net, public utility network documentation and planning;
- SICAD-Tele, documentation of telecom lines;
- SICAD-Sewer, digital and video documentation of sewage lines;
- SICAD-Thea, thematical cartography;
- SICAD-Area, area intersection and buffering;
- SICAD-RBS/RIM, raster data processing;
- SICAD-Plot, output to many standard plot devices; and
- SICAD-MM, user interface generator;

Various calculation programs and interfaces are available for linking the SICAD system to existing programs and data: digital terrain model; network calculations; raster/vector conversion; and homogenisation.

• SMALLWORLD

SMALLWORLD is a UNIX-based GIS written using the object-oriented language Magik (algol-like), which makes the system highly customisable. It includes integrated raster and vector data, version managed database (which allows the storage of multiple versions of all aspects of data, even raster, in a single database), a "virtual database" concept which allows the integration of commercial databases such as Oracle and Ingres), a seamless mapbase and a CASE tool to aid design and development of new applications. In detail, SMALLWORLD GIS models the world as a collection of objects which have relationships and attributes. Complex object relationships are modelled by one-to-one, one-to-many or many-to-many relations. Object attributes are aspatial (represented as numbers or characters) and spatial (represented as points, chains and areas). The SMALLWORLD GIS Version Managed Datastore holds all of its spatial and aspatial data seamlessly; there are no tiles or layers. Fast spatial retrievals are achieved by means of an improved quadtree indexing technique. The structure of SMALLWORLD can be described by means of the following modules:

- SMALLWORLD GIS Core;
- SMALLWORLD GIS Update;
- SMALLWORLD GIS Drafting;
- SMALLWORLD GIS 2D Analysis;
- SMALLWORLD GIS Output;
- SMALLWORLD GIS 3D/Grid Analysis;
- SMALLWORLD GIS Query/View;
- SMALLWORLD ACE;
- SMALLWORLD Magik; and
- SMALLWORLD Case.

One approach to providing GIS services to a user who has a PC or Macintosh is to configure it as an X-terminal. Although all the capabilities of the GIS are available, it is especially suited for simpler applications such as querying, viewing, reporting and plotting.SMALLWORLD has also started implementation of a port to Windows NT.

• SPANS

SPANS (SPatial ANalysis System) is a modular system, with products available for the Windows, OS/2 and UNIX operating system environments. Three submodels make up the SPANS integrated geographic model:

- Data model, which includes capabilities for: digitizing paper maps, importing a wide variety of data types from many sources, projection transformation, geometric registration, and data editing;
- Analysis model, supporting both vector and grid data structures; for added convenience, and to facilitate analysis and modeling, SPANS provides rapid routines for transforming data from one form to the other; and
- Visualization model, providing a rich set of capabilities for visualizing data and information.

The complete suite of software modules is the SPANS GIS system. The standard package is GIS Builder, a set of basic tools for building databases, constructing analytical models and visualizing and querying data. Additional modules are available to expand the functions of GIS Builder, such as:

- 3D view;
- Contouring;
- Surface generator;
- Point aggregation;
- Neighborhood analysis;
- Topological analysis;
- Visibility analysis;
- Interaction modeling;
- Multicriteria modeling;
- Map modeling;
- Table modeling;
- Point modeling; and
- Application developer program.

• System 9

This package shows a topological vector structure. It is based "object oriented" modular GIS based on a toolbox approach. System 9 is available on IBM AIX, HP UX, Solaris 2 and Solaris 1 (SunOS).

System 9 operates at the UNIX shell level, Korn shell preferably. Commands can be put into scripts (or C code) similar to AMLs but much faster. The database Empress is far superior to many imbedded GIS databases (INFO comes to mind). It is primarily a vector based system with an ability to display geo-referenced raster imagery as a background.

On the down side, output graphics are weak in System 9 and importation of standard data formats is laborious. This is primarily because System 9 requires all feature classes to be fully defined before data capture/input can occur. In many ways it is an advanced product with a robust data model but it has some definite rough edges. The product is very expensive but that may change now that Unisys owns it. It is used more in Europe than the US. Another drawback is that there is an embedded archaic user interface that hampers some of the crucial operations of the product. This interface is not major impediment to the product's functionality or operation, just an annoyance.

• TNTmips

TNTmips is an integrated system for advanced image processing, GIS, CAD, desktop mapping and other spatial database management and visualization applications. With TNTmips can be created complex multi-layer displays from project materials including raster, vector, CAD, and relational database objects. It permits moreover to process multi-band images to produce classification rasters, draw interpretations, make annotations, overlay CAD and vector materials, and produce large-format color maps and posters. TNTmips works on PCs (Intel 80386, 486) with Windows 3.1 and NT, DEC Alpha AXP series with Windows NT, Apple Macintosh with A/UX, DECstation 3000 and 5000 series with Ultrix, Sun SPARC-station series with Solaris 1.x or 2.x, HP 9000 series with HP-UX, Data General

AViiON series with DG-UX, Silicon Graphics IRIS series with IRIX and IBM RS6000 with AIX. The features of this package, such as:

— Airvideo capture;
— Photo interpretation;
— Image classification;
— Mosaicking;
— Raster-to-vector conversion;
— Orthoimage and DEM from stereo;
— Relational database query;
— Geographic analysis (GIS);
— Computer Aided Design (CAD);
— Map and poster layout;
— Spatial manipulation language;
— Widest peripheral support;
— Internationalization and localization translation tools; and
— Dozens more;

can be integrated with the Microimages'TNT products:

— TNTview for visualization and analysis;
— TNTatlas for publishing spatial information; and
— TNTsdk (Software Development Kit).

• WINGIS

This software, for MS-Windows and MS-Windows/NT, consist of a number of modules of a well devised system which can link geographical elements with data. This package has three modules: WINMAP, WINVIEW, WINSAT and WIN3D.

WINGIS permits interrelating of the most varied streams even from different databases and works through SQL standard even with mainframes as well as with network databases. The graphic background can be provided by digitizer, imported coordinate data or by satellite or aerial photographs. The minimum hardware configuration for WINGIS is a 486 PC with at least 16MB of RAM, 120 MB free HD-space and Windows or Windows/NT running.

The theoretical limitations of WINGIS program are mainly in function of the disk space. Project limitations are 42,000 x 42,000 kilometers with a resolution of 0.01 meter, a distance function of maximally 21,000 km, a maximum of 16354 layers and 16354 points per graphic object and a maximum of 2,1 billion objects, polygons to a maximum of 5000 points per polygon. The database used is the GUPTA SQL-Windows, which can handle DB@, Informix, Ingres, Sybase, HPAIIBase, AS/400 as well as Oracle, dBase, Lotus, DIF, CSV, SQL or ASCII data.

More detailed information concerning the GIS-Company addresses are shown in Tables 16.1 and 16.2.

TABLE 16.1. GIS-Company addresses.

SOFTWARE	COMPANY ADDRESS	PHONE	FAX
AllyMAP	Allyson Lawless (Pty) Ltd. P O Box 73285, Fairland 2195 Johannesburg, South Africa	(+27)(11) 476-4100	(+27)(11) 678-7518
Atlas GIS	Strategic Mapping, Inc. 3135 Kifer Road, Santa Clara, CA 95051 U.S.A.	(408) 970-9600	(408) 970-9999
ARC/INFO ArcView ArcCAD	ESRI, Inc. 380 New York Street, Redlands, CA 92373 U.S.A.	(909) 793-2853 ext.1375	(909) 793-5953
CARIS	Universal Systems Ltd. 270 Rookwood Avenue, P.O. Box 3391, Station B Fredericton, New Brunswick, Canada	(506) 458-8533	(506) 459-3849
ERDAS	ERDAS, Inc. 2801 Buford Highway, Suite 300, Atlanta, Georgia 30329-2137 U.S.A.	(404) 248-9000	(404) 248-9400
ER Mapper	Earth Resource Mapping Level 2, 87 Colin Street, West Perth Western Australia 6005	(+61) (9) 388-2900	(+61) (9) 388-2901
GENASYS	Genasys II Inc. 2629 Redwing Rd., Suite 330, Fort Collins, CO 80526, U.S.A.	(303) 226-3283 (800) 447-0265	(303) 226-0869
GISPlus	Caliper Corporation 1172 Beacon Street, Newton, MA 02161 U.S.A.	(617) 527-4700	(617) 527-5113
GRASS	U.S. Army Corps of Engineers, CERL P.O. Box 9005, Champaign, Illinois 61826-9005, U.S.A.	(217) 373-7220	************
IDRISI	Clark University, Graduate School of Geography 950 Main St. Worcester, MA 01610 U.S.A.	(508) 793-7526	(508) 793-8842
ILWIS	Internat. Inst. for Aerosp. Survey and Earth Sciences ITC 350 Boulevard 1945, P.O. Box 6, 7500 AA Enschede, The Netherlands	(+31)(53)874-337	(+31)(53)874-484
INTERGRAPH MGE	Intergraph Corp. Mapping Sciences Division, Hunstville, AL 35894-0001, U.S.A.	(205) 730-2700 (800) 345-4856	************
MacMap	Klik Developpements La Faisanderie, 10, route des Aubris, F-78490 GALLUIS, France	(+33)(1)34.86.74.44	(+33)(1)34.86.89.11
MAPII	Machine Geography Laboratory, Dept. of Geography Università of Manitoba, Winnipeg, Manitoba R3T-2N2, Canada	(204) 474-6602	************

TABLE 16.2. GIS-Company addresses.

SOFTWARE	COMPANY ADDRESS	PHONE	FAX
MapBox	Decision Images, Inc. 196 Tamarack Circle, Skillman, NJ 08558, U.S.A.	(609) 683-0234	(609) 683-4068
MapGrafix	ComGrafix, Inc. 620 E Street, Clearwater, Fla 34616 U.S.A.	(813) 443-6807 (800) 448-MAPS	(813) 443-7585
MapInfo	MapInfo Corporation One Global View, Troy, NY 12180 U.S.A.	(518) 285-6000 (800) FAST-MAP	(518) 285-6060
MOSS	Bureau of Land Management, Service Center Denver Federal Center, Denver, CO 80225-0047, U.S.A.	(303) 236-0990	************
OSU MAP	Prof. Duane F. Marble, Department of Geography The Ohio State University, Columbus, Ohio 43210-1361, U.S.A.	(616) 292-2250	(616) 292-9180
Oz GIS	Clever Company QMDD box 6108, Queanbeyan, Australia 2620.	Internet: [192.88.110.20] [128.214.87.1]	************
pMAP	Spatial Information Systems, Inc. 19 Old Town Square, Fort Collins, CO 80524, U.S.A.	(303) 490-2155	(303) 482-0251
SICAD	Siemens Nixdorf, Informationssysteme AG International Business Development Otto-Hahn-Ring 6, 81730 Muenchen, Germany	(+49)(89)636-41988	(+49)(89)636-45202
SMALLWORLD	Smallworld Systems Brunswick House, 61-69 Newmarket Road, Cambridge, U.K.	(+44)(223)460-199	(+44)(223)460-210
SPANS	TYDAC Technologies Inc. 2 Gurdwara Road, Suite 210, Nepean, ON K2E 1A2, Canada	(613) 226-5525	(613) 226-3819
System 9	GIS Division, UNISYS Canada Inc. 61 Middlefield Road, Scarborough, Ontario, M1S 5A9, Canada	(416) 609-7788	(416) 297-2520
TNTmips	MicroImages, Inc. 201 N. 8th Street Suite 15, Lincoln, NE 68508-1347, U.S.A.	(402) 477-9554	(402) 477-9559
WINGIS	PROGIS W.H.M. Corporation 114 W. Magnolia Street, Suite 423, Bellingham, WA 98225, U.S.A.	(206) 738-2449	************

TABLE 16.3. GIS-Functions.

SOFTWARE	FUNCTIONS								STRUCTURE		
	A	B	C	D	E	F	G	H	NTV	TV	R
AllyMAP	Y	N	N	Y	Y	Y	N	Y		•	
Atlas GIS	Y	N	N	Y	N	Y	N	Y	•		
ARC/INFO	Y	Y	Y	Y	Y	Y	Y	Y		•	•
CARIS	Y	Y	Y	Y	Y	Y	Y	Y		•	•
ERDAS	Y	Y	Y	Y	N	Y	Y	Y			•
ER Mapper	Y	Y	Y	Y	N	Y	Y	Y			•
GENASYS	Y	N	Y	Y	Y	Y	Y	Y		•	
GISPlus	Y	N	Y	Y	Y	Y	N	Y		•	
GRASS	Y	Y	Y	Y	Y	N	Y	Y			•
IDRISI	Y	Y	Y	Y	N	Y	Y	Y			•
ILWIS	Y	Y	Y	Y	Y	Y	Y	Y		•	•
INTERGR.MGE	Y	Y	Y	Y	Y	Y	Y	Y		•	•
MacMap	Y	N	N	Y	Y	Y	N	Y		•	
MAPII	Y	Y	Y	Y	N	N	N	Y			•
MapBox	Y	N	Y	Y	Y	N	N	Y			•
MapGrafix	Y	Y	Y	Y	N	Y	N	Y	•		•
MapInfo	Y	N	N	Y	N	Y	N	Y	•		
MOSS-PC	Y	N	N	Y	N	Y	N	N	•		
OSU MAP	Y	N	Y	Y	N	N	N	Y			•
Oz GIS	Y	N	N	Y	N	Y	N	Y	•		
pMAP	Y	N	Y	Y	Y	Y	Y	Y	•		•
SICAD	Y	N	Y	Y	Y	Y	Y	Y		•	•
SMALLWORLD	Y	N	Y	Y	Y	Y	N	Y		•	•
SPANS	Y	N	Y	Y	Y	Y	Y	Y		•	•
System 9	Y	N	N	Y	Y	Y	N	Y		•	
TNTmips	Y	Y	Y	Y	Y	Y	Y	Y		•	•
WINGIS	Y	N	Y	Y	Y	Y	Y	Y		•	•

16.4. GIS Comparison

After having analysed some of the main commercially available GIS packages, in this section a comparison in terms of data structure (NonTopological Vector, Topological Vector, and Raster) and functions will be done. The main purpose of a GIS is to process spatial information, and the spatial relationships of the data can be summarized (database inquiries) or manipulated (analytic processing). All GIS packages contain hardware and software for data input, storage, processing, and display of computerized maps. In particular, the comparison among the analysed GIS packages has been done considering the processing functions grouped into eight functional areas:

A) Map algebra (reclassify map categories, overlay two or more maps, measure distance and connectivity, and characterize cartographic neighborhoods);
B) Image Processing (scene and sub-scene extraction, color enhancements, image registration, terrain correction, automated classification, user directed classification, filtering, histograms, etc.);
C) Surface analysis (slope angle, compass aspect, interpolate elevation at any point, generate contours, calculate drainage network/optimal path, and generate cross-sections);
D) Buffering (build buffer zones around routes, line segments, points, and areas);
E) Network analysis (shortest path along network, accumulation of attribute values, spatial adjacency search, and nearest neighbor search);
F) Polygon operations (polygon overlay, point-in-polygon, line-in-polygon, and merge/dissolve on basis of attribute);
G) Data structure conversion (raster to vector, vector to raster, and register vectors over raster; and
H) Data format conversion (import-export of ARC, DXF, ERDAS, ASCII, TIFF, GRASS, IDRISI formats).

Finally, in Table 16.3 the comparison among the GIS packages is synthesized, using "Y" when the GIS package has the function, and "N" when the GIS package does not have the function.

ACKNOWLEDGEMENT

The author would like to thank Professor V.P. Singh for the suggestions given and all those who have collaborated directly (vendors and GIS companies) and through GIS-L mailing list (listserv@ubvm.cc.buffalo.edu) to the listing of the GIS packages above mentioned.

References

Colosimo, C. and G. Mendicino, 1989. "Sistemi Informativi e Ingegneria Ambientale". Editoriale Bios, Cosenza, Italy.
Mark, R.L., Kemp, K.K. and H.A. Loaiciga, 1993. "Implementation of GIS for Water Resources Planning and Management". Journal of Water Resources Planning and Management, Vol.119, n. 2, pp.184-205.
Nyerger, T.L., 1993. "Understanding the Scope of GIS: Its Relationship to Environmental Modeling". In M.F. Goodchild, B.O. Parks and L.T. Steyaert, eds, Environmental Modeling with GIS, Oxford University Press, Chapter 8.

Contributors

Dr. Claude Collet
Institut de Geographie
Université de Fribourg
CH-1700 Fribourg
Switzerland

Professor C. Colosimo
Dipartimento di Difesa del Suolo
University of Calabria
C/da S. Antonello
87040 Montalto Uffugo Scalo
(Cosenza), Italy

Mr. J. Connolly
CH2M HILL
2510 Red Hill Avenue
Santa Ana, CA 92705, U.S.A.

Dr. D. Consuegra
Department of Rural Engineering
Soil and Water Management Institute
Swiss Federal Institute of Technology
CH-1015, Lausanne
Switzerland

Dr. A. Dalton
Prairie Geomatics
P.O. Box 417
Birtle, Manitoba,
Canada R0M 0C0

Dr. A.P.J. De Roo
Department of Physical Geography
University of Utrecht
Heidelberglaan 2
P.O. Box 80.115
3508 CS Utrecht
The Netherlands

Dr. T.P. Gostelow
Engineering Geology Group
British Geological Survey
Keyworth, Nottingham NG12 5GG
U.K.

Dr. S.K. Gupta
CH2 M HILL
2510 Red Hill Avenue
Santa Ana, CA 92705, U.S.A.

Professor C.T. Haan
Dept. of Biosystems and
Agricultural Engineering
Oklahoma State University
Stillwater, OK 74078, U.S.A.

Professor S.S. Iyengar
Department of Computer Science
Louisiana State University
Baton Rouge, LA 70803, U.S.A.

Dr. F. Joerin
Department of Rural Engineering
Soil and Water Management Institute
Swiss Federal Institute of Technology
CH-1015, Lausanne
Switzerland

Dr. G.W. Kite
Hydrometeorological Research Division
Atmospheric Environment Service
National Hydrology Research Institute
11 Innovation Boulevard
Saskatoon, Saskatchewan S7N 3H5
Canada

Dr. N. Lam
Dept. of Geography and Anthropology
Louisiana State University
Baton Rouge, LA 70803, U.S.A.

Dr. E. Ellehoj
Ellehoj Consulting
11456 43 Ave
Edmonton, Alberta
Canada T6J 0Y4

Professor M. Fiorentino
Department of Engineering
and Environmental Physics
University of Basilicata at Potenza
Via della Tecnica 3
85100 Potenza
Italy

Dr. E.B. Moser
Dept. of Experimental Statistics
149-A Agricultural Admin. Bldg.
Louisiana State University
Baton Rouge, LA 70803-5606, U.S.A.

Professor I. Muzik
Dept. of Civil Engineering
The University of Calgary
Calgary, Alberta T2N 1N4
Canada

Mr. C.A. Quiroga
Dept. of Civil & Environmental Engineering
Louisiana State University
Baton Rouge, LA 70803-6405, U.S.A.

Professor S.F. Shih
Remote Sensing Applications Laboratory
Institute of Food and Agricultural sciences
University of Florida
Gainesville, FL 32611-0570, U.S.A.

Professor V.P. Singh
Department of Civil & Environmental
Engineering
Louisiana State University
Baton Rouge, LA 70803-6405, U.S.A.

Dr. R.E. Macchiavelli
Dept. of Experimental Statistics
Louisiana State University
Baton Rouge, LA 70803-5606, U

Dr. G. Mendicino
Dipartimento di Difesa del Suolo
Università della Calabria
87040 Montalto Uffugo Scalo,
(Cosenza), Italy

Dr. A. Sole
Dipartimento di Ingegneria e
Fisica dell Ambiente
Univ. of Basilicata at Potenza
via della Tecnia 3
85100 Potenza, Italy

Dr. D.E. Storm
Biosystems and Agricultural
Engineering Department
Oklahoma State University
Stillwater, Oklahoma 74078
U.S.A.

Dr. A. Valanzano
Centro Interfacoltà Servizi
Informatici e Telematici
Via della Tecnica n.3
85100 Potenza, Italy

Dr. M.L. Wolfe
Biological Systems
Engineering Department
Virginia Polytechnic Institute
and State University
Blacksburg, VA 24061-0303
U.S.A.

Mr. G. Woodside
CH2M HILL
2510 Red Hill Avenue
Santa Ana, CA 92705, U.S.A.

438

Subject Index

Water Science and Technology Library

1. A.S. Eikum and R.W. Seabloom (eds.): *Alternative Wastewater Treatment. Low-Cost Small Systems, Research and Development. Proceedings of the Conference held in Oslo, Norway (7–10 September 1981).* 1982
   ISBN 90-277-1430-4
2. W. Brutsaert and G.H. Jirka (eds.): *Gas Transfer at Water Surfaces.* 1984
   ISBN 90-277-1697-8
3. D.A. Kraijenhoff and J.R. Moll (eds.): *River Flow Modelling and Forecasting.* 1986
   ISBN 90-277-2082-7
4. World Meteorological Organization (ed.): *Microprocessors in Operational Hydrology. Proceedings of a Conference held in Geneva (4–5 September 1984).* 1986
   ISBN 90-277-2156-4
5. J. Němec: *Hydrological Forecasting. Design and Operation of Hydrological Forecasting Systems.* 1986
   ISBN 90-277-2259-5
6. V.K. Gupta, I. Rodríguez-Iturbe and E.F. Wood (eds.): *Scale Problems in Hydrology. Runoff Generation and Basin Response.* 1986
   ISBN 90-277-2258-7
7. D.C. Major and H.E. Schwarz: *Large-Scale Regional Water Resources Planning. The North Atlantic Regional Study.* 1990   ISBN 0-7923-0711-9
8. W.H. Hager: *Energy Dissipators and Hydraulic Jump.* 1992
   ISBN 0-7923-1508-1
9. V.P. Singh and M. Fiorentino (eds.): *Entropy and Energy Dissipation in Water Resources.* 1992   ISBN 0-7923-1696-7
10. K.W. Hipel (ed.): *Stochastic and Statistical Methods in Hydrology and Environmental Engineering.* A Four Volume Work Resulting from the International Conference in Honour of Professor T. E. Unny (21–23 June 1993). 1994
    10/1: Extreme values: floods and droughts   ISBN 0-7923-2756-X
    10/2: Stochastic and statistical modelling with groundwater and surface water applications   ISBN 0-7923-2757-8
    10/3: Time series analysis in hydrology and environmental engineering
    ISBN 0-7923-2758-6
    10/4: Effective environmental management for sustainable development
    ISBN 0-7923-2759-4
    Set 10/1–10/4: ISBN 0-7923-2760-8
11. S.N. Rodionov: *Global and Regional Climate Interaction: The Caspian Sea Experience.* 1994   ISBN 0-7923-2784-5
12. A. Peters, G. Wittum, B. Herrling, U. Meissner, C.A. Brebbia, W.G. Gray and G.F. Pinder (eds.): *Computational Methods in Water Resources X.* 1994
    Set 12/1–12/2: ISBN 0-7923-2937-6
13. C.B. Vreugdenhil: *Numerical Methods for Shallow-Water Flow.* 1994
    ISBN 0-7923-3164-8
14. E. Cabrera and A.F. Vela (eds.): *Improving Efficiency and Reliability in Water Distribution Systems.* 1995   ISBN 0-7923-3536-8

Water Science and Technology Library

Kluwer Academic Publishers – Dordrecht / Boston / London